T0146288

Elephant Slaves and Pampered Parrots

Animals, History, Culture

Harriet Ritvo, Series Editor

Louise E. Robbins

Elephant Slaves and Pampered Parrots

Exotic Animals in Eighteenth-Century Paris

The Johns Hopkins University Press

Baltimore and London

© 2002 The Johns Hopkins University Press
All rights reserved. Published 2002
Printed in the United States of America on acid-free paper

2   4   6   8   9   7   5   3   1

The Johns Hopkins University Press
2715 North Charles Street
Baltimore, Maryland 21218-4363
www.press.jhu.edu

Library of Congress Cataloging-in-Publication Data
Robbins, Louise E.
Elephant slaves and pampered parrots : exotic animals in
eighteenth-century Paris / Louise E. Robbins.
p. cm. — (Animals, history, culture)
Includes bibliographical references (p.   ) and index.
ISBN 0-8018-6753-3 (hardcover : alk. paper)
1. Wild animals as pets—France—Paris—History—18th century.
2. Human-animal relationships—France—Paris—History—18th century.
3. Exotic animals—France—Paris—History—18th century. I. Title. II. Series.
SF411.36.F8 R63 2002
636.088′7′09443609033—dc21
2001000243

A catalog record for this book is available from the British Library.

To my family

CONTENTS

List of Illustrations   ix

Preface   xi

Introduction   1

1 · *Live Cargo*   9

2 · *The Royal Menagerie*   37

3 · *Fairs and Fights*   68

4 · *The Oiseleurs' Guild*   100

5 · *Pampered Parrots*   122

6 · *Animals in Print*   156

7 · *Elephant Slaves*   186

8 · *Vive la Liberté*   206

Epilogue   231

Notes   237

Primary Material   311

Note on Secondary Sources   323

Index   337

ILLUSTRATIONS

1.1 Some overseas destinations for French eighteenth-century traders and explorers   14

1.2 Number of ships per year arriving in Marseille from the Antilles   15

1.3 Drawing a bird, frontispiece from Sonnerat, *Voyage à la Nouvelle Guinée* (1776)   18

2.1 The Versailles menagerie: *a*, front view; *b*, aerial plan   41, 42

2.2 Panther at the Versailles menagerie (1739)   46

2.3 *Ostrich Hunt*, painting by Charles Vanloo (1738)   48

2.4 Pair of zebras, from *Histoire naturelle* (1764)   53

2.5 Quagga, from *Histoire naturelle* (1782)   59

2.6 Head and penis of the Versailles rhinoceros, drawing by Petrus Camper (1777)   63

3.1 Wild "leopard" in a fair booth, illustration by Jean-Baptiste Oudry   77

3.2 Performing monkey in a fair booth, illustration by Jean-Baptiste Oudry   78

3.3 Lion-tailed macaque *(ouanderou)*, from *Histoire naturelle* (1766)   81

3.4 Poster for Paris animal fights (1805)   82

3.5 *African Lion Being Attacked by Dogs*, painting by Jean-Jacques Bachelier (1757)   84

3.6 Black wolf *(loup noir)*, from *Histoire naturelle* (1761)   90

3.7 Bronze and ormolu rhinoceros clock (ca. 1749)   96

4.1 Bird merchant selling canaries (1774)   102

4.2 Oiseleurs releasing birds for Charles VII's entry into Paris   105

5.1 Mme Louise, Louis XV's daughter, with a showy parrot, painting by Adelaide Labille-Guiard (1788)   127

5.2 Conversational parrot   133

5.3 White-faced capuchin monkey *(saï à gorge blanche)*, from *Histoire naturelle* (1767)   135

5.4 "Unexpected fire in the modern hairdo of Mme Rise-to-the-sky"   144

5.5 "The modish monkey"   147

5.6 Trumpeter *(agami)*, from *Histoire naturelle* (1778)   154

6.1 Buffon   171

6.2 Elephant-human hybrids, from Restif de la Bretonne, *La découverte australe* (1781)   178

7.1 "The Gévaudan beast devouring a woman"   189

7.2 Elephant *(éléphant)*, from *Histoire naturelle* (1764)   194

8.1 "Vive la liberté!"   208

8.2 Provisional menagerie at Jardin des plantes (1794)   223

8.3 Mating elephants, from Houel, *Histoire naturelle des deux éléphans* (1803)   227

PREFACE

From the first idea to the last comma, many people had a part in the production of this book. At the University of Wisconsin–Madison, I thank especially the faculty who guided me in creating its initial incarnation as a dissertation for the Department of the History of Science. Tom Broman introduced me to the Enlightenment, put up with my stubbornness, and helped me find out what it was I wanted to write. The gentle musings of Hal Cook pushed me to consider big issues of science and natural history. Lynn Nyhart read with great care and encouraged me to get to the point. Suzanne Desan shared her knowledge of French history, communicated her revolutionary enthusiasm, and always raised my energy level. And from Anne Vila I gained an entrée to eighteenth-century French literature and a greater understanding of *sensibilité*.

Other kinds of aid came from a variety of sources. For financial sustenance, I am grateful for two years of fellowship support from the University of Wisconsin–Madison, travel money from the Department of the History of Science, teaching assistantships from the Department of the History of Science and from Integrated Liberal Studies, and project assistantships with the *Isis* editorial office and with Professor Ronald L. Numbers. My parents also contributed generously.

For assistance with research in the United States, I thank the staffs of the University of Wisconsin–Madison Special Collections Department, the Microforms and Media Center, and the Interlibrary Loan Office. John Neu, former history of science bibliographer at UW-Madison, deserves special gratitude for keeping an eagle eye out for relevant references. Several other institutions and libraries provided research facilities and supplied material for illustrations.

In Paris, I appreciated the facilities and thank the personnel at the

Archives nationales, the Bibliothèque nationale de France, the Bibliothèque historique de la ville de Paris, and the Bibliothèque centrale du Muséum national d'histoire naturelle. Mme Françoise Serre, at the Muséum, was particularly helpful, and I also thank Yves Laissus for meeting with me. In Montbard, Pierre Ickowicz very kindly gave me a personal tour of the Musée des anciennes écuries and the parc Buffon, where I was able to contemplate the "grandes vues d'un génie ardent" from the window of Buffon's study. Bernard Rignault and Luc Dunias at the Musée de la sidérurgie showed me around the Grande forge de Buffon and humored me in my search for fan letters written to Buffon.

Many colleagues read and commented on chapters, informed me of relevant sources, or noted down monkey stories. I might not have discovered some of my most important sources had Michael Lynn not pointed them out to me; he also passed along innumerable references, very generously sharing the fruits of his hard work. Richard W. Burkhardt Jr. dug through his notes to answer my questions about the Paris menagerie and carefully read two chapters, Jonathan Lamb gave me insightful suggestions concerning pets and animal slaves, and Patrice Higonnet made some recommendations for improving chapter 8. Morag Martin provided many useful references, and Mary Salzman guided me through the labyrinth of the BnF Estampes collection. Members of the History of Science Dissertation Group, the French History Dissertators' Group, and the Madison Area French History Discussion Group offered encouraging comments and great company—thanks to all: Franca Barricelli, Jessica Coulbury, Suzanne Desan, Susan Dinan, Ralph Drayton, Steve Eardley, Fa-ti Fan, Hae-Gyung Geong, Judy Houck, Ted Ingham, Tomomi Kinukawa, Alan Krinsky, Jamie Lee, Jody LePage, Mike Lynn, Craig McConnell, Sarah Pfatteicher, David Reid, Michael Robinson, John Rudolph, Alison Sandman, Karen Walloch, Hsiu-Yun Wang, and Deirdre Weaver. Friends who kindly made time to read drafts of chapters include Jane Huth, Joanna Inglot, Dan Simberloff, and Mary Tebo.

During the last stages, I especially appreciated Harriet Ritvo's enthusiasm for including the project in her new series. The manuscript benefited from her constructive criticism as well as that of an anonymous reviewer. Dennis Marshall spied many errors that had been lurking unseen for years. And I thank Robert J. Brugger at the Johns Hopkins University Press for his persistence and patience, as well as Melody Herr, Julie McCarthy, and everyone else who helped to turn electronic files into a beautiful book.

I end this brief preface by defining and explaining a few terms. First, *exotic.* Products that originated elsewhere began to appear in France in abundance in the eighteenth century, but the word used to describe such objects was almost always *étranger* (foreign), not *exotique.* In 1787, a popular guidebook defined the word for its readers, which suggests that it was still not common, although it had been used as early as the sixteenth century. Referring to a garden containing "exotic or indigenous" trees, the author explained that the word came from the Greek and meant "foreign *[étranger]*, that which is not a production of the country in which one lives."[1]

Second, *exotic animals.* I use the term for any species not native to western Europe. In some places, particularly chapter 5 (pets), the boundary I have drawn between exotic and native species is rather artificial: parrots and monkeys joined cats and dogs in the household, and some of the literature I discuss relates to both native and exotic pets. One reason I have avoided treating native animals is that I would have had to write at least an additional volume; I hope that someone will, one day. More important, exotic animals did, in many cases, hold a different place in the culture. Their imported status distinguished them and often increased their value; their numbers were constantly increasing; unlike cats or dogs, they usually had to be chained or caged and were not bred domestically; and many were geographically linked with France's colonies. As for the word *animal,* I use it to refer only to nonhuman animals. I often use *man* rather than *human,* especially in translations, to accord with eighteenth-century usage and to underscore the fact that the word *homme,* even when intended to refer to both men and women, would have carried a certain male flavor.

Third, *science* and *natural history.* I use the word *science* in its modern English-language sense; it had a much broader meaning in eighteenth-century France. Almost any subject, from finance to theology, could be a science if approached with rigor. Instead of dividing knowledge into "sciences" and "humanities," Enlightenment scholars used the categories "philosophy," "history," and "poetry." In the tree of knowledge that accompanied the *Encyclopédie,* Diderot and d'Alembert included physics, chemistry, anatomy, and physiology as varieties of philosophy, whereas natural history fell under the heading of history.[2] Studying the natural history of a species involved much the same process as studying the history of nations or languages: classifying, verifying facts, describing characteristics and behavior, and drawing generalizations through comparison. Natural objects (along with the occasional human artifact) were often displayed in *natural history cabinets,* which could vary in size from a glass-fronted case to a several-room exhibit.

Finally, a note on translation. All translations are mine unless otherwise indicated. Punctuation and capitalization are sometimes modified. I have attempted to translate most animal names into present-day English-language common names. In cases where I could not identify an animal, I either left the name in French or translated it directly into a descriptive but not specific English name: a bird called a *perroquet vert*—a green parrot—for instance, could have been any of a variety of different parrot species.[3]

Elephant Slaves and Pampered Parrots

Introduction

IF YOU HAD been walking down the rue Dauphine in central Paris in mid-January 1771, you would have encountered a large crowd of people clustered on the street. After elbowing a way in, and probably having to pay for a closer look, you would have been treated to a rare sight: a young male elephant was entertaining the crowd with tricks. According to a report in a contemporary periodical, the *Avant-coureur*,

> It takes grains of rice with its trunk from ladies' hands; it uncorks a bottle of beer and gulps it down. It's a remarkable sight to see it perform that task. . . . A bottle is put before it with the cork slightly loosened. The elephant grabs the bottle with its trunk, turns it over and puts the bottom in its jaws; then it brings its trunk under the neck of the bottle, pinches the cork, and removes it; the cork falls and the liquid pours into its trunk. It lets the bottle go when it's empty, then puts its trunk, which acted like a funnel, in its mouth, and pours in the beer.[1]

The fuss over the elephant suggests that exotic animals were an uncommon spectacle in eighteenth-century France; indeed, the *Avant-coureur* account began, "It is a rare and interesting sight to see a living elephant in Paris." And in fact no elephant had appeared in Paris for more than a century, since 1668. If elephants were unusual, though, exotic animals as a whole were not, as I discovered when I began looking for their traces. Parisians and visitors to the city had plenty of opportunities to observe creatures from all parts of the world. Just across the river from the rue Dauphine, on the right bank of the Seine, at the quai de la Mégisserie, was the shop of

Ange-Auguste Chateau, oiseleur du roi (the king's bird seller). After entering Chateau's shop and being kicked at playfully by the resident crowned crane, customers searching for an interesting pet would find cages containing parrots and parakeets from Africa and South America, cockatoos from Australasia, cardinals, painted buntings, blue jays, and flying squirrels. If sieur Chateau had no capuchin or green monkeys for sale, then surely one of the other bird-sellers' shops in the area would.

During annual fairs, roving show people displayed menageries of exotic animals containing up to several dozen different species. Animal fights, which took place in an amphitheater on the edge of town and were advertised by posters and in newssheets, provided another location for viewing creatures from overseas. Ordinary fights usually involved bulls, dogs, bears, and deer, but special fights that took place during religious festivals featured lions, tigers, and once even a mandrill. Some distance outside of Paris, at the king's palace at Versailles, visitors could tour the menagerie that had been built by Louis XIV. There, in the walled-in enclosures, they could observe a wide range of birds, carnivores, grazing animals, and monkeys, and—the favorite of many visitors—a bizarre, wrinkled rhinoceros.

Not only were many species of animals present physically, but they also made frequent appearances in all kinds of written works, from satires and fables to scholarly tomes. Some people in the crowd on rue Dauphine would already have known something about elephants before seeing one: the announcement in the *Avant-coureur* and another newssheet even gave recommendations about where to read up about them. One of the suggested works, a best-selling natural history encyclopedia by the renowned naturalist Georges-Louis Leclerc, comte de Buffon, was available in a variety of formats, including a relatively inexpensive pocket-size edition. Of course, without journals or letters describing their experiences, we can't know what the observers had read or were thinking when they looked at the young elephant, but at least a few of them who were familiar with Buffon's description (which was widely reproduced) might have felt sorry for it, thinking that it would probably never mate with another of its kind. According to Buffon, "the disgust for [the captive elephant's] situation lodges in the bottom of its heart," and, unlike domestic animals, those "born slaves" that man can manipulate and propagate at will, the elephant "consistently refuses to reproduce for the profit of the tyrant."[2]

Stumbling onto the subject of exotic animals in France was something like stumbling across an elephant on a city street. I could not ignore it, and once I had discovered it I became more and more intrigued. I had started out planning to study Buffon and to analyze the reasons for the popularity of his natural history encyclopedia. That topic remained a part of my inquiry, but it became secondary when I realized that there was a whole world of real animals to explore.

Exotic animals were a major presence in eighteenth-century Paris—not only materially but also culturally. Although individual animals had been present in earlier times, their numbers swelled during this period. They showed up on the streets and in private homes, as well as in jokes, poems, stories, posters and paintings, and in works of natural history. In the transformation to literature or art, they often took on metaphorical meanings. Two sets of separate but interconnected questions called out for answers: How did these exotic animals get to France, who brought them and why, and where did they reside in Paris? What kinds of meanings did people ascribe to them, and how were those meanings related to other aspects of French culture? This book gives my answers to those questions.

One rationale for rummaging around in a lost corner of history is simply the pleasure of spying on the past, finding out what it was like to be alive in a different time. A great deal is known about everyday life in eighteenth-century Paris, but, curiously, very little about the city's animal inhabitants. To imagine what people would have seen and experienced there would be incomplete without the exotic animals that were an ever-increasing presence.

Once the animals are in place, we can begin to look at their cultural meanings. As a biologist-turned-historian, I am particularly fascinated by how attitudes toward nature have changed over time. What I "see" when I look at an elephant or a patch of woods is determined in part simply by when and where I was born. Exploring shifts in such attitudes has been popular among scholars, especially since the growth of the environmental movement in the 1970s and, in more recent decades, since the interest in animal rights and in cultural studies of science. Many authors—for example, Clarence Glacken, in *Traces on the Rhodian Shore* (1967), Keith Thomas, in *Man and the Natural World* (1983), and Richard Grove, in *Green Imperialism* (1995)—have traced very broad transformations in ideas about nature, linking them to changes in economic structures, biological theories, urbanization, and European expansion. Analysts of animal-human relations have also proposed broad schemes to describe changing attitudes toward animals in the West. Many of these studies contrast a past golden age when humans respected their animal companions with a present-day culture of exploitation brought on by the mechanistic outlook of the scientific revolution, by the domineering Judeo-Christian ethic, or by the transition to pastoralism.[3]

There may be some truth to these overarching visions, but their wide-angle views smooth over much varied topography. More nuanced works on specific times and places, such as Harriet Ritvo's *The Animal Estate*, about Victorian Britain, have shown how closely animals become interwoven with issues peculiar to that culture. Some animals may become cultural symbols—the independent bald eagle as a symbol of the United States, for example. In an often-repeated phrase, anthropologist Claude Lévi-Strauss remarked that societies accord animals special status not because they are "good to eat" but because they are "good to think."[4] According to Lévi-Strauss and other anthropologists, the meanings cultures ascribe to animals are "not installed in nature": symbolic significance, although based on the animals' features and behavior, is not determined by those traits.[5] Considering the number of characteristics each species possesses (color, shape, diet, habitat, mode of reproduction, call or song), the possibilities are immense. For instance, where Europeans valued parrots for their mimicking abilities, the Asmat in New Guinea esteemed them as brothers during head-hunting expeditions, because parrots eat fruit, and fruit looks like the human head.[6]

Eighteenth-century France is especially interesting for looking at exotic animals because of their growing presence, because of their connection with the popularity of natural history, and because of the way they became linked with political and social issues. Comparisons could be made concerning the presence and meanings of exotic animals among different cultures during this period, but I have chosen to focus on France and specifically on Paris, the largest city. Readers should keep in mind that the French were not the only ones importing exotic animals or incorporating them into their culture.

To understand the meanings of animals in eighteenth-century Paris, I have looked at all kinds of writings. The audience for such texts would have included few peasants or day laborers, but it expanded considerably in the eighteenth century, along with literacy, publishing, and consumption. The wealthy could buy deluxe editions of new books; those with moderate means could buy cheaper editions, subscribe to journals that printed excerpts and book reviews, go to lending libraries, or read periodicals in cafés; illiterate people could listen to someone reading aloud, attend a lecture, or enjoy the presentations of exhibitors at the fair. As knowledge of exotic species grew and natural history became popular, the public bought books that provided them with authenticated facts about animal behavior. I suggest, however, that many readers wanted to read about animals not so much to learn about the animals themselves, but rather as a way to think about human behavior and society: that is why the most popular natural history texts, especially Buffon's *Histoire naturelle,* were those that provided moral, social, and po-litical lessons in a manner similar to that of fables. This connection between natural history and fables is important for understanding why people cared to look at elephants or read about them, and it highlights a side of eight-eenth-century natural history that has long lingered in the shadows.

This neglect is particularly evident in the case of Buffon, a major En-lightenment scientific figure. Through much of the nineteenth century, Buf-fon was acclaimed more for his literary than for his scientific accomplish-ments. Since 1950, however, the pendulum has swung back again and many scholars have highlighted his scientific achievements. Historians have ana-lyzed his novel ideas about geological change, species transformation (some have identified him as a precursor to Darwin), and reproduction. But for Buffon and many other Enlightenment thinkers, science and style were allies, not enemies. "The principles of the sciences would be repellent if literature did not lend its charms," read a passage under the heading "Sciences" in the *Encyclopédie:* "Truths become more accessible *[sensibles]* through the clear style, the pleasant images, and the clever turns of phrase by which they are presented to the mind."[7] For most readers of the time, the stylistic pleasure of Buffon's work enhanced its scientific value, a value that included draw-ing conclusions about human society from observations about animals.

The writings of Buffon and other authors in eighteenth-century France were often infused with metaphorical language linking animals with debates that were going on in the social and political realm concerning the pros and cons of colonial holdings, the ethics and economy of slaveholding, the proper role for women, and the legitimacy of the monarchy and the extent of its power. For instance, I believe that the proliferation of sympathy toward "animal slaves" was most likely connected with the widespread critical movement that arose in late-eighteenth-century France. Social critics portrayed wild animals as symbols of freedom and independence, contrasting them to the enslaved creatures that lived under a tyrannous regime. A similar, but less widespread, strain existed in England; although Enlightenment criticism was a broad, cross-cultural phenomenon, it flowered especially profusely in France, where it was richly fertilized by the absolutist monarchy. Criticisms of tyranny were much stronger in France than in England; likewise, oppressed animals seem to have received more sympathy (at least rhetorically). The case of eighteenth-century France provides an instructive contrast to cultures that have already been studied, where concern for "enslaved" animals remained a minority viewpoint. Such concern does not necessarily, however, reflect a shift toward new concern for the environment or even newfound sympathy for animals. As we will see, animal meanings were complex and shifting.

History rarely hands us answers to present-day dilemmas, but it does often give us a step stool that we can use to see over the walls of our cultural habits and assumptions. Today, animals are a particularly prominent part of popular culture, and new books or documentaries keep appearing on topics such as devoted animal fathers, mind-reading dogs, or noble elephants.[8] *Slavery* is not an uncommon term in some animal-rights literature, and there seems to be growing popular support for the idea that humans should no longer set themselves above other animals.[9] "We may, in fact, be present at the dawn of a fundamental shift in the human-animal relationship, one that does not put the human animal at the center of all creation," predicts one author.[10] My own response to these issues, as well as my understanding of how humans make meaning of animals, has changed considerably since I immersed myself in the world of eighteenth-century parrots and elephants. I have become much more aware of how hard it is to see animals without seeing ourselves, how complicated the history of human-animal relations is, and how careful we have to be with our metaphors, especially those that link animals with slaves or other oppressed human groups.

In the first chapter I examine how exotic animals got to France. Few of them walked or flew, or traveled overland—they came in ships, ships that were primarily engaged in trade. Sugar, coffee, and indigo were the primary goods traded, along with African slaves; I trace the tracks of the animals that often shared space with slaves and trade goods, looking at who transported them and why they bothered to do so. Understanding the connection between animals and commerce helps to understand both why particular species such as African grey parrots became relatively common in the eighteenth century and why they became clothed in language connecting them with issues such as luxury, colonialism, and slavery.

In chapters 2 through 5, I explore the sites that exotic animals inhabited once they reached Paris: the king's menagerie, fairs and fights, bird-sellers' shops, and private homes. The sequence begins with the most traditional site for exotic animals, turns, then, to two groups of people (show people and merchants) who made their livelihoods from the animals they acquired (an increasing number of them from overseas), and ends with private individuals. In all of these venues, changes occurred as exotic animals became more common and the popular interest in natural history grew. Each site, however, had its own dynamic and characteristics.

Of these four sites, only the royal menagerie at Versailles has up to now received much attention from scholars. My perspective on the Versailles animal collection (chap. 2) is somewhat different from that of previous historians, who have examined it either as a scientific location or as a site for kings to show off their power. In addition to these aspects of the menagerie, I explore how the animals were acquired and the ways in which spectators reacted to them. One result of this shift in focus is that we see the ragged edges of royal power: desired animals slipped out of the monarch's grasp, and people started to interpret the menagerie in a very different way than its owners intended.

Chapter 3, on fairs and fights, surveys the exotic animals that were obtained and displayed by entrepreneurs hoping to cash in on the public's interest in the exotic and in science. One of the fascinating aspects of this side of Parisian animal life is how entertainment and education merged and conflicted. Fair entrepreneurs took advantage of the fad for natural history by pitching their announcements to naturalists, who indeed visited their booths. Although the naturalists complained about exaggerated claims and faked rarities, they obtained valuable information from the animal handlers and often borrowed their sensational animal-behavior stories, too.

Discovering the papers of the bird-sellers' guild, the *oiseleurs*, gave me the opportunity to explore a specific commercial outlet for exotic birds and some small animals (chap. 4). Bringing to life the bird shops, which were in the center of Paris, resurrects what was probably one of the primary spots where city people would have encountered unusual species. Naturally, naturalists were among the visitors to these shops, and, as at the fair, they acquired useful data from the shopkeepers, one of whom (sieur Chateau) even made it into Buffon's *Histoire naturelle* in a few places. As the eighteenth century progressed, though, the oiseleurs had more and more difficulty maintaining their monopoly in the face of increased trade and upheavals in the guild system.

In chapter 5 I follow exotic animals into private homes. From advertising newspapers and other sources, I have been able to add exotic pets to the picture of households that historians have previously filled with furniture, clothes, and books. The popularity of these pets depended on the same commercial growth that brought the animals to France in the first place: larger numbers of people, especially in cities, were accumulating enough money to be able to purchase expensive pets and other luxury goods. I also look at how pets became incorporated into discussions of luxury, gender, domestication, and the nature of the human-animal boundary.

People not only displayed and observed, and bought and sold animals, they also read and wrote about them. Where the first five chapters use texts mostly as tools to recover the presence of "real" animals (with some excursions into meanings), the next two look in depth at the texts themselves in order to understand cultural representations of exotic animals. In chapter 6, I survey the category that I call "animal books," which includes fables, works of natural theology, and encyclopedias. I explore the reasons for their popularity and analyze the ways in which they incorporated moral lessons in the guise of animal tales. In chapter 7, I analyze the metaphor of animal slavery and try to understand why authors chose to compare—or, in a few cases, not to compare—mistreated animals to enslaved humans.

In chapter 8, I explore the fate of animals and of animal rhetoric during the Revolution. I discuss why the symbols associated with wild animals shifted around (such that they were regarded as victims at some times and as enemies at others) and compare various utopian schemes for the liberation of animal slaves with the reality of the revolutionary government's decision to establish a national menagerie.

CHAPTER I

Live Cargo

IN 1764, just after France lost Canada to the English in the Seven Years War, the crew of a ship transporting French Canadians to a new colony in the Malouines Islands (now known as the Falkland or Malvinas Islands) were ordered by their captain, Louis-Antoine de Bougainville, to kill a young *"tigre"*—probably a jaguar. It had been on board for only a week, having been loaded when the ship stopped for provisions at Montevideo. According to the Dominican priest Dom Pernetty, who wrote an account of the trip, the local governor had given the animal to Bougainville as a present. It had been raised from cubhood in the courtyard of the governor's palace, where the servants played with it as though it were a pet cat. After a week on board in a made-to-order cage, however, it began to roar, especially during the night, and there was no more fresh meat to feed it. "These considerations determined M. de Bougainville to have it strangled."[1]

That was not the end of the ship's exotic fauna; several parrots were also on board, the most stunning of which had been a gift to Bougainville from the governor of Brazil. Its feathers were a gaudy mixture of jonquil and lemon yellow, carmine red, crimson, dark green, and bright blue (a pattern created by plucking individual feathers from the young bird and injecting a liquid potion at the root). Merchants had sold several more parrots, two of which could speak Portuguese, to crew members. Of seven parrots, only two made it back to France alive. Bougainville's and Pernetty's both died of a cold in the head, followed by asthma, and M. de Belcourt's fell overboard and drowned. One of the surviving parrots, Pernetty reported, "was of the small kind, [and] had no tail, because it pulled out its feathers as soon as they appeared. The sailor to whom it belonged didn't take nearly as good care of it as we did of ours, yet he preserved his. It spoke quite well and imitated perfectly the cries of the children on board, those of the ships' boys when they are whipped for having made mistakes, those of the chickens, and the varied languages of all the animals on the frigate."[2]

Parrots, it turns out, were passengers on Bougainville's later and more famous voyage, as well. Historians' accounts of Bougainville's three-year voyage around the world always mention one exotic passenger, a young Tahitian man, Aotourou, who created a sensation when he arrived in Paris, but they rarely mention the exotic birds that were also on board.[3] According to Bougainville, indigenous people in present-day Malaysia and Indonesia would paddle out to the French ships in canoes, bearing parrots along with food items like pigs, chickens, bananas, and coconuts, which they traded for red handkerchiefs (the going rate was one handkerchief for one cockatoo or several chickens). Most readers probably would not envision the exchange of more than a few parrots. But in his journal of the voyage, Louis Caro, first lieutenant on the *Étoile*, noted offhandedly that after one such trading episode there were now more than four hundred parrots on the ship.[4]

This practice of collecting jaguars, parrots, and other live animals is difficult to trace because of sparse documentation. For most of the merchant ships, which made up the bulk of sea voyaging in the eighteenth century, bare-bones logs preserve only navigational data. For voyages on which memoirs or journals were kept, acquiring birds here and there may have been too routine to warrant notice, even when numbers ran into the hundreds; only one of the four chroniclers of Bougainville's voyage found occasion to mention the multitudes of parrots.[5] Reconstructing this trade requires scavenging bits and pieces from a variety of sources.

Exotic animal passengers may often have escaped mention in part because any traveler on a long voyage would have been used to the company of bleating and cackling barnyard animals. Ships normally embarked with a full set of domestic animals as a supply of fresh meat: each of the ships on the attempted circumnavigation by Jean-François de Galaup, comte de La Pérouse, in the 1780s carried five cows, thirty to forty each of sheep and pigs, and two hundred poultry (ducks, geese, chickens, and turkeys).[6] Exotic animals, like Bougainville's *tigre* and parrots, were often more difficult to transport because of fierceness or fragility, but their edibility, too, put them at risk.

Although not everyone liked their flavor, many travelers found parrots to be particularly tasty: "multicolored parrots, red, grey, green, yellow, and mixed, are the best and the most exquisite of all: their flesh is tender, rich *[courte]*, and melts in the mouth," reported Robert Challe during a stop in the East Indies.[7] Appetizing or not, any animal protein was fair game when provisions ran low; the line separating food from nonfood was neither distinct nor fixed. In a frequently reprinted early travel account, Jean de Léry described how, after several weeks becalmed in the middle of the Atlantic on the way back to France from Brazil in 1558, he and his compatriots finally ate the monkeys and parrots they had been hoping to take home alive.[8] Extreme famine was rare, but periods of hunger afflicted most long voyages. When they ran short of food near New Guinea, Bougainville's crew ate a dog they had taken on board in South America ("it was young and plump . . . we found it excellent"), and La Pérouse devoured a scrawny curlew that had made the mistake of landing on the ship ("[it] tasted hardly better than the sharks.")[9] Such incidents entered the popular imagination enough to become the subject of an eighteenth-century fable, in which the famished crew of a becalmed ship eats first the parakeets, then the cardinals and the cockatoos, and finally the overconfident parrot, which had been saying all along, mimicking the captain, that everything would turn out fine (and so chose not to escape through the hole in its cage).[10]

Considering both chronic food shortages and lax food prohibitions, it is amazing that any animals made it back alive. Those that did, obviously had a value that placed them above consumable creatures, as is evident in the reluctance to eat them even in extreme circumstances. Léry reported that before they sacrificed the parrots and monkeys, he and his shipmates ate every last bit of the bitter, black mash of biscuit crumbs, maggots, and rat droppings that remained in the hold. Those who had been teaching their parrots a new language held out the longest.[11] In exploring the circumstances under which French travelers took the trouble to collect animals and keep them alive, I have found several distinct (but always intertwined) motivations. Contrary to what one might expect, scientific study was rarely the principal one. Most of the time animals were taken on board as gifts, as commissions for the king's menagerie, as commercial items, or as shipboard companions. Similar reasons probably motivated seafarers of other nationalities, who transported exotic species into Dutch, British, and other ports as well.[12]

Nobody counted the numbers of parrots being disembarked from incoming vessels, so chronological trends in the exotic animal trade cannot be quantified. But evidence from a variety of sources suggests that it increased dramatically during the eighteenth century, along with the increase in colonial trade. Nobody counted the myriad parrot deaths, either, but the vulnerability of imported creatures seems to have added to the value of those that survived. Many, indeed, did survive, though the percentage was probably low. As we will see, travelers responded to their losses in different ways—sometimes with indifference or resignation, sometimes with regret and sharpened desire, and sometimes with expressions of respect for resistant animals. As in so much writing about animals, these passages often reveal conflicting sentiments: the same animal being sincerely mourned fades into a two-dimensional prop when it provides a metaphorical entry point for musings on the limits of power or the morality of slavery. Before turning to the animals, though, we'll take a brief look at the types of overseas voyages that carried them to France.

Trading and Exploring

The exotic animals that arrived in France during the ancien régime rode in on ships returning from Africa, the Americas, and the East Indies. Major, well-documented expeditions, such as those of Columbus, Cook, and Bou-

gainville, have provided the primary fodder for intellectual historians interested in the encounter of the Old World with the New and for historians of
science interested in the importation of new scientific data and specimens to
Europe. Not these, however, but the much more frequent commercial voyages were responsible for the importation of most live animals. Anyone unfamiliar with economic history might be surprised at the extent of this traffic:
for instance, a voyager who arrived in the port of Concepción (now in Chile)
in the 1710s encountered fifteen French ships already there, containing more
than fifteen hundred men—and this in a port that could be reached only by
braving brutal Cape Horn.[13] The number of ships sailing annually to the
West Indies just from the port of Bordeaux mushroomed from 23 in 1682 to
310 in 1782, and commerce with East and South Asia also grew. Overall, the
value of foreign trade in France quintupled from the 1710s to the 1780s.[14]
By the late eighteenth century, hundreds of French ships were flying to and
from the Caribbean, Africa, India, and the Far East (see fig. 1.1).

Apart from Mediterranean trade, which had flourished since the medieval period, the earliest French trading voyages were to the west coast of
Africa (fifteenth century), the northern Atlantic, and Brazil (early sixteenth
century). Trade in the north Atlantic first centered on cod and later expanded as the fur trade developed. After several failed attempts, colonization of French Canada began in the early seventeenth century and advanced rapidly, but the colony passed into English hands in 1763—the same
year that put an end to Louisiana as a French colony. La Louisiane, which
encompassed much of what is now the central United States, had started
out strong with financier John Law's popular investment and colonization
scheme, but this "Mississippi bubble" soon burst, partly because of resistance to forced deportation of prisoners and beggars from France (vividly
portrayed in Prévost's *Manon Lescaut* [1731]). Although the colony never
flourished, ships stopping there on the way back to France from the West
Indies had an opportunity to take on board North American species such as
cardinals.

France's Caribbean colonies, especially Saint Domingue (now Haiti),
attracted more settlers and fared much better economically than did those
in North America. Beginning in the late seventeenth century, settlements
developed on Saint Domingue, Martinique, and Guadeloupe, and in Guyana, South America. Whites and, later, people of mixed race established

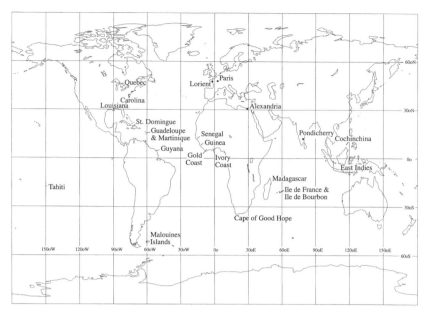

FIGURE 1.1. Some of the major locations to which French traders and explorers traveled in the eighteenth century. Map by Gail Ambrosius.

sugar plantations and imported astonishingly large numbers of enslaved Africans to labor in them.[15] The plantations were very productive, and for most of the century France was the world's leading sugar producer. Slaves were transported to the French islands at a rate of ten thousand to fifteen thousand per year in midcentury and about thirty thousand per year in the 1780s, for an estimated total (from the late seventeenth century) of 1.5 million, out of 6 million for the eighteenth-century Atlantic slave trade as a whole.[16] Most of this trade was carried on by independent merchants, although at various times the government accorded exclusive rights to certain parts of the African coast to companies such as the Compagnie du Sénégal, the Compagnie de la Guyane, and for a brief period the Compagnie des Indes. Colonial trade skyrocketed during the century, becoming a significant element of the national economy. Hundreds of ships left every year from the increasingly prosperous ports of Nantes, La Rochelle, Le Havre, Bordeaux, and others: in the last quarter of the century, five hundred to six hundred ships a year sailed to the Antilles, and fifty to one hundred a year to Africa (fig. 1.2).[17] These numbers not only help to explain where the Pa-

Number of Ships

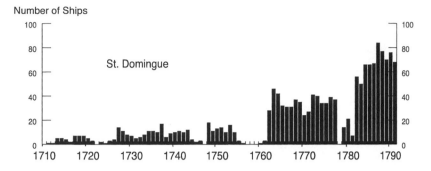

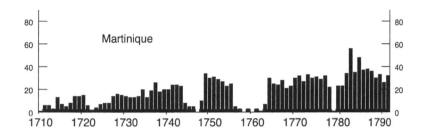

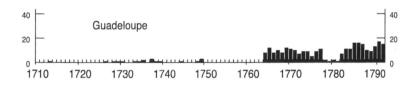

FIGURE 1.2. Number of ships per year arriving in Marseille from the
Antilles. The dips correspond to the plague outbreak (1720),
the War of the Austrian Succession (1744–48), the Seven Years War (1756–
63), and the American Revolution (1778–82). Adapted from Charles
Carrière, *Négociants marseillais au XVIIIe siècle*, 2 vols. ([Marseille]:
Institut historique de Provence, 1973), 1:332. Chart by Gail Ambrosius.

risian parrots came from, but suggest why such turbulent debates swirled around the issues of colonization and slavery toward the end of the century.

East Indian trade played a smaller role economically than did the Caribbean trade but brought prized animals such as cockatoos, elephants, and zebras to France. The Compagnie des Indes orientales, modeled after the prosperous Dutch and English East India companies, was created in 1664 by Louis XIV and his chief minister Colbert. Suffering from wars, financial troubles, and repeated fruitless attempts to set up a colony and refreshment station at Madagascar, the company fizzled out by 1708. After a period when the Indies trade was accorded to merchants from Saint Malo, the Compagnie des Indes revived in 1719 and began to flourish: trading factories and forts were built in India, principally at Pondicherry; slave colonies were founded on the Mascarene Islands—île de France and île de Bourbon, now Mauritius and Réunion—which served as way stations; and the company's port at Lorient in western France, with a flotilla of more than one hundred ships, became a center for imports from Asia. Heavy ships, with crews averaging 100 to 150 men, would embark with holds full of precious metals (mostly silver from the Spanish colonies) and wine. One and a half to two years later, they would return with textiles, porcelain, spices, tea, and coffee. In 1769, suffering from financial trouble following the Seven Years War and public pressure for free trade, the company was dissolved and the Asia trade opened up to private merchants, who every year continued to send dozens of ships eastward around the tip of Africa. From 1730 to 1785, 479 French ships sailed to the Indies and China and 400 to the Mascarenes; another 400 or so plied the Eastern trade during the busy period from 1786 to 1793.[18]

Voyages of discovery and exploration involved far fewer vessels and crew than did commercial trading ventures and brought back few live animals, but the publicity surrounding them fueled public interest in the geography and natural history of newly described lands. Voyage accounts sold well and were collected into popular, multivolume anthologies, where readers could find out about everything from Iroquois torture techniques to African monkeys that lusted after human females. When naturalists like Buffon put together synthetic works, they turned to these accounts for information about plants, animals, and people in different parts of the world—always with the impossible task of culling reliable reports from imaginary and exaggerated ones.

Although these voyages invariably had economic or political goals or both, botanists, zoologists, mineralogists, and astronomers sometimes went along—more frequently later in the century—and brought back specimens, drawings, descriptions, and observations to add to the store of scientific knowledge at home.[19] Several important expeditions took place before Bougainville's circumnavigation from 1768 to 1771, but it was not until then that the inclusion of scientists became routine. Bougainville took along a botanist and an astronomer; La Pérouse, commander of the next circumnavigation attempt (1785–89), had on board two astronomers, four naturalists, a botanist, a gardener, three artists, and two engineers; and naturalists also accompanied d'Entrecasteaux's expedition (1791–94) to search for the missing La Pérouse.

Scientific Specimens

It might seem curious that explorers and naturalists returned with few live specimens and that, in fact, scientific study was not an important motivation for collecting exotic animals in the eighteenth century. When naturalists began traveling overseas, however, what they most often collected were shells, dried plants, seeds, and pickled or stuffed animals. Although they occasionally tried to bring back a living zoological curiosity, they did not embark on expeditions expecting to do so. Shipboard animals might even be detrimental to the scientific mission. A botanist on La Pérouse's expedition wrote to the head gardener at the Jardin du roi (the king's garden in Paris) that the seeds he had collected on the island of Madeira and put out on the deck to dry had been eaten—by a monkey.[20] The collections of live animals in menageries belonging to royalty, wealthy individuals, and traveling show people were not acquired for scientific purposes. Once in Europe, however, the animals became the object of naturalists' observations, and they further stimulated the curiosity about exotic nature that natural history cabinets and travel accounts had already aroused.

Naturalists were expected to describe the behavior and appearance of new species as best they could and then procure and preserve specimens. Official instructions to the naturalists who accompanied the expeditions of Bougainville, La Pérouse, and d'Entrecasteaux advised them to do so, and they took along with them manuals and materials necessary for stuffing and

FIGURE 1.3. A naturalist drawing a bird with help from the local people.
Note preserved specimens in the foreground. Frontispiece from [Pierre]
Sonnerat, *Voyage à la Nouvelle Guinée* (Paris, 1776). Reproduced
by courtesy of the Department of Special Collections, General
Library System, University of Wisconsin–Madison.

pickling.[21] The spoils could be enormous: upon returning from his trip to the East Indies and China in 1782, for example, Sonnerat deposited in the Cabinet du roi (the king's natural history cabinet) three hundred birds and fifty quadrupeds, along with fish, reptiles, insects, and plant specimens (fig. 1.3).[22] The first significant effort to collect live animals on a scientific expedition did not occur until Nicolas Baudin's voyage (1800–1804).[23]

Occasionally, eighteenth-century travelers and naturalists overseas did try to send or bring back live animals. The chevalier Chastellux, a military officer who was in North America during the American Revolution, sent an opossum to Buffon via a captain who agreed to look after it.[24] A Compagnie des Indes ship took on a baby elephant to deliver to the naturalist Réaumur, but the timing was bad: the ship was returning to France in 1755 just when the Seven Years War was beginning and the English were commandeering every French ship in the Atlantic. The elephant ended up not in Paris but in Portsmouth, where it died. Its death seemed to make little difference to Réaumur, who was mainly concerned that the animal be well stuffed and carefully shipped back to him. Thanks to friends' efforts and to the animal's small size, he was able to retrieve it and place it in his natural history cabinet.[25]

For the most part, animals weren't collected alive because it was so much easier to collect them dead. Pernetty, the chronicler of Bougainville's voyage to South America, would have liked to trap a live hummingbird, but, not knowing how, he shot at it instead. Although the bullet only grazed the bird, it dropped dead. Its skin was probably among those that Pernetty donated to the natural history cabinet of the Abbey of Saint Germain des Prés in Paris.[26] Sailors on d'Entrecasteaux's expedition caught a live kangaroo, but they attached the rope around its neck too tightly and it ended up as dinner; the meat tasted like hare, they reported.[27] Le Vaillant wanted to bring back a gazelle from South Africa, but it was too fearful to be restrained and it, too, ended up being eaten.[28] Most wild animals are fragile or ornery—indeed, the majority of animals successfully collected and transported alive had already been captured and tamed by someone else.[29] Bougainville's *tigre* did not last long when it lost its veneer of tractability.

Dead animals might make fine museum specimens, but as gifts they were decidedly inferior to living creatures. Many animals suffered the long, expensive, often fatal trip to France as tributes to the king, commissions for the royal menagerie, or personal gifts. In all of these cases it was their symbolic value as exotic, hard-to-acquire items that made the animals worth troubling with.[30]

Many foreign dignitaries presented visitors with live animals, often with the intention that they be delivered to the king. These were usually distinctive animals, large and beautiful, like the *tigre* given to Bougainville or two eagles for Louis XVI that Governor Kasloff of Kamchatka presented to La Pérouse.[31] Smaller animals often formed part of gift exchanges during trading missions. When negotiating with African rulers, French merchants gave their hosts items such as mirrors, spices, and eau-de-vie and came away with monkeys and birds, among other things.[32] When the director of the Compagnie du Sénégal was scouting West Africa for places to buy slaves, gum arabic, ivory, and gold in the late seventeenth century, one of the gifts he received was a tame eagle. He fed it raw meat and kept it in his cabin at first, but because it made such a mess—it shot its excrement out with the force of a syringe, he wrote—he let it loose on the deck "to play with the sailors." He was upset when someone carelessly knocked over a heavy barrel and killed it.[33] In Brazil, Captain de Gennes received a large tortoise, a fine gift that required little care: it lived for the rest of the trip under one of the cannons, eating and drinking nothing.[34]

Because monarchs could not rely on gifts to fill their menageries, they frequently commissioned agents to supply animals or sent collectors to bring back desired species. During Louis XIV's reign, when the new menagerie at Versailles had just been built, the governors of French colonies were instructed to send rare animals, the Compagnie des Indes had a standing order for exotic species, and Colbert sent an animal purveyor to the Levant every year. Between 1687 and 1694 the purveyor brought at least nine hundred animals to Versailles, including more than one hundred ostriches and five hundred purple swamphens.[35] Louis XV showed less interest in the menagerie and did little active searching for rarities, but he continued to receive interesting contributions as tributes from naval and colonial officers:

highlights included tigers, a rhinoceros, and an Asian elephant. Under Louis XVI the animal park was revived, especially after the end of the American war in 1782, during which disruptions of trade had hindered the importation of animals. The king and his ministers sent directives around the world requesting particular species that were lacking. Each animal in the menagerie represented obstacles overcome and could be viewed by the monarch and his subjects as a symbol both of his global reach and of the potency of his will (chap. 2 describes these acquisitions in more detail).[36]

Even the king's orders (and funds) could not prevent mishaps on the way to the menagerie. Consuls in Cairo apologized for the deaths of fourteen ostriches due to a delay in one leg of their journey and the loss of many purple swamphens: one hundred on a ship that was seized by a pirate, and another sixty to eighty on a boat that overturned in the Nile. Delicate animals had to be shipped during warm weather; a captain sailing from Alexandria to Marseille was urged to wait until spring because on a winter voyage "everything would die en route."[37]

Few captains rushed at the chance to take strange animals on board. One can understand why. The animals took up room, might be noisy or troublesome, and demanded individual care. Detailed instructions, for example, accompanied two porcupines embarked from the Cape of Good Hope: "The two porcupines eat wheat, all kinds of pasturage, fruit, and bread. These animals must not be exposed to the sun, and the darker and less exposed the place they are kept, the better they will survive. One must never give them anything to drink."[38] Big animals (except tortoises) required enormous amounts of food, which, to a captain, meant space. The two tigers brought from India in 1770 devoured approximately four hundred sheep during the voyage to France.[39] Unwilling captains caused continual problems for shipments to the menagerie. In 1711, in response to complaints by French agents in the Levant that ships' captains would not transport animals, the king ordered that legal proceedings be undertaken against those who refused to comply.[40] Uncooperative captains still roused complaints in the 1780s, when an agent at the Cape of Good Hope had to send animals by a merchant ship because no captains of the royal fleet would accept them.[41]

Once the animals got to France, there was still a long trip to Versailles. By water from Le Havre took ten to twelve days; by land from Lorient was about three weeks.[42] Letters among bureaucrats involved in naval affairs are full of instructions and worries about who would transport a particular animal, by what route, how much they would be paid, and so on. A letter from the navy minister to an official in Marseille about a recently arrived animal gives a sense for the sorts of details that had to be attended to: the minister ordered the person arranging the transport of the *caracolak* to find someone trustworthy to conduct the animal and pay him in advance (a *passeport* for the conductor was included) and also to ask the captain of the ship that brought the animal from Alexandria what sort of food and care it needed; all the minister knew was that it disliked heat and ate meat. He apparently had no idea what the animal was, and asked that a drawing of it be sent to him.[43]

On a personal rather than an official level, exotic animals—not tigers and zebras, but smaller creatures, most often birds or monkeys—were often given as gifts or were commissioned from overseas. Tame guineafowl from West Africa were often taken to France as presents, remarked La Courbe, director of the Compagnie du Sénégal, who himself returned with yet rarer offerings for two different navy ministers: two beautiful birds with black-and-white tails.[44] Other travelers returned with talkative parrots and Java sparrows from Cochinchina.[45] Jacques-Henri Bernardin de Saint-Pierre's host at the Cape of Good Hope gave him three parakeets from Madagascar as a going-away gift.[46] Letters from officials of the Compagnie des Indes mention the arrival at the port of Lorient of a small parrot and monkey for the comte de Brionne in 1711; two parakeets for a royal commissioner in 1736; and, in 1751, two cockatoos for Mme la comtesse de la Guiche and a parakeet for Monsieur Sainte-Catherine.[47] In 1750, a group order for birds came to the controller-general from some courtiers, which he passed on to the local director of the Compagnie des Indes in Lorient: "Monseigneur the controller-general, desiring to attain from Senegal a certain quantity of parrots, parakeets, and small birds to satisfy the requests many members of the court have made of him, entreats you to have made four cages to contain 200 small birds each, four other cages that could hold about 25 parakeets each, and two other cages for 12 parrots each. Please be so kind as to prepare these cages to be sent to Senegal by the first ship that can be dispatched from Lorient for this commission."[48]

Traveling friends often served as sources for special gifts. An army captain, Douin, wanting a gift for his wife, asked his friend Bossu, a navy captain, to bring a parrot back from Saint Domingue. Bossu readily complied, writing to Douin that he had passed the parrot on to a coachman whom he paid 12 livres to deliver it "dead or alive." He apologized that the parrot spoke Spanish and that he had only had time to teach it to say *"bonjour"* to Mme Douin.[49]

So many arriving animals were spoken for that they might be difficult to acquire unless one had connections. When Académie des sciences member Le Monnier asked his friend Cossigny to buy him a bird in Lorient to give to his benefactress "la Princesse," he received disappointing news: "I still haven't found any kind of bird to send you," wrote Cossigny. "I contacted several people, who all assured me that there were no exotic *[étrangers]* birds here. When they arrive, they are snatched up immediately, or they belong to some ship's officer who intends them for his relatives or friends."[50]

The bestowal of many such gifts by the chevalier de Boufflers, governor of Senegal from 1785 to 1787, shows the important role living animals could play as means for cementing personal and professional ties.[51] In the spring of 1787, the chevalier sent a vital reminder of himself and his surroundings to his lover, the comtesse de Sabran. Sabran wrote back to him: "An officer arrived this afternoon well roasted, well burned, straight from Senegal, to bring three birds from you . . . [they] arrived in the best condition, and I thank you with all my heart; they are very beautiful to my eyes, and to those of the connoisseurs, and even if they weren't, they would still have the first of their merits, that of coming from you." She added them to the menagerie of other birds he had sent her. A week later, however, the countess was no longer so pleased. After a dinner she gave in honor of a seven-year-old prince, the son of the king of Cochinchina, the prince's cousin—"the ugliest creature existing in the four corners of the world"—jumped from the table and opened the birdcage, which had been put by the window for air. Two birds escaped immediately; the third stayed behind, exhibiting a constancy, the countess remarked in her letter to the chevalier, of which "you would surely not be capable." She told him to send no more: seven had already died in one day, they evoked melancholy memories of him, and she disliked the bad grace with which he offered them. She was exhausted after chasing the escaped birds all day. "I hoped . . . that I would be able to get them back; but it's madness. When one has once broken one's chains, one doesn't put them back on, and liberty alone balances all the evils of life.

Happy he who can enjoy it without fear of losing it, or can regain it when he has lost it!"[52]

When he returned from Africa, the chevalier embarked with a menagerie of presents for everyone except Sabran. But the winds dropped, food ran out, and "everything I brought, and it's not much, is dying around me":

> I lost a green parrot with a red head, destined for Elzéar [the countess's son], two little monkeys that I had reserved for M. de Poix [the duc de Noailles], the spoonbill for the bishop of Laon, five or six parakeets; finally, yesterday evening, I saw expire a poor little yellow parrot, the first of its species that had ever been seen in Africa, and which I counted on linking with someone who is unique in the human species, who is to the human species what the human species is to parrots. I had the stupidity to regret it terribly, because it was the best child of Africa. I am left with a parakeet for the queen, a horse for the maréchal de Castries [navy minister], a little captive for M. de Beauvau [his uncle], a purple swamphen for the duc de Laon, an ostrich for M. de Nivernois, and a husband for you.[53]

Imagine the ship crowded with parrots, parakeets, monkeys, a spoonbill, an ostrich, and a horse—only someone with power and the right connections could obtain such creatures and find room and shipboard care for them. The chevalier, who was angling for a job better than that of governor of Senegal, no doubt knew that his intended recipients would come to the same conclusion. The ability to get and keep alive exotic animals indicated thoughtfulness, effort, and connections and so seems to have conveyed particularly strong messages of devoted friendship or sincere fealty. An exotic animal's symbolism could be nudged in different directions, though: a piqued recipient like Sabran, for example, might focus on the oppression of chains and cages rather than the gentle ties of a bird-mediated bond. Even bird and man could switch places. When Boufflers returned to France, rather than offering a bird in place of himself, he offered himself in the image of a bedraggled bird seeking refuge, and he begged Sabran not to be so cruel as to roast the poor creature or to sell it to a bird merchant.[54]

While travelers like the chevalier collected animals to present to specific people, sailors gathered them, sometimes by the dozen, with an eye to profit. The importation of quantities of small animals had begun with the earliest voyages to the New World. Only two years after Columbus's first voyage, Antonio de Torres took sixty American parrots back with him to Cadiz.[55] The French quickly followed suit. "What an intimacy," Lévi-Strauss wrote about long-standing French-Brazilian relations, "can we read into the fact that in 1531 the frigate *La Pélérine* brought back to France, along with three thousand leopard skins and three hundred monkeys, 'six hundred parrots that already know a few words of French'!"[56]

Even if these numbers are exaggerated, which seems likely, they signify a different style of animal acquisition than picking up a few animals for companionship or as gifts. African animals, although lacking the novelty that attached to those from the New World, were also popular and became more available as European trading posts became firmly established and the slave trade boomed. Trading ventures farther afield, to the Indian Ocean and the East Indies, sometimes yielded bonanzas such as the 260 monkeys from Madagascar that arrived in Paris in 1670.[57] The demand for exotic pets spurred this piggyback trade, which must have resulted in the transfer of thousands of animals and the deaths of many thousands more. As the eighteenth century progressed, naturalists and philosophes interested in agricultural reform began to call for more attention to importing edible and draft animals, but pets and ornamental species remained the most popular animal commodities.

Evidence of this trade comes indirectly from the popularity of exotic pets in Paris (see chaps. 4 and 5) and from descriptions of the goods available in port towns. By the late sixteenth century, exotic species were for sale in Dieppe and Marseille. King Henri III bought quantities of parrots and monkeys in Dieppe when he visited the Norman town in 1576.[58] A traveler to Marseille in 1597 watched foreign animals being unloaded onto the crowded docks and reported seeing a pet ostrich at a country house; a chained leopard (which had supposedly killed seven people a few weeks before) in a town square; a lion cub at an inn; a porcupine in a private home; and, at the duc de Guise's mansion, a large monkey named Bertram.[59] At the end of the eighteenth century, the baron de Wimpffen described the port of Le Havre as brimming with birds: "The city of Havre consists of hardly more than one road, but it is so animated, so noisy, that you don't have to see the ocean to know you are in a port. Legions of parrots from all over the world, in all sizes and colors, hanging in doors, in shops, in windows on every floor, talk, whistle, sing, scream, babble like . . . magpies. . . . *Those nasty birds!* said my hostess, whom I had been listening to for an hour; *I wish they were all at the bottom of the ocean!*—Ah, madame, I thought, as long as you were with them in the cage."[60] From natural history texts and travel memoirs, it is possible to put together a rough picture of where these animals came from, how they got from their native lands to French ports, and why people took the trouble to import them.

Pacific islanders seem to have been reliable suppliers of birds to visiting Europeans. They had been trading with foreigners for a long time and must have been accustomed to seeing the visitors' eyes light up at the sight of a pretty bird and bargaining for useful metal and cloth items in exchange. The naturalist Sonnerat, in his account of a voyage to New Guinea, remarked that while the chief on an island they visited was talking with the captain, the inhabitants were trading with the sailors—beautiful red parrots and other goods in exchange for "a few nails or some inferior knives."[61] Accounts of Bougainville's, La Pérouse's, and d'Entrecasteaux's voyages all tell of trading for parrots in the Pacific, although the report of four hundred parrots on one of Bougainville's ships is the only one, of the accounts I have seen, to mention a number.[62] Compagnie des Indes and private merchant ships probably also filled their cages with birds in the Pacific and China. Although private trade frequently drew complaints or even outright ban, poorly paid sailors expected to stash away exotic booty to be redeemed at the end of the journey.[63] One parrot would have earned a sailor the equivalent of three to six months' wages, if he got the average Paris price for it.[64]

French traders did little business in South America, but when they did they apparently loaded up with Amazonian parrots and intriguing New World monkeys. A casual remark in a 1779 memoir suggests that there were few restrictions on their import, since they were considered ornaments rather than trade goods. In this memoir, the abbé Beliardy proposed a method for spiriting vicuñas out of the Andes in order to try to domesticate them in France. Beliardy suggested that, because the Spanish king would never give permission for the export of useful animals, a few vicuñas be brought down the Rio La Plata and picked up by French ships as "curiosities," "the way we now bring back from those countries monkeys and parrots."[65]

Although trade with North America was more active than with South America, the fauna were less exciting. One traveler to Louisiana noticed pretty green-and-blue parrots there, but he suggested that they had not been imported because they had difficulty learning to talk. Even when they did learn, they tended to be taciturn, "similar in this way to the Natives [Naturels], who speak little. It is doubtless because a silent parrot would not be popular with our women that one does not see them in France." By contrast, he praised the cardinal's song, but complained that it was so loud that it would deafen you if kept in a house.[66] Others did not seem to mind the cardinal's resonant voice. The Jesuit father Charlevoix thought they would make a fortune in France, especially if they could be bred like canaries, because of their gentle song, striking plumage, and little crest "that bears a resemblance to those crowns that painters give to Indian and American kings."[67] Another traveler, writing toward the end of the century, reported that cardinals "are easy to maintain, and sing in their cage; as a result, all the ships take away a large number of them." Other popular species he mentioned were the violet pape, which was rarer and thus much more "recherché" than the cardinal, and the remarkable flying squirrel, which he rated especially suitable for presents.[68]

Most small birds and animals were imported into France from Africa and the Caribbean, reflecting the predominance of these trade routes. While they were in Africa buying and loading slaves (a process that took several weeks), sailors would buy or trade for parrots, parakeets, finches, and monkeys.[69] The French also traded tools and cloth for parrots in Guadeloupe and bought live birds from native people in Guyana.[70] Paralleling the population dynamics of native populations, however, Antillean parrots quickly declined in numbers (they were reportedly gone from Martinique by the 1750s),[71] while African birds kept being imported to the New World and from there back to France. Writing about the rose-ringed parakeet, Buffon noted that "the ships that leave from Senegal or Guinea, where this parakeet is commonly found, bring quantities of them along with the Negroes to our islands in America [the Antilles]."[72] Lovebirds reportedly arrived in France via slave ships, on which they died in great numbers.[73] A resident of Saint Domingue in the 1780s admired a physician's collection of two hundred Senegalese birds and noted that since 1783 it had become popular to keep aviaries with birds from Senegal, Guyana, Mississippi, and the Spanish part of the island.[74]

An account of two slaving expeditions in the 1690s written by Jean Barbot, a commercial agent from La Rochelle, portrays the routine of buying animals along the west coast of Africa. Different areas apparently had reputations among traders as good places to buy particular kinds of monkeys, parrots, and parakeets. Barbot mentions "grey and green parrots"—along with slaves, millet, poultry, and cattle—as readily available from the Bissago Islands, off modern-day Guinea; fine monkeys and parakeets along the Gold Coast (Ghana); many grey parrots and monkeys to be had from Príncipe Island; and blue parrots and monkeys at Calabar (between Nigeria and Cameroon). Although on one occasion he received some parrots as a gift, he usually paid for them in cash or in kind. The natives of Calabar "give three or four monkeys for an old hat or coat, taking much pride to dress themselves in our sailors' old rags."[75] According to a Dutchman, William Bosman, who traveled in Guinea at about the same time, the old clothes were beginning to wear thin, however. Concerning the region around the Gabon River, he wrote,

> Formerly a great Trade was driven here in old Perukes [wigs] by our Sailors. For these they got whatever they pleased of these People, as Wax, Hony, Parrots, Monkeys and all sorts of Refreshments.
>
> But for these four Years so many Merchants of these sorts of Goods, have been here, that the Sailor swears the Trade is utterly spoiled; and tho' this prime Stock costs him nothing, yet it doth not at present turn to account.[76]

Even though Barbot gave away some of his animals as gifts, one of his primary concerns seems to have been financial gain. He noted, for example, that a beautiful monkey he had acquired in Boutroe would fetch 20 louis d'or (480 livres) in Paris. His anticipation of profit probably made it easier to laugh when the monkey took from his cabin a case containing a silver-handled knife, fork, and spoon and "threw each of them, one after another, into the sea, which was then very calm, skipping and dancing about very merrily, as each of them went overboard."[77]

Barbot had a difficult time keeping his purchases alive, but it was probably still worth acquiring them because they were so cheap. On one voyage all fifty of his parakeets died; elsewhere he remarked that only one in twenty green-and-red parakeets generally survived the trip. The few that he did manage to bring back alive, he presented to the French royalty. Of more than fifty birds and one hundred monkeys acquired at Príncipe Island, again, only a few survived all the way to France. The structure of the slave-trading route made the import of African animals difficult because they had to cross the ocean twice, first to the Antilles with the slaves, and then back again to France. Barbot noted that he would have liked to take some gentle monkeys with him, but "it is a very difficult matter to preserve them alive in so long a passage, as it is from Guinea to Europe, especially considering that our carrying slaves over from thence to America lengthens it considerably." Barbot's propensity for picking up a variety of species also caused difficulties. His servant having forgotten to feed his civet one day, it broke into his cabin, where it tore off the head of his "curious talking parrot of the Amazons river." Other hazards that plagued Barbot included the ships' guns, which, when fired, made some of his parakeets drop dead, and the pervasive rats, which killed some of his grey-and-blue parrots during the night.[78] Seamen attempting to transport African finches later in the eighteenth century also claimed that they lost several with each report of the ship's cannon. The ornithologist who reported this phenomenon guessed that it was caused not by the disturbance of the air but because the birds, in their fright, slammed their heads into the sides of their cages.[79]

By the latter half of the eighteenth century, in concert with the growth of interest in utility and practical agriculture, some commentators began to suggest that animals be imported more systematically. This viewpoint was strongly argued in an anonymous, unpublished memoir (probably written by Buffon's collaborator Daubenton) that set out suggestions for exporting birds from the neglected and unlucky colony of Guyana, as well as in the volumes of the *Encyclopédie méthodique* on ornithology, written by the physician and ornithologist Pierre-Jean-Claude Mauduyt de la Varenne, who owned the largest collection of preserved birds in Paris.[80]

The memoir, written some time in the 1760s or 1770s, meshes with a rash of contemporaneous attempts to build up the colony of Guyana.[81] It begins with a summary of the Atlantic bird trade and a pitch for the avifauna of Cayenne, the principal settlement:

> There is no country more abundant in birds than Cayenne, where the species are more varied and their plumage more brilliant. So far only skins have been sent, in large numbers it is true, but living birds have very rarely been sent. However the thing is no more difficult than to stop by Brazil, where they take live birds daily to Lisbon, [or to] travel from Carolina and Virginia to London where they frequently receive living birds from English colonies [or] in going back from Senegal to France where live birds from that part of Africa are starting to become common due to the large number that are brought. The transport of these animals is at the same time an object of curiosity and amusement for those who receive them, [and] an object of profit for those who take charge of them during the crossing. . . . It could also become an object of utility in the case of several species that could become acclimatized and become an additional food source.[82]

Beginning with those that would enchant Europeans and enrich their importers, the author listed a number of beautiful or interesting birds, some of which were popular cage birds in Cayenne. He recommended in particular the troupial, a kind of oriole, because of its tameness and mimicking abilities, and suggested that the parrot species already common in Europe be bypassed in favor of more unusual varieties. Another list included species that might be useful, presumably to eat, although the first on the list, the trumpeter, was known especially for its abilities as a house guard.

Detailed recommendations followed concerning proper shipping procedures. Through such methods, the author declared, he once saw seventy-two small birds successfully transported from Senegal in a cage two feet by one foot by one foot. Be sure to send pairs, he advised; young, tame birds do better than grown, wild ones; and teach them to eat in the cage before sending them off. Cages should be constructed to minimize injury when birds become scared: they should be round, with wide slats, few perches inside, with cloth or leather covering, and a double floor that can be removed for cleaning. Two containers should be attached to the inside for food and for water that can be filled through a pipe, without opening the cage. One "window" in the front should be covered at night or at times when the birds might become frightened. If you know what the birds eat, take enough of that grain for the trip; otherwise take millet for small birds or wheat for large birds. Feed fruit- or insect-eating species bread moistened with honey and water; if possible, include chopped fruit or meat.

An optimistic supplement to the memoir proposed steps to be taken to tame the intransigent hummingbird, which "transported live would surely be a great pleasure."[83] First it must be determined if they live only on flower nectar or if they also eat insects; then they should be conveyed with care in cages large enough to allow them to fly around a bit in their "prison." One should be sure that they arrive in France during warm weather, and during the final leg of the trip the cage should be suspended from the ceiling of the carriage to reduce the effect of jolting on rough roads.

Mauduyt's treatment of ornithology in the *Encyclopédie méthodique* similarly urged the importation of birds for both pleasure and use and gave recommendations for transporting them. Mauduyt, too, stressed the commercial gain that could come from such endeavors: he claimed that toucans, for instance, would be very easy to import, and because of their beauty and uniqueness, the person who did so would be well compensated for his pains. Above all, Mauduyt urged importation and domestication of the species that had come first on the list of useful birds in the anonymous memoir: the trumpeter. This large bird, native to Guyana, had been highly praised by residents of the colony and in Buffon's *Histoire naturelle*. Because they were so tame, loving, and faithful, trumpeters could even replace dogs as companions and guardians, he suggested, and they would be superior because one would not fear catching rabies from them.[84]

This wishful thinking about naturalizing trumpeters was repeated for scores of animals by naturalists and others interested in the utilitarian aspect of exotic species. Llamas, pacas (tasty South American rodents), and zebras were among those that, it was thought, could be transported to France, bred, and used for food or as sources of labor. With the exception of flocks of Spanish sheep, however, most of these proposals came to naught, and animal importations continued to consist largely of ornamental species.[85]

As gifts, commercial or utilitarian objects, or scientific specimens, animals earned their value only after they had arrived at their destination. But during the voyage they could supply a different sort of value. There are enough accounts like that of the saucy parrot on Bougainville's ship and the seed-eating monkey on La Pérouse's to suggest that many shipboard animals, especially parrots and monkeys, earned their keep by amusing the crew. Work on an eighteenth-century sailing vessel could be backbreaking and frenzied, but during their free time (most crews followed a schedule of four hours on, four hours off) and during calm weather, sailors told stories, sang songs, and probably played with pets and taught them tricks. More than distraction, animals must also have satisfied a need for affection for men separated from their families, crammed into tiny, smelly, dirty quarters for months or years at a time, fearing injury, sickness, or death. Mortality rates of sailors on slaving routes, mostly from disease, were around 15 percent, about the same as that of slaves (whose circumstances were of course incomparably worse).[86] Manuals of advice for captains stressed how important it was to keep the sailors from becoming bored or sad and recommended taking along musical instruments and encouraging games and dancing.[87] Captains may thus have tolerated or even invited animal companions, despite the trouble they sometimes caused.

Birds could bring pleasure on board: La Courbe enjoyed his eagle until its accidental death, and he wrote affectionately of tame guineafowl that drank out of his milk bowl.[88] The parrot on Bougainville's trip to the Malouines Islands apparently entertained the crew with its prattle. Remarks about shipboard birds are rare in travel accounts, however, perhaps because birds were such common companions. Offhand references to monkeys suggest that they, too, were frequent passengers that only entered the written record when something unusual happened. The monkey that ate the seeds collected by La Pérouse's botanist never entered the account again. La Courbe said nothing about acquiring monkeys, but he revealed their presence on the ship in remarking that when tropic birds flew overhead "the monkeys we had on board saw them first and started to scream."[89] Officers on d'Entrecasteaux's ship mentioned taking a monkey ashore in Tasmania to see how the native people would react to it.[90]

Monkeys do, however, scamper into view more often than birds, perhaps because of their mischievous behavior and their special status as a link between humans and other animals. Monkeys and apes that earned mention for their engaging antics include some chimpanzees that ate at the table and bit the cabin boys, and monkeys from Malacca that did acrobatics in the rigging.[91] The baron de Wimpffen wrote that when the ship he was sailing on lost its masts and was foundering in a storm, he suddenly felt someone tightly hugging his legs: "That *someone* was a large monkey who, having managed to break . . . the cord that attached him to the poop deck, had taken refuge next to me. Thus, I served . . . as an asylum for a poor wretch when I saw none for myself."[92]

Two extended accounts of shipboard monkeys give us a taste of the strong emotions such animals sometimes aroused. Although I believe the accounts can be read quite literally to find out about monkeys' experiences on ships, their authors no doubt lingered over them—and probably ornamented them—because the stories tapped into an issue of particular interest at the time.[93] Many cultures are fascinated by monkeys, but, as we will see later, they sat at a pivotal reference point for eighteenth-century debates about what it means to be human. The first tale is told by Robert Challe, the connoisseur of parrot flesh. Somewhere off the coast of Africa, a wounded monkey was brought on board with her infant, which she caressed tenderly until it died three days later. Challe wrote that she wept, sighed, and gave the dead body a final kiss when a sailor took it away. She didn't fuss when the surgeon removed bullets from her flesh and bled her, although she kept pointing to a spot beneath her left breast. When she died, the surgeon, who thought he had located all of the bullets, found one more, just where she had been pointing. The crew member who rescued her was in fact one of the ship's hunters, who had shot her himself; had his aim been better, she would no doubt have gone directly to the stew pot, not the sick bay.[94]

Another heart-rending tale comes to us from François Froger, who traveled to Africa and South America in the 1690s. He tells the story of Mango, an old monkey who had belonged to the governor of Gambia and who "gave us [the crew] from time to time several quarter hours of pleasure." He was kept on a chain, but he broke free almost once a week. "His only concern was to look for food, and when he had taken advantage of *[déniaisé]* some sailor, it was a pleasure to watch him climb to the top of the mast and jump hand over hand in the rigging, a plate of rice or a large hunk of lard between his paws." One day Mango threw overboard the ivory gears of a clock that the captain was having made. The captain did not laugh. He sentenced Mango to death, and the monkey was taken ashore to be executed. According to Froger, Mango, who knew what was happening, broke his cord just as the executioner's shots fired and escaped, injured but alive. The ship must have stayed at anchor for several days, for Froger concludes: "Every day we saw the animal, wounded as he was, running along the shore looking for a way to come back on board; and if he had regret for having left us, we had no less for having been deprived of his dear face."[95]

Stubborn Animals

People trying to transport live animals constantly had to face the deaths of their charges, which usually occurred in much less dramatic circumstances than in the case of Mango. If the animal slipped away, though, the story (if not the skin) remained, and accounts of losses often make emotional high points in travel narratives. Thus La Condamine writes of a little monkey he was given while in South America on an expedition to measure the earth's diameter. The only individual he had seen of this species, its fur was silver, "the color of the most beautiful blond hair," the tail shone glossy brown, and its ears, cheeks, and muzzle were a vermillion so bright it looked fake. He kept it alive for a year, taking great pains to shelter it from the cold, only to have it die almost within sight of the French coast. He preserved the monkey in alcohol, hoping to be able to demonstrate that his description was not fanciful.[96]

The dullness brought on by death often irked travelers. The comte de Maudave could only describe in words his remarkable catch, "the most unusual [bird] of Brazil," whose color changed first from grey to white, then from white to "the most beautiful color of fire." The young bird he sent back to France died from "the rigors of the crossing" when still half white.[97] La Pérouse also mourned the loss of a marvel. During a stop at an island in the South Pacific, the crew accepted as a gift a charming tame turtledove, white with a most beautiful violet head, green wings, and a red-and-white dotted front. "It was unlikely that it could arrive alive in Europe; indeed, its death allowed us to keep only its skin, which soon lost all of its brilliance."[98]

The cases of the ones that almost made it were the hardest to bear. Bertrand Bajon, a surgeon in Guyana, tried to take home a tame crested guan, a gift from the governor, who had gotten it from the native people. Bajon carefully kept it alive during a "long and unpleasant crossing," only to experience "the displeasure of losing it" just at the entrance to the port of Marseille.[99] Jean de Léry described with regret how, during the weeks-long famine on his becalmed ship, he hid his talking parrot for as long as he could, but finally, not able to find even two or three nuts to feed it, and afraid that someone would steal it, he killed and ate it—guts, feet, beak, and all— only a few days before the wind returned.[100]

Frustration, of course, just enhanced desire; and hummingbirds aroused by far the most expressions of thwarted desire. Everyone who saw one wanted to cage it and take it home. Skins would not do: "only dead ones have been brought to Europe," wrote one voyager; "the brilliance of their magnificent plumage is considerably dulled."[101] The first dilemma was how to nab one. The native people on Guadeloupe were said to know how to catch hummingbirds alive, but Europeans either gave up or got them by luck.[102] One person trapped one when it happened to be blown through an open window by the wind; another saw one enter the funnel of a large flower and quickly pinched the flower closed, cut the stalk, and put the flower-enclosed bird in a cage.[103] Bernardin de Saint-Pierre saw a beautiful captive hummingbird at the Cape of Good Hope, iridescent green, ruby, and brown, but ants ate it during the night after it had bathed in a bowl of sugar water.[104] None ever lived more than a few days, and death made them "ugly in comparison to how they appeared when alive."[105]

Some writers, like the author of the memorandum about importing birds from Cayenne, believed that with the right techniques, hummingbirds would eventually be successfully transported to Europe. A correspondent to the Jardin du roi, who lamented that the effort to establish a colony in Cayenne left no time for pursuing scientific activities, maintained that, with enough patience and skill, hummingbirds could be tamed.[106] The authors of a popular natural history book thought that hummingbird eggs could be imported, and they chided travelers for being more interested in military conquests than in conquests of nature's beauty.[107] Others were more skeptical. A natural history dictionary claimed that the birds were easily tamed if taken young, but because they were accustomed (fixé) to their native climate, they could never be enjoyed in France.[108] Buffon doubted that they could ever be domesticated.[109] Their elusiveness was sometimes attributed to a resistance to imprisonment. One author claimed that his friend's bird died "from sadness, no doubt, at having lost its liberty."[110] Another, writing from Massachusetts during the American Revolution, claimed that "it is impossible for them to live without the enjoyment of the most unrestrained liberty."[111]

Stubborn animals that refused to thrive in captivity provided metaphorical food for musings about patriotism and slavery. Buffon related stories of a captive chimp that died from melancholy, and of the solitaire, a bird that, although tamed easily, cried and refused to eat when captured.[112] One species that seemed to be utterly unattainable achieved almost mythical status during the century, after Labat wrote about it in his widely read description of Africa. Natural historians and collectors of voyagers' narratives repeatedly reprinted this story.[113] According to Labat, the famous white monkeys of the kingdom of Galam were the most beautiful, pleasant, and amusing animals in the world, but foiled all attempts at removal. Although some had been conveyed as far as Fort Saint-Louis, they refused to eat and would die in a few days. It was not simply a dislike of captivity, for when they were in their own country they survived perfectly well on a chain. Could it be, Labat wondered, that their strong love for their country caused them to languish when they were exiled from it?[114]

Some animals continued to evade importation attempts; others, like rhinoceroses and elephants, arrived in only very small numbers. But because of active colonial trade and because so many small animals were taken on board by travelers who valued them as gifts, tributes, profitable goods, curiosities, and companions, a great many exotic animals arrived in France during the eighteenth century, despite high mortality rates. Once there, they spread out into a variety of habitats, from the king's menagerie to street fairs and private homes, where they served as sources of profit for entrepreneurs, knowledge for naturalists, pleasure for spectators and owners, symbolic display for the monarch, and vehicles for thinking about colonialism, luxury, and other cultural concerns.

CHAPTER 2

The Royal Menagerie

MANY OF THE spectacular or rare animals that arrived in France became inhabitants of the royal menagerie. Until the late seventeenth century, the modest royal collection moved around from palace to palace according to the preference of the monarch. Exotic birds and small animals provided diverting ornaments for the court; lions and other large animals were kept primarily to be brought out for staged fights. The collection grew and attained more permanent lodgings in the 1660s, when Louis XIV constructed two new menageries: one at Vincennes, next to a palace on the eastern edge of Paris, and a more elaborate one, which became a model for menageries throughout Europe, at Versailles, the site of a royal hunting lodge two hours (by carriage) west of Paris. At Vincennes, lions, tigers, and leopards were kept in cages around an amphitheater where the king could entertain courtiers and visiting dignitaries with bloody battles. In 1682, for instance, the ambassador of Persia enjoyed the spectacle of a fight to the death between a royal tiger and an elephant. In between fights, the animals were carefully looked after: when one of the lions lost his appetite, his keepers perked him up by feeding him only baby animals and sheep that had been skinned alive. (He vomited blood and died shortly thereafter—the anatomists who dissected him concluded that this diet had been too rich.)[1]

The menagerie at Versailles was to be something very different from the one at Vincennes. Louis XIV and his architect Louis Le Vau included a compound for the display of peaceful animals as an integral part of the grandiose new palace and grounds. The menagerie centered on a small chateau with an octagonal observation room, from which visitors looked out at the animals in walled, wedge-shaped enclosures encircling the building. From the end of the seventeenth century through the end of the eighteenth, this was the destination for the exotic creatures presented to or acquired by French kings.

In writing about the gardens and animal collections of early modern rulers, historians have shifted from a documentary approach to analyses that stress the symbolic importance of these public displays. They have shown how rulers metaphorically demonstrated their control of domestic and international affairs through demonstrating their control of native and exotic plants and animals. A recent review article, for example, describes menageries as places where "animals were caged for human amusement and as symbols of status and power."[2] The Versailles menagerie is a good place to see this sort of symbolism.

It also makes a good place for exploring the limits to royal power. As any parent, teacher, or police officer knows, attempts to brandish authority are not always successful: resources may be lacking or the group to be controlled may resist or misinterpret the orders and symbols of authority. Authority itself can be exerted in various ways. The history of the menagerie becomes more complicated and more connected with other political and cultural changes when we look more closely at the varied intentions of its masters and at two areas over which the monarchs had less than complete control: the networks of transportation and personnel necessary to transfer the animals, and the responses of spectators and commentators.[3]

Kingly intentions shifted considerably from the 1660s to the 1780s, paralleling changes in monarchical styles and concerns that could be characterized very broadly as theatrical under Louis XIV, recreational/passive under Louis XV, and utilitarian/reformist under Louis XVI. As we will see, these changes affected the amount of time and money spent on the menagerie buildings and on the types of animals sought. Early in his reign, Louis XIV lavished attention on his new menagerie and provided a powerful show of absolutist power. He frequently conducted visitors there, and he invited members of his newly established Académie des sciences to use it for scientific study. Louis XV gladly accepted a host of spectacular animals that arrived during his reign, but otherwise paid little attention to the menagerie. By the 1780s, the monarchy was trying to shore up its crumbling reputation, and Louis XVI responded to criticism of the menagerie as an ostentatious prison by trying—unsuccessfully—to represent it as a site for both spectacle and utility in the areas of natural history and rural economy.

In order to show off African lions or Brazilian parrots, a monarch had to arrange for them to be acquired, transported, fed, and housed. These processes involved large numbers of people whose motivations varied widely, and who could choose not to cooperate. Although this behind-the-scenes activity has not been much discussed in historical works on menageries, which have emphasized the finished display, the complexity of these steps in fact constituted part of the symbolic value of exotic animals. Procuring such animals required at the least forays to other parts of the world.

During the century and a quarter of the Versailles menagerie's existence, the provenance of its inhabitants changed along with shifting trade routes and the vagaries of foreign relations. North African species predominated in the late seventeenth and early eighteenth centuries, when Mediterranean trade flourished; animals from India and southern Africa arrived from the 1760s through the 1780s because of both the growing East Indian trade and the political alignments that characterized the period after the end of the Seven Years War. Importations stopped during wars. The connection between diplomatic relations and the makeup of the menagerie emphasizes the fact that importing animals was as much a political as a logistical undertaking.

If kings could not completely control the acquisition of animals for the menagerie, neither could they control the responses of the spectators. Unfortunately, there is no way to recover the experiences or thoughts of most visitors. All we have are sporadic accounts by literate and usually elite visitors for the early part of the century, supplemented for the latter part with commentaries in periodicals and natural history books. Among elite visitors, however, a decided change in attitude occurred: at its inception, the menagerie received much praise as a spectacular display, but beginning in the 1750s it began to come in for attack. This criticism was part of a general crisis of the French monarchy—a "desacralization" that involved critiques and caricatures of previously revered symbols.[4] By the 1780s opinion was divided. Some commentators saw the menagerie as an important institution for promoting useful domestication experiments and generating knowledge about natural history; critics continued to portray it as a wasteful luxury and cruel prison. Opinions about the menagerie were constantly knotted up with opinions about politics and society.

Naturalists' attitudes toward the menagerie also transformed, although in an unexpected way. Louis XIV's menagerie served as an important resource for early studies of comparative anatomy, where the animals' insides became material for display and description. Anatomical work continued, half-heartedly, in the eighteenth century, but was supplemented by a new emphasis on animal behavior and breeding experiments. The menagerie, however, was a less important site than might be expected for naturalists, and Buffon, even while using it as a resource, questioned the menagerie's value as a place for observing behavior because the animals were artificially constrained. The menagerie's importance also dwindled as exotic animals became more common. By the end of the eighteenth century, naturalists did not have to go to Versailles; they could view animals at fairs or in shops and private homes.

The Menagerie to 1750: Splendor and Decline

Le Vau designed the menagerie for Versailles to maximize the spectacle of its exotic inhabitants (figs. 2.1a and 2.1b). The little chateau with its radiating courtyards provided diversion for visitors and insistently reminded them of the king's imperial ambitions. Indeed, as Chandra Mukerji has shown, the entire layout of the grounds demonstrated royal command of territory: "in the gardens of Versailles, no one needed to provide evidence for the state's power; no one had to persuade foreign dignitaries about the grandeur of French accomplishments. The glory of France was a truth presented in the landscape itself."[5] The menagerie also presented animals in a

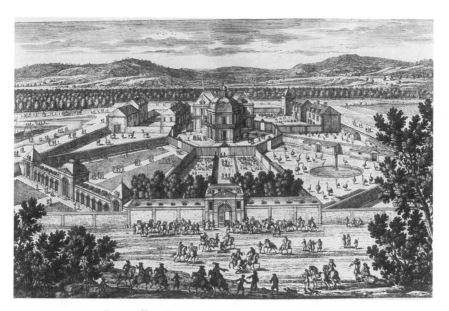

FIGURE 2.1a. "Versailles: Vue et perspective du derriere de la Ménagerie,"
engraving by Pérelle. This illustration (though not entirely accurate)
shows the central mini-chateau and radiating courtyards of the menagerie
built at Versailles during Louis XIV's reign. BnF Département
d'Estampes, va78f. Photo: Bibliothèque nationale de France, Paris.

setting of sculptures and paintings that emphasized their function as living
fables, telling moral as well as political tales.[6]

The visitor entered the menagerie along a walled pathway that obscured
sight of the animals.[7] You would enter a courtyard and then go into the
small chateau—still no animals in sight—and climb up a flight of stairs to
the second floor, where you would find three doors. The doors to the south
and north led into wings containing two sumptuously decorated suites of
rooms; the third door, to the west, took you into a gallery from which windows along one side afforded a first glimpse of the fauna. The full glory of
the king's animal collection came into view when you entered a single, large
octagonal room at the end of the gallery: from the seven walls of the octagon (the eighth led to the gallery) opened seven French doors onto bal-
conies from which you could gaze out into the seven enclosures below, full
of ostriches, crested cranes, camels, and sometimes even an elephant. In an

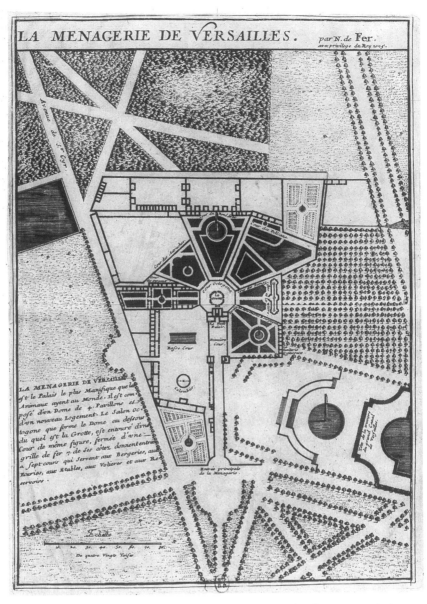

FIGURE 2.1b. "La ménagerie de Versailles," by Nicolas de Fer. Aerial plan of the Versailles menagerie. BnF Département d'Estampes, va78f, vol. 2. Photo: Bibliothèque nationale de France, Paris.

ingenious design that must have heightened the effect of the animals as decorative performers, the walls of the gallery and the salon were covered with several dozen paintings of the very animals that appeared out the windows—like instant still lifes. Here the king, and later his granddaughter the duchesse de Bourgogne, to whom he turned over the menagerie in 1698, hosted dinners and card parties.

Viewing the animals from the octagonal salon provided a prelude for the rest of the merriment in store for the menagerie visitor. Underneath this salon lay a grotto, encrusted with shell designs and tricked out with hidden pipes that could be turned on to douse people with jets of water. Outside, an octagonal courtyard around the observatory had more hidden water jets and also allowed closer access to the animals. Each enclosure communicated with the courtyard by an ornamental gate; on the portals were sculpted animal heads based on Ovid's *Metamorphoses*. The enclosures' occupants and names changed over the years, but around 1700 they consisted of (1) the Cour des Belles Poules, or Quartier des Cigognes, containing storks and sheep; (2) the Quartier des Demoiselles, for the demoiselle cranes (which later housed a vast aviary with more than forty species); (3) the Cour des Pélicans, containing large birds from Asia and Africa; (4) the Cour du Rondeau, which featured a large round basin for wading birds; (5) the Cour des Autruches, for the ostriches (later containing cages for eagles, porcupines, and other small animals); (6) the Cour des Oiseaux, containing various birds and small animals (later, cassowaries, an elephant, and camels); and (7) the Basse-cour, where animals for the king's table were raised.

The menagerie provided quarters for royal tributes such as an elephant from the king of Portugal and three crocodiles from the king of Siam, but to stock it properly took deliberate, sustained, and expensive effort. In the last two decades of the seventeenth century, under the orders of chief minister Colbert, an animal purveyor made forty-one trips to North Africa and the Near East to acquire ostriches, gazelles, goats, sheep, guineafowl, purple swamphens, and demoiselle cranes. Spectators could vicariously appreciate Louis XIV's domination of the Mediterranean as they watched the ostriches strut in a simulated desert of sand and rocks.[8] On special occasions, visitors were treated to a pretend Mediterranean voyage along the grand canal that bisected the Versailles park; they would travel to the menagerie in a miniature thirty-two-gun frigate manned by ornately costumed sailors and escorted by six skiffs full of musicians. In this grand outdoor theater, the king portrayed his commanding role on the world stage.[9]

Visitors' descriptions attest to the king's success at eliciting both awe at his ability to gather species from all corners of the world and enjoyment at the sight of nature's delights. La Fontaine described the menagerie as a "place full of all sorts of birds and quadrupeds, most of them very rare and from very distant countries," and an obscure poet wrote of it, "Abundance is everywhere; this royal mansion / Shows that in everything, nothing is its equal."[10] According to the architect and anatomist Claude Perrault, the graceful demoiselle cranes made perfect ornaments:

> No one who has seen the demoiselle cranes in the park at Versailles has failed to note that their gait, their movements, and their jumps have much in common with those of the gypsy dancers *[Bohemiennes]*, whose dance they seem to imitate. One could even say that they love to have people admire their grace and skillful jumps, and that they follow people around not for the bits of food that people throw, as most tame wild animals do, but because they want to be noticed; never failing, when they see that they are being watched, to begin dancing and singing.[11]

What more pleasing sight for courtiers fond of fables and mirror games than birds that, like them, loved to show off their elegant beauty for admirers?

The menagerie was widely accessible to a larger public, whose reactions, however, we can only guess at.[12] The trip from Paris would have been time-consuming but within almost anyone's means. In 1727, a guidebook noted that "the Coach for Versailles, as it is called, is not for persons of quality. It is usually full of fifteen or as many as twenty people seated pell-mell, and there is very little room. In addition to that, it takes a day to make the four-hour journey, since the coachmen stop at every inn for a drink." Those with more means could get there in two hours by renting a horse or sharing a four-horse carriage with three other people for 3 livres each. Near the end of Louis XIV's reign, forty to fifty carriages per day made the trip.[13] During the reigns of Louis XV and Louis XVI, a concierge who guided visitors may have described something about the animals and where they came from.[14]

The menagerie animals were of particular interest to fellows of the Académie royale des sciences, which was established in 1666, two years after construction of the menagerie. Academy fellows—fifteen at its founding—received pensions and conducted research in mathematics, physics, chemistry, anatomy, and botany. The academicians were mainly concerned with the animals after they died, when they opened them up, pulled them apart, observed and described their innards. It was an expedient division of labor, from royal spectacle in life to anatomical exhibit in death. The dissections took place in genteel surroundings—the Bibliothèque du roi (king's library) or the rooms of one of the fellows—but the operation was messy and smelly: masking the stench of a bear's entrails took three pints of eau-de-vie, and even then the onlookers had to hold alcohol-soaked handkerchiefs to their noses.[15] Their biggest task, at which the king himself made an appearance, was dissecting the elephant that had been a present from the king of Portugal; the academicians were surprised to discover that the brawny beast was a female.[16] The animal dissections resulted in some of the most impressive publications from the Académie's early years, the *Mémoires pour servir à l'histoire naturelle des animaux*, a collective effort guided by physician-architect Perrault that described close to fifty different species. The work concentrated almost entirely on details of structure and anatomy, paying little attention to behavior, although the academicians did test the chameleon's color-changing ability before dissecting it, and they described the dance of the favorites of Versailles, the demoiselle cranes.[17]

Louis XIV lost interest in the menagerie, spending time instead with his aviary at the Trianon, the small palace at the other end of the huge Versailles park, but his granddaughter the duchesse de Bourgogne had the minichateau completely remodeled when she took command of it in 1698, and she entertained there extensively. According to Masumi Iriye, during this period the menagerie lost its function as an observatory and became primarily a pastoral "pleasure pavilion."[18] Gentle pleasure within would have contrasted with the change in fauna outside; at about this time, the lions, leopards, and tigers from the menagerie at Vincennes were transferred to Versailles, where they were housed in newly built enclosures fronted with iron bars. It would not be long before some commentators started to see prisoners when they looked at these caged animals (fig. 2.2).

During the Regency (when the court moved from Versailles to Paris) and the early part of Louis XV's reign, the menagerie went out of fashion. Animal presents kept arriving, but the king seems to have taken little active

FIGURE 2.2. A panther behind bars in the royal menagerie at Versailles, 1739. Engraving by Basan, after a painting of Jean-Baptiste Oudry. BnF Département d'Estampes.

interest in it. A guidebook from 1727 praised the magnificent little chateau, especially the lifelike animal paintings within, but remarked that all sorts of rare and wild animals which *used* to be kept there had been sold or given away after Louis XIV's death.[19] Royal funding for dissections stopped, and stories circulated of the concierge raising turkeys there for his own consumption.[20]

Louis XV spent less time at the menagerie than at the Trianon, where he had greenhouses with rare plants and a small menagerie of poultry, and near which his mistress, Mme de Pompadour, built her Hermitage, with its own collection of golden pheasants.[21] The king was an avid hunter, however, and his interest in exotic animals manifested itself in imaginary hunting scenes. Between 1735 and 1739 he commissioned a series of paintings, *Chasses exotiques,* from several of the most prominent artists of the day, a few of whom may have used menagerie animals or drawings of the animals as models. These imposing, gory works (about six feet by four feet), which hung until the late 1760s in the Petite Galerie of the upper apartments of Versailles, depicted wild-eyed dogs and spear-wielding, turbaned men attacking a different fierce animal in each painting: a lion, an elephant, a tiger, a leopard, a bear, a wild bull, a crocodile, and an ostrich (fig. 2.3).[22] The muscularly aggressive exoticism of these paintings seems a fitting theme for a period when France and England were fighting each other abroad.

The Menagerie from 1750 to 1793: Revival and Criticism

Despite Louis XV's preference for hunting over collecting, the menagerie sustained itself under bureaucratic inertia. When the duc de Luynes visited in 1750, he found a respectable collection of animals, including a seal, a pelican, two tigers, two or three lions, and a camel. He went there with a group of people expressly to see a newly arrived condor, courtesy of the navy minister, Rouillé. The duke remarked that visits to the menagerie had gone completely out of fashion, but that he thought it was worth paying attention to again.[23] Its symbolic role was still being touted to the public: a 1755 guidebook commented about the menagerie animals, "One would say that Africa has paid a tribute of her productions, and that the other parts of the world gave as homage to the king their most rare and unusual animals and birds."[24] About this time the menagerie began to regain importance, probably for several reasons: increased colonial trade resulting in an influx of animals, growing interest in natural history, and the beginning of the publication of Buffon's *Histoire naturelle*.

In the second half of the eighteenth century, the contents of the menagerie changed and so did the ways people used and responded to it. It became filled with more, and more diverse, species, primarily from India in the 1760s and 1770s, and then from southern Africa in the 1780s. The two waves differed in that the first was mostly unsolicited whereas the second resulted from the king's direct commands. These changes in geographical origin and in royal interest are related to shifts in both external and internal politics. Externally, wars and shifting trade routes determined where and how animals could be acquired. Internally, a growing crisis in the monarchy urged Louis XVI (who acceded to the throne in 1775) to actively groom his public image in a way that Louis XV had not done, and the menagerie was an important part of this image. Visitors under the influence of the fashionable quest for useful knowledge began to see it as a site for natural history observation, even as they also continued to find visceral awe and

FIGURE 2.3. Charles André Vanloo (1705–65), *La chasse de l'autruche* (Ostrich hunt), 1738. One of a series of eight paintings commissioned by Louis XV and displayed in the Versailles palace. Note the hunters' costumes and the palm tree, evocative of the "exotic" Near East. Réunion des Musées Nationaux / Art Resource, N.Y. Courtesy of Musée de Picardie, Amiens, France.

laughter there. Increased visibility came at a price, however; the menagerie became a target for critics of absolutist pomp, who urged it to become more utilitarian. Even Buffon, who occasionally observed animals at Versailles, joined the circle of critics.

Animal Acquisitions

The menagerie acquired several spectacular new residents during Louis XV's reign with little effort on the part of the king or his ministers, since most of them were gifts. Despite the disruption of shipping during the Seven Years War, one animal, a zebra, made it through right in the middle of the war. The young zebra, the first to be seen in France, arrived in 1760 as a gift from the Dutch governor of the Cape of Good Hope, who supposedly had paid 30,000 livres for it.[25] During the one year it survived, it drew huge crowds to the menagerie.

The 1763 Treaty of Paris brought temporary peace to the seas and an influx of new animals. Most of these were from India (tigers in 1765 and 1770, a rhinoceros in 1770, and an elephant and a Tibetan musk deer in 1772)—a puzzling fact, since the treaty had forced France to relinquish her colonial ventures in India. The major French settlement, Pondicherry, had been razed by the English, and the commander of the French troops in India, Lally, had been publicly beheaded for his ineptitude. France still maintained a presence in the subcontinent, however, at the handful of trading posts allowed by the treaty. The substantial animal gifts offered up to the king by the governors of those posts, Jean Law de Lauriston and Jean-Baptiste Chevalier, may well have expressed in material form the elephantine ambitions of these two men. Both constantly tried to destabilize the English hold in India by making alliances with rebellious Indian princes. They sent to Paris numerous memoranda and proposals for political and military ventures and chafed at the lack of response. Chevalier in particular was known as aggressive and crafty. Perhaps he hoped that by positioning impossible-to-overlook, giant-sized animals right in front of the king's nose, he was improving the chances that the king would give a favorable hearing to his requests. Jean-Bernard Lacroix, who has documented the travels of these animals, suggests that Indian species dominated menagerie acquisitions because Indians, unlike Africans, had a menagerie tradition of their own.[26]

Although this may help explain availability, it does not take into account the fact that procuring and shipping the animals required the active involvement of people who had to have strong reasons for undertaking such projects.

These "gifts" entailed considerable expense and effort. The cost of fresh meat for the ten-month sea voyage and twenty-four-day overland journey to Versailles for the tigers sent in 1770 came to about 4,500 livres. The rhinoceros, being vegetarian, posed fewer provisioning problems, and the six-month voyage from India seems to have been uneventful; the captain's log for the day it was loaded, for example, tersely notes, "north wind, cool, good weather, in the afternoon we loaded a rhinoceros for the king." During the trip, however, the young (probably two-year-old) rhinoceros had grown big and bad *(fort et méchant),* and it was not unloaded from the ship until a made-to-order cage was ready. It then took two and a half months to prepare the animal for the trip to Versailles. Preparations included fashioning a horsehair-stuffed leather collar studded with four strong iron rings and constructing a sturdy cart. Despite a cumulative effort of seventy-two days of labor by wheelwrights, fifty-seven by blacksmiths, thirty-six by locksmiths, and two by carpenters, the cart and wheels still collapsed and had to be repaired en route. During the twenty-day trip to Versailles, the rhinoceros was tended by two butchers and a "chief conductor," who fed it and kept its skin supple by rubbing it with fish oil. Total cost for sea and land transport: 5,388 livres, 10 sous, 10 deniers.[27]

Two years later an elephant and an Indian mahout named Joumone followed in the footsteps of the rhinoceros. This time, however, the navy minister ordered that the animal be transported to Versailles from Lorient "in the most economical way possible." Joumone and the elephant walked the whole way, thus sparing the expense of a cart; the trip took somewhere between four and six weeks.[28]

Louis XVI's reign, which began in 1775, brought many political and economic reforms, including revival of the menagerie. Despite the recent influx of Indian animals, physical maintenance had been neglected. In 1774, a list of needed repairs warned of numerous impending disasters: the elephant wouldn't last through another winter unless its enclosure was repaired; the aquatic birds' pond was almost filled in with silt; the rhinoceros was about to knock over a crumbling wall in its pen. By 1785 the animal pictures in the chateau had faded so much that they could hardly be made out.[29] The state of the menagerie must have been an embarrassment, especially when visiting dignitaries asked for tours. More crucial than decent menagerie lodgings, however, were the animals to fill them; this became something of a crisis in the late 1770s and early 1780s, for the American war disturbed regular trade and prevented replenishment of the constantly dwindling stock of animals. The menagerie suffered a serious loss in 1782 when the elephant, the menagerie's "most beautiful ornament," who "continually attracted crowds," broke out of her enclosure at night and died after falling into a canal.[30]

That the king was not keeping up with other sovereigns is suggested by the observations of foreign visitors. In 1782, the baronne d'Oberkirch remarked on the dearth of rare animals, and two years later an English visitor sniffed that, despite the presence of the rhinoceros and other curious birds and animals, the menagerie was "not very well stocked."[31] The Dutch were far ahead: in the 1776 and 1782 supplements to the *Histoire naturelle,* Buffon often quoted Dutch naturalists' observations of animals from the prince of Orange's menagerie that had never been seen alive in France.[32]

The 1780s was a decade of economic crisis and reassessment. The controller-general from 1783 to 1787, Calonne, responded to the financial troubles by spending lavishly, taking out loans, financing public works, and trying to create the impression of a robust state. The fact that the menagerie received high priority rather than being left to deteriorate attests to its important symbolic role. The king himself accorded the menagerie a central role in reviving the tarnished monarchy and asked the governor of Versailles (the duc de Noailles, prince de Poix) to oversee its replenishment. The prince wrote to the navy minister, the maréchal de Castries, in January 1783,

> Please allow me, Monsieur le Marquis, to contact you on a matter of concern for the king's menagerie, where for several years many animals, both quadrupeds and birds, have not been able to be replaced because of the war. Since His Majesty has decided that the menagerie will be maintained as a manifestation of royal magnificence *[comme tenant à la magnificence Royale]* I wish it to be in good order. Because some of the animals that are lacking are found in India and because you send a convoy there every year, I am attaching a list

of what has been requested; you will do me great pleasure to give the necessary orders [to obtain them]. I am very grateful for any attention that you should give to my wishes.

A small piece of paper that probably went along with this letter contains the following list: "an elephant; 2 zebras, male and female; mandrill and baboon monkeys; 6 guineafowl."[33]

The menagerie did acquire many interesting animals in the 1780s, including a panther, a couple of hyenas, a young American *"tigre,"* two badgers, some ostriches, a mandrill, and several other monkeys.[34] But the king never got his wish for an elephant; in fact, the next elephants to be seen in France arrived from Holland in the 1790s as spoils of war. And the request for a pair of zebras—which Buffon had described as "perhaps the handsomest and most elegantly attired of all the quadrupeds" (fig. 2.4)—was only halfway fulfilled: only one zebra arrived at Versailles in July 1786.[35]

The arrival of the zebra was the end result of a surprisingly protracted and avid quest that I have been able to partially reconstruct by means of the letters that slowly sailed back and forth from the navy minister and the prince to officials in île de France and the Cape of Good Hope, a correspondence in which the juxtaposition of state affairs with the zebra affair attests to the menagerie's significance. This episode reveals the negotiations and personal connections necessary for acquiring an exotic animal and suggests why a king concerned with bolstering his reputation might do well to invest in a menagerie: each success demonstrated his command over overseas transportation networks, political allies, financial resources, and subordinates. Each failure, however, betrayed bad links in the chain of command: uncooperative captains, self-serving agents, budget limits, or stubborn animals.

The Zebra Quest

From the maréchal de Castries, the order for menagerie animals went to the next in line, the vicomte de Souillac, governor, and Chevreau, *intendant,* of île de France, headquarters for the East Indian branch of the navy. "I am sending along," wrote Castries, "the note that the prince de Poix addressed to me concerning procuring for the king's menagerie the foreign *[étrangers]*

FIGURE 2.4. A pair of zebras, which Buffon called "perhaps the handsomest and most elegantly attired of all the quadrupeds." From Buffon et al., *Histoire naturelle,* vol. 12 (1764), plate 2. Reproduced by courtesy of the Department of Special Collections, General Library System, University of Wisconsin–Madison.

quadrupeds and birds that are lacking. Please take advantage of favorable occasions to entrust several reliable people to buy these animals. Nothing is to be neglected in fulfilling the prince de Poix's wishes. Please send me the list of expenses incurred by these commissions. I urge *[recommande]* you to take the most economical measures in this regard."[36] Souillac and Chevreau in turn passed the word to the French agent closest to the African habitat of zebras, M. Percheron, at the Cape of Good Hope.

Percheron became the man in the front lines of the zebra hunt—recipient of commands and reproaches, responsible for acquisition and transport. The task required diplomatic rather than animal-finding skills, for Percheron had little direct authority: he was simply a French agent in a Dutch colony. The Dutch East India Company controlled the Cape of Good Hope, having set up a colony at Table Bay in 1652 as a provisioning spot for ships involved in the spice trade, a venture requiring almost year-long voyages (one-way) between Europe and Indonesia. Cape Town became known as "the tavern of the two seas." The vegetable garden that the colonists had originally established for feeding hungry sailors was later supplemented by a botanic garden and menagerie, where travelers in the 1770s reported seeing gazelles, elands, zebras, ostriches, and cassowaries. The animal park served as a reservoir for animals to be sent to the prince of Orange for his menagerie at Honselaarsdijk.

The 1780s, it turns out, was a propitious period for a French agent to ask for favors from a Dutch governor. In 1780, Holland joined France in alliance with the rebellious American colonists, and French and English ships raced toward the strategic southern tip of Africa. The French won the race and subsequent battle. They then fortified Cape Town with troops brought over from India and settled in for a several-year occupation, during which the town earned the nickname Little Paris. The war brought temporary prosperity to local merchants and to the governors of the Cape Colony, and friendly relations between the French and Dutch; more than two hundred French ships stopped at Table Bay between 1785 and 1789.[37]

In early 1784, Percheron received the letter from Souillac and Chevreau requesting that he supply two zebras, male and female, for the king's menagerie. He replied that it was almost impossible to get zebras because the people *(paysans)* who lived near their mountain habitat had neglected capturing them, despite being paid by the government to do so. However there was one in the governor's menagerie that he would try to obtain, and he had been able to buy a wild horse with zebra-like stripes that he would send as soon as possible.[38]

Percheron succeeded in getting the zebra, but then he had to find a way to transport it and the striped horse to France. Because captains of French navy ships refused to take the animals aboard, pleading lack of space, he arranged for them to travel on a private merchant ship. He confidently wrote to Castries that he had loaded them along with supplies of hay and wheat for the trip, that the captain had promised to look after them, and that a passenger would care for them as well.[39]

After the animals were loaded, the ship was retained in the harbor for three days by strong southeast winds. Just after it set sail, Percheron received a note sent ashore by the captain: the zebra had died on the third day. With the ship already under way and no immediate means to send a message to France, Percheron was unable even to warn Versailles to expect a dead zebra, not a live one. After the ship docked in France four months later, Castries penned an instant rebuke (which took six months to get to Percheron). The zebra had died, reported Castries—unaware that Percheron already knew that—and the striped horse was in bad shape. The horse, no matter how unusual, was obviously no substitute for a zebra. Castries repeated the original orders: procure two zebras, male and female, and send them as soon as possible.[40]

Percheron, a good bureaucrat, made sure to shift the blame away from himself: next time, he wrote to Castries, he would be careful not to trust precious animals to such a negligent captain. It was most surely the captain's fault that the zebra had died, because he (Percheron) had kept the animal for a year in a stable, and it had been in excellent shape when loaded on the ship.[41] Percheron also wrote to Souillac and Chevreau, who complained that he had not been doing his duty. He blamed the "criminal negligence" of the captains who had not transmitted his correspondence and immediately had copies of all of his letters made to prove that he had been following orders.[42] In his letter explaining the death of the zebra, Percheron stated that the loss disturbed him greatly because it might take another ten years to find a replacement for it.[43] Souillac and Chevreau passed this on to Castries with their own snide commentary: the loss of the zebra was really not so grave, since it did not fulfill the prince de Poix's orders anyway; he had requested not one, but *two* zebras, one male and one female.[44]

Anticipating potential transport problems to come, Percheron complained to Castries that he would have no way to ship all of the wonderful animals he was about to procure if the navy captains refused to accept them. He requested that Castries issue an order requiring the captains to transport animals destined for the menagerie, since these ships generally had more room than did the merchant ships, and the animals' chances of survival would be better.[45]

To procure the zebras, Percheron again went to the governor of the Cape. Luckily, he told Castries, the new governor (Cornelis Jacob van de Graaff) was very attentive to the interests of the French state. Not only had he given out orders for zebras across an area of two hundred lieux (about five hundred miles), but he let Percheron know that he would pass on other curiosities from the Cape menagerie.[46]

Success. November 1785: Percheron writes to Castries, "I have at the menagerie of the [Dutch East India] company two zebras, male and female, and a large number of rare birds destined for the king's menagerie." He won't send them right away, he explains, because they would arrive in Europe in the winter; he will wait for ships leaving in April. Oh, and by the way, there is a small matter of compensation for the governor, who paid 120 piasters each for the zebras. The prince of Orange, to whom the governor was supposed to send them, usually recognizes such consignments with a valuable gift; Percheron thinks it "indispensable" that Castries send something. Here, perhaps, is an explanation of how the zebras slipped from Dutch to French hands; Percheron may have promised the governor, a notorious lover of luxury, a generous bonus in return for his aid.[47]

Three months later, Percheron had found a willing captain to take to France a number of birds, a pair of axis deer, and one zebra, a male "of the greatest beauty." The captain apparently agreed to take good care of the animals in return for a promise that he could present them in person to Castries.[48] Perhaps the only way Percheron could find someone to take on the onerous task of watching over tender animals at sea for several months was to let them take credit for the delivery.

But why only a male zebra? What happened to the female? In his letter to Castries, Percheron said that the ship did not have room for all of the animals, and he would send the others—the female zebra, a wild horse, two large ostriches, and several other birds—later. But he must have been counting zebras before they materialized, since two weeks later he remarked to the vicomte de Souillac that he was still waiting for a female zebra that had been promised to him by the governor.[49] He did not mention the animal again.

In the meantime, the prince de Poix was getting impatient. He reminded Castries again of the menagerie's need for an elephant and several zebras, because these animals "are the most beautiful ornament of the king's menagerie." Castries replied that he had reiterated his orders that nothing be neglected in fulfilling the prince's wishes. Two months later, the prince heard of the male zebra's arrival in France, along with other animals shipped by Percheron. He gushed that, thanks to Castries, the menagerie had never been so beautiful: *now*, if only he could get an elephant and a lion, the menagerie would be complete.[50]

The zebra and his retinue arrived in Lorient in June 1786, where the superintendent of the port arranged for their transport to Versailles. He used the usual means for conveying strange fauna, the regular Paris-Lorient coach service. The escorts were to be the navy officer Lardy and a butcher, Pierre Baras. Both were generously paid, and Lardy was instructed to encourage the coachmen, by means of tips, to be careful of the animals and to choose the smoothest route. Cages containing the deer, a heron, a pelican, and an "extraordinary starling" occupied the front of the coach, to which the zebra was "solidly attached" (it must have been tethered alongside). Lardy was to make sure that the cages were scrubbed and washed twice a day, that the animals received adequate food and water, and that the zebra was lodged separately in the inns along the route, with good fresh litter and a companion: the butcher's duties included sleeping near the zebra at night. Baras and the zebra bedded down together must have made a memorable sight for fellow travelers along the Paris-Lorient road that summer. About three weeks after they left Lorient, the director of the Versailles menagerie signed a receipt declaring that Lardy had successfully deposited the zebra and the other animals, all in good condition.[51] That is the last trace I have found of the long-desired zebra, which never acquired a mate. Presumably the king came to gaze on the prize, but it must have expired some time in the next few years; it was not among the menagerie animals that survived the Revolution. In May 1789 Percheron's successor bought another zebra (its sex not indicated), but there is no evidence that it ever reached France.[52]

One of Percheron's acquisitions that did survive, however, was the striped horse. It became one of the founding specimens in the new menagerie at the Jardin des plantes in 1794, and today its skin can be seen in the Muséum national d'histoire naturelle—in the hall of extinct animals. It turns out to have been a quagga, a relative of the zebra, extinct since 1883 when the last known individual died in a zoo in Amsterdam.[53] Sturdier and less strikingly striped than the zebra, the quagga generated little excitement among those trying to revive the image of the menagerie (fig. 2.5).

Percheron left the Cape in March 1787 without having been able to obtain a mate for the male zebra. He continued to acquire other rare specimens, but the ordeal seems to have cost him much trouble and earned him little credit. His last acquisition was a leopard, for which, he explained to Castries, he had had to pay a high price because he was competing with the English for it. It traveled from Mozambique to Saint Domingue on a slave ship, from there to Lorient, and finally by coach to Versailles. It provoked, however, only a scolding from Castries, who wrote that the orders Percheron had received to procure rare animals did not absolve him of the necessity to economize: "You should not always try to outbid [couvrir] the high price that other foreigners are willing to squander [sacrifier] on these sorts of things. Moreover, you have not even indicated how much the leopard cost."[54]

At the time that Castries was drafting his rebuke, Percheron was recuperating in Lorient. He had personally accompanied on a merchant ship from the Cape two porcupines, three purple swamphens, and seventeen crates of rare plants, not trusting navy captains to deliver them safely—"but such are the thorns attached to my position," he grumbled. He had hoped to escort the treasures to Versailles, but eventually sent them on alone (minus two of the swamphens, which died during the summer) because his health was so poor that he did not want to risk the rigors of the coach trip.[55]

Animals continued to arrive from Africa and the Indies in the late 1780s: a cassowary, a gift to the menagerie from the vicomte de Souillac, in July 1788; a serval in July 1789.[56] The prince de Poix, sensitive to the embarrassing deficiency at the menagerie, continued to thank Castries and his

FIGURE 2.5. A quagga. The hardy quagga that arrived in France in 1784 was one of the few animals from the Versailles menagerie to survive the Revolution. From Buffon et al., *Histoire naturelle, supp.*, vol. 6 (1782), plate 7. Reproduced by courtesy of the Department of Special Collections, General Library System, University of Wisconsin–Madison.

successor (the comte de la Luzerne) for the latest acquisitions, all the while urging more attention to its proper embellishment. After the arrival of Percheron's porcupines, he repeated that he would not really be happy until he received a lion, which had long been promised.[57] What is a royal menagerie without the king of the beasts, after all? One was finally sent from Senegal in 1788, with a dog as its devoted friend. The two would become potent but unstable symbols during and after the Revolution. Still, the prince de Poix kept looking for animals. In June 1789 he reminded the navy minister that the commanders of troops in India should be reminded to forward any animals they came across that would be worthy to be placed in the menagerie.[58] Despite the prince's efforts, the reform of the menagerie, like other reforms, came too late.

Visions of the Menagerie

Turning from acquisition to observation takes us back in time again to mid-century. While Louis XV was blithely accumulating unsolicited animals for the menagerie, it was becoming a focus of criticism for some of his subjects. Here, for example, is the entire entry under the word *ménagerie* in the volume of the *Encyclopédie* published in 1765: "A building where large numbers of animals are maintained as a curiosity. Only sovereigns have menageries. Menageries should be destroyed when people are short of bread; it would be shameful to spend large sums to feed animals when all around them people are dying of hunger."[59] Scholars who have quoted this passage have sometimes portrayed it as a straightforward prelude to the dismantling of the menagerie in the 1790s, when very similar rhetoric ruled the day.[60] But the Revolution was still a generation away; many of its leaders had not even been born yet. Just as the criticisms of the philosophes did not lead directly or inevitably to revolution, neither did snipes at the menagerie foretell its doom.

The 1750s was a particularly turbulent decade, when Louis XV's exiling of magistrates of the Paris Parlement, his involvement in the costly Seven Years War (1756–63), and his extravagant expenditures turned pens against the king who had once been known as *bien-aimé* (well-loved). The philosophes derided all kinds of ostentation; the article on Versailles in the *Encyclopédie* opined that the millions wasted on construction of the palace could have been better spent on "several works that would be useful and necessary to the kingdom."[61] The early 1750s had also seen renewal of the smoldering rumors that the king and his henchmen hoarded flour during periods of famine—thus, perhaps, the reference to starving people.[62] Less biting but equally derisive commentators turned *menagerie* into a metaphor for the petty brutishness of elite society, as in a satirist's definition of a menagerie as "a name that designates a place where all kinds of animals are collected, and which would aptly describe many houses where one finds nothing but individuals who babble and peck."[63]

Criticism had its reform side, however, and many believed that the menagerie had the potential to be a useful institution. In his imaginary South Seas utopia (*New Atlantis* [1624]), Francis Bacon had described a park full of wild birds and beasts used for breeding and experimentation, rather than "for view or rareness."[64] The *Avant-coureur,* a journal oriented toward the interests of amateur naturalists, put forward such a vision for the Versailles menagerie in 1767:

> Why is not more being done [to try new unions among different species of animals]? Do we flatter ourselves that we know all of the immense variety and infinity of nature's operations? Nothing could be more interesting /*curieux*/ and perhaps easier than such experiments, and it is to be hoped that the menageries in which we assemble at great expense the rarest animals of all species should be more than useless lairs meant only to please the eyes concerning the form and appearance of these animals. All of Paris saw the zebra at the menagerie of Versailles; the coupling of this animal with asses, mares, and other different species would have taken only the least trouble, and we wouldn't find ourselves reduced to the simple and frivolous advantage of being able to say *we have seen a zebra.*[65]

A year later, a utopian, best-selling (and censored) novel by Louis-Sébastien Mercier, *L'an 2440,* described the menagerie of the future. Here an inhabitant of Paris in 2440 is giving the time-traveler protagonist a tour of the former Cabinet du roi, now a "palace of nature" belonging to the state:

> We have vast menageries for all sorts of animals. We have found in the depths of the wilderness several species that were absolutely unknown to

you. We combine these races to see what different results are produced. We have made extraordinary and very useful discoveries. We mix these tribes to see the effects they will produce. The discoveries we have made here are astonishing, and highly useful, and the species has [sic] become more than twice as big; we have finally realized that the efforts one devotes to nature are rarely unfruitful.[66]

Utilitarianism was the byword of the day, and menageries could do their part by being centers for creating new animal servants.

The philosophes' criticism did not stop visitors from coming to Versailles to see the exciting new acquisitions from India and Africa. As the report from the *Avant-coureur* noted, the 1760 zebra was especially popular. The Hungarian count Teleki declared it to be the most interesting of the animals at the menagerie, its stripes looking like they had been painted on.[67] The *Affiches de Province* announced its arrival and reported that zebras were so expensive because they were so rare and difficult to catch.[68] The irresistible combination of beauty, rarity, elusiveness, and high value brought coaches full of zebra gawkers. (Zebra stripes even became the rage for men's suits in the 1780s, supposedly modeled on the 1760 Versailles zebra, which after its death was stuffed and put on display in a glass case in the Cabinet du roi.)[69] The idea that such a stunning creature could also be useful may well have been behind the vain pursuit of a zebra pair. Everyone believed that zebras could be tamed if trained properly—Buffon declared that they could eventually replace horses—and rumors circulated that the Dutch had succeeded in hitching a team of zebras to a carriage.[70] Perhaps the king and the prince de Poix envisioned stylish zebras pulling coaches throughout France.

After the zebra died, the crowds did not have to wait long before more marvels arrived. The elephant and rhinoceros appear in many contemporary guidebooks and memoirs, the absurdly wrinkled rhinoceros most often being cited as the high point of the menagerie. Writing about his visit in 1771, the duc de Croÿ marveled at the toucan, "but," he continued, "what drew my attention . . . was a rhinoceros that M. Bertin had brought a year ago from the Cape of Good Hope; it is the first male rhinoceros that has been seen in Europe. About twenty years ago a full-sized one was seen in Paris, but it was a female"[71] (fig. 2.6). During another visit six years later,

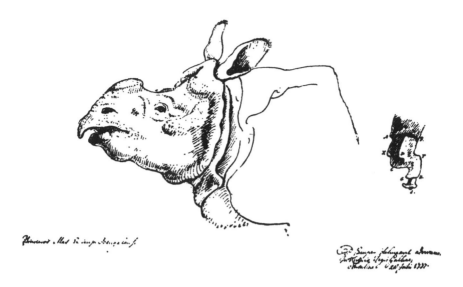

FIGURE 2.6. The rhinoceros that came to the Versailles menagerie from the Cape of Good Hope in 1770 was a crowd pleaser; it also interested naturalists. Petrus Camper made this drawing of its head and penis on 28 July 1777. University of Amsterdam library, department of manuscripts.

he judged the rhinoceros, which had grown even larger, "a menagerie item unique in Europe and of the greatest interest."[72] The public loved watching the elephant uncork bottles of wine with her trunk and accept tobacco from their snuffboxes.[73] The *Correspondance secrète,* an underground newsletter mostly filled with tales of political intrigue, devoted a long article to the elephant's demise.[74]

Guidebooks continued to portray the menagerie as a place of pure spectacle, a place for the curious to see rare animals and exotic birds. It seems to have been a regular stop not only for Parisians but for tourists from the provinces and other countries.[75] The concierge of the menagerie may have regulated the entrance and acted as a guide. No record remains of his patter, but since he is consistently described as vulgar, ignorant, and on the constant lookout for tips, it is unlikely that he gave a very erudite tour.[76]

Enlightened visitors, however, could apply what they had learned from natural history books or courses to their observations. Such a visitor, who has left us a record of his impressions of the menagerie over a stretch of many years, is the duc de Croÿ, courtier and naturalist "en amateur."[77] The duke was at once a student of natural history, a connoisseur of luxury, and a fond friend of his old favorites in the menagerie. His multiple, sometimes conflicting, responses bring out the particular tensions in the menagerie's meaning during this time. Was it entertainment or education? Extravagance or necessity?

The duke's admiration for the menagerie was damped by his concern that the king was spending too much on his many small chateaus and gardens (1:135, 148).[78] But the desire for exotic display seduced him as well: when he put his own financial affairs in order in 1773 and decided to economize, the duke vowed to cease all projects "except for a menagerie that my daughter-in-law is making, a work more pleasurable than expensive, for which we just sent from Paris . . . a carriage full of rare birds" (3:51).[79]

Observers like the duke distinguished themselves from common menagerie observers, as we see in the duke's description of the Holy Roman Emperor Joseph II's tour of the menagerie in 1777. When the emperor, who conducted much of his visit incognito, arrived at the menagerie in the morning, the guide would not admit him as a sole individual. So he waited until, shortly after, a large group arrived, then he followed them in. The guide later told the duke that the single gentleman had stood out from the crowd because of the discerning comments he made about the elephant and the rhinoceros and because he paid attention only to "very rare, large objects." The guide's admiration was no doubt improved retrospectively when the mysterious man slipped 15 louis into his palm on his way out (4:16–17).

After talking to the emperor about the menagerie, the duke paid his own visit, concluding that it was more worthy of a visit than one would think. "I examined with the greatest pleasure the elephant which, although female, is very large, being more than seven feet tall. . . . The rhinoceros looked larger [he had last seen it six years before] . . . the camels, dromedaries, lions, tigers, etcetera, make a fine collection, and I made note of everything that is lacking" (4:16). Three years later he went to the menagerie in the early morning, "which always gives me pleasure." He approved of the collection, noted that the rhinoceros, the sight of which gave him "greatest pleasure," had grown, and although he again regretted that the elephant was a female, he marveled that she did things with her trunk that one could hardly believe. The thirty-five-year-old pelican also earned his praise ("remarkable"), but he dismissed the rest of the collection as "ordinary" (4:216).

In 1782, two years before his death, the duke visited his old friends one last time. "All in all, [the menagerie] is most beautiful, and the rhinoceros, as well as the elephant, both bigger, gave me great pleasure. I learned that the elephant, though female, who was almost nine feet tall, could no longer get up after lying down, which confirms what was thought to be a fable. It rests by leaning against trees" (4:263).

The duke's musings suggest that the king had succeeded in turning the menagerie into a place for both study and contemplation, and some commentators have indeed described it as an important scientific site: a former page in Louis XVI's court remarked that the barbaric animal fights of the past had been superseded by a collection of animals useful for the progress of natural history, and a nineteenth-century history of menageries claimed that "it is to [the menagerie at Versailles] that we owe the Natural History of Buffon and his coadjutor Daubenton."[80] Naturalists, however, continued to criticize the menagerie both for its attention to display rather than utility and for its unsuitability as a place to study animal behavior. In his 1784 volume on birds in the *Encyclopédie méthodique*, for example, Mauduyt urged travelers to bring back to France species that could be bred for food, for their hides, or as beasts of burden; they should then be acclimatized in suitable habitats rather than being shut up in menageries, which are too confining.[81] Buffon and Daubenton did observe about a dozen species at Versailles, but, as Jacques Roger has noted, Buffon preferred to stay at Montbard, where he had his own small menagerie.[82] Moreover, they criticized the menagerie as an insalubrious spot where the animals were kept in conditions unsuited for observation. The caracal and the serval lunged so fiercely at the bars of their somber cages that the naturalists could not even get close enough to accurately describe the exact shade of their fur.[83]

Deaths supplied corpses to Daubenton's measuring implements and dissecting tools, but these only confirmed the pernicious effects of menagerie life. Buffon contended that the tiger skeleton they examined was not as large as a full-grown one from the wild would have been because the animal had been kept since cubhood "enclosed in a narrow cage *[loge]* at the menagerie, where the lack of movement and space, the boredom of prison, the constraint of the body, [and] unsuitable food shortened its life and retarded its development, or even reduced the growth of its body." The caracal, which died six months after baring its teeth at the naturalists and so appeared in an

addendum to a later volume, displayed even greater deformities. It was ema-
ciated (only twelve pounds), partially bald, and had a large bone spur on its
femur and extensive decay in the jaw. Daubenton concluded that "bone
spurs and decay often affect animals enclosed in menageries; lack of fresh
air and exercise cause other diseases as well, from which many of them die."
Buffon, in the article on the tiger, had similarly generalized that animals kept
in menageries were reduced in size and often had undeveloped genitals.[84]

Buffon expressed his reservations about the usefulness of menageries for
observing animal behavior more strongly in an article near the beginning of
the second volume of the *Histoire naturelle des oiseaux* (1771). This eloquent
passage is in the history of the bustard *(outarde)*, a large goose-like bird,
which begins with a discussion of method. Discovering the history of an
animal, Buffon writes, is a project of several generations. Most animals are
afraid of man: they resist his approach with aggression or by escaping into
burrows, deserts, oceans, or the sky.

> How could we, then, in a short time, see all of the animals in all of the sit-
> uations necessary in order to thoroughly understand their natures, their hab-
> its, their instincts; in short, the principal facts of their history? In vain does
> one collect at great expense extensive series of these animals, conserving
> their skins with care and mounting them artistically on their skeletons, giv-
> ing each one the correct bearing and a natural appearance: all of that repre-
> sents nothing more than a still life, inanimate and superficial. And if some
> sovereign came up with the grand idea to advance this enchanting science by
> creating vast menageries, gathering together for observers a large number
> of living species, one would still gain only imperfect knowledge of nature;
> most animals, intimidated by man's presence, bothered by his observations,
> tormented by the constant anxiety of captivity, would exhibit only distorted
> and constrained behavior, unworthy of the scrutiny of a philosopher, for
> whom the only beautiful nature is the nature that is free, independent, and,
> if you will, wild.[85]

This is a remarkable passage: it seems to be at once disparaging both the
very volume in which it is printed and the pretensions of its patron, the
king. Buffon's *Histoire naturelle* was, after all, conceived as a description of
the contents of that (useless?) storehouse of bones and skins, the Cabinet
du roi. Buffon may have been gambling, quite safely, that the king, although

presented with handsomely bound volumes of the *Histoire naturelle*, was not likely to spend his spare time reading about bustards. Even if the king were to hear of it, Buffon could count on the shield of his renown to protect him. The passage could also be seen as a self-serving justification for Buffon's failure to visit the menagerie more often; he relied heavily, especially in the *Histoire naturelle des oiseaux*, on travel narratives and reports by correspondents abroad as sources for information about animal behavior, and he hated to go to Paris.

Two broader issues are involved, though. First, Buffon was genuinely interested in epistemological problems, and the question of how to observe animals without affecting their behavior is a still sticky and ultimately unsolvable problem.[86] Second, it does not seem that Buffon was as opposed to menageries as the language in this passage would suggest. He collected many animals at his own estate, sometimes recording the lingering deaths of those that refused to eat in captivity.[87] In the 1770s and 1780s he angled for the transfer of the Versailles animals to the Jardin du roi, and according to the architect Verniquet he had approved of a plan to create a vast menagerie reproducing all of the habitats of the world, where indigenous people would be displayed along with the flora and fauna of each region.[88] Buffon's references to imprisoned animals probably had as much to do with the political issues of the day as with the animals themselves. The portrait of imprisoned, deformed, starved animals meshes with other passages in the *Histoire naturelle* in which Buffon praises liberty and freedom and denounces tyranny, imprisonment, and slavery.

The menagerie animals, which, at the beginning of the century, had reflected the glory of the sun king, could not escape the shadow of increasing disaffection falling on their owner. At the same time, their distinctiveness was fading. With growing commercial traffic and the increasing flow of exotic animals into the city, urban menageries began to rival the collection at Versailles.

CHAPTER 3

Fairs and Fights

PARISIANS had to make the trip out to Versailles in 1760 if they wanted to admire a zebra, but right at their own doorsteps they could see lions, elephants, baboons, cockatoos, and many other species on display in booths at the city fairgrounds or along encircling boulevards, or sometimes even engaged in fights at the *combat d'animaux*. Bears and monkeys had been dancing and doing tricks since at least the beginning of the seventeenth century, and more unusual animals had made occasional appearances; in the early 1600s, Parisians could have seen a troop of ostriches whose master sold their feathers to onlookers, or an elephant that traveled from city to city with its Dutch owner.[1] During the eighteenth century, however, the number and variety of animals that visited Paris expanded immensely. By the end of the century, some of these animal shows rivaled the king's collection: in 1775, Laurent Spinacuta's Grande Ménagerie at the Saint Germain fair featured two tigers, several kinds of monkeys, an armadillo, an ocelot, and a condor—in all, forty-two live animals.[2] The variety of animals that could be seen in the French capital was certainly greater than it had ever been before, owing to increased trade, and the general public may have had greater access to them than do people in today's large metropolises. A large proportion of Parisians of all classes probably had some experience of species other than those used for transportation and food, such as horses, hogs, and chickens.

Laurent Spinacuta and other animal exhibitors acquired animals, carted them around in cages, obtained permission to show them, and competed for the spectators' cash. Like monarchs, they had to think about pleasing the public—to maintain their incomes rather than their images. By emphasizing the bizarre and exotic, they capitalized on the public interest in marvels; and by displaying their charges either as fierce monsters locked safely behind bars or as tamed beasts showing off tricks and incredible talents, they played to the public's dual desire for danger and for domestication.

With the midcentury fad for natural history (instigated in part by the presence of the animals themselves), animal exhibitors began to pitch their shows to amateur naturalists. This transition was generated in part by the writings of naturalists, who gathered material from the stage animals and from their proprietors. Just as they had faulted the king's menagerie for poor accommodations and useless ostentation, naturalists and others criticized private entrepreneurs for sensationalism, imprecision, and cruelty. Still, people flocked to the fairs, and naturalists happily reproduced in their writings the lurid anecdotes passed on to them by the exhibitors. This high visibility of wild and exotic animals helps to explain the widespread interest in natural history as well as the frequent appearance of animals in social satires and political metaphors.[3]

Fairs, Boulevards, and Fights

Entrepreneurs of animal shows set up shop primarily at the annual Paris fairs. The word *fair (foire)* has lost some of its effervescence and now seems too flat a term for the hugely popular sites of entertainment that were such a distinctive feature of the city. The major fair, the Saint Germain, the Saint Laurent fair, and the smaller Saint Ovide and Saint Claire fairs all originated in privileges accorded by the City of Paris to religious orders, who rented booths out to merchants and show people. By the eighteenth century, the Saint Germain and Saint Laurent fairs featured elaborate covered boutiques laid out in a grid pattern, as well as theaters, coffee shops and tea gardens, and designated areas for parking carriages. "Like a town in miniature," wrote an English visitor of the Saint Germain fair.[4]

The Saint Germain fair opened on February 3, around the time of Carnival, and ran until Palm Sunday, a week before Easter. What a delicious opportunity it offered for getting around the gloomy strictures of Lent, when meat-eating, marriages, merrymaking, and theatergoing were officially forbidden. The fairgrounds were near Saint Germain des Prés, in the heart of the city, where the sedate marché Saint Germain stands now. The fair stayed popular and successful through the revolutionary period, despite a disastrous fire in 1762 and competition in the 1780s with the spectacles of the boulevards, where year-round entertainment became available on the wide avenues bordering the city.

Summer ushered in the Saint Laurent fair, located on the north side of Paris between the rue du Faubourg Saint Denis and the rue du Faubourg Saint Martin, which was usually open from the end of June until the end of August. The Saint Laurent was something of a poor cousin to the Saint Germain fair until 1663, when the temporary, muddy stalls were replaced by 260 covered boutiques in the large open square. The crowds of visitors came especially for the Opéra Comique and to relax in the outdoor gardens, eating and drinking while flames from hundreds of candles outshone the slowly fading light of long summer evenings. "It deserves to be put among the greatest pleasures of Paris," declared the author of a guidebook from the 1720s, who described it as packed shoulder to shoulder with all kinds of people and advised visitors that although the fair was open every day but Sunday, it was enough to go twice a week.[5] The Saint Laurent fair declined in the 1760s when the Opéra Comique moved and fairgoers patronized the more accessible Saint Ovide and Saint Claire fairs, but came back to life for the last decade before the Revolution after the Saint Ovide fairgrounds burned down.

The Saint Claire fair took place for a week at the end of July on the rue Saint Victor, and the Saint Ovide from August 14 to September 15 at the place Louis XV, now the place de la Concorde. Both attracted animal exhibitors, whose displays added to the lively atmosphere. An English visitor described the booths at the Saint Ovide fair as "form[ing] a Circus of the

gayest Appearance I ever saw and perfectly singular—[they] are temporary, . . . but adorned with a sort of Frippery Finery, Ribbons, Looking-
Glasses, Cutlery, Pastry, every thing one can imagine that is at once brilliant
& worthless—but which when illuminated with numberless Lights gives an
Air of Festivity."[6]

Another place to see strange animals was along the boulevards, wide,
tree-lined streets built atop the old ramparts on the north side of the city.
The boulevards became especially popular in the 1780s after they were paved
and provided with benches and street lamps, and one could find any sort of
entertainment there. Nicolet's Grands danseurs du roi, who put on humorous pantomimes, and Audinet's Ambigu-Comique, featuring spectacular
productions with child actors, drew large crowds, and throngs came to
promenade while breathing healthy air, carouse in the numerous taverns and
cafés, gamble, hustle, attend plays, marvel at roving menageries, or witness
performances of physics experiments.

One of the popular attractions at an arena just off the boulevard was the
combat d'animaux, a bloody show featuring bulls, deer, bears, boars, and
wolves. Lions also appeared regularly, and tigers, leopards, a polar bear,
and even a mandrill took their turns facing the dogs. Although the staged
fights that had been put on by the French royalty had largely disappeared by
the early eighteenth century, these shows for the public, run by private (but
licensed) entrepreneurs, thrived from at least the late sixteenth century
through the early nineteenth century.[7]

An announcement from 1713 informs us that the fights took place at the
porte Saint-Martin, on the northeastern edge of Paris, by permission of the
king and the lieutenant general of the Paris police.[8] Little documentation
exists for the next several decades. From the 1750s until 1776, sieur de Saint
Martin presented the spectacles at the barrière de Sève, or Sèvre, on the
southwest side of Paris. Sieur Leleu took over in 1776, and the location
changed to a hill above the Saint Louis Hospital along the boulevard's north
side. Admission was less than at government-authorized theaters, but on
the high end for popular entertainment: normal prices ranged from 3 livres
down to 24 sous, with double the price on important feast days.[9] The spectacle was officially permitted by the authorities, but it was closed down temporarily in the mid-1780s when sieur Leleu tried to introduce a Spanish-
style bullfight (combat du taureau), with human toreadors. The Paris police

attempted to extinguish it again in 1790, after it had been reestablished by a sieur Martin (probably the same person who had run it in midcentury, or his son), on the grounds that the bloody spectacle would corrupt people's morals (see chap. 8). Martin avoided the ban by moving the arena just outside the city limits. After several more incarnations, it disappeared for good in 1833.

Entrepreneurs and Their Animals

Animal exhibitors ranged from well-known entrepreneurs like Laurent Spinacuta and sieur Leleu to the obscure Martin Endric, who toured with a performing monkey and is known to us today only because his name ended up in a police report. Celebrated or unknown, they all had to manage the practical concerns of life with wild animals as well as the business side, attracting customers. Most of them have left behind few traces, but it is possible to sketch out some aspects of their lives from references in natural history texts, transcription of handbills, and police records (collected and published by Émile Campardon), which often include detailed testimonies.

Unlike the king, animal exhibitors did not have their own armies and agents to get animals for them, but they managed to acquire an impressive number through commercial channels. Unfortunately, we have little information on how they did so. It appears that, rather than setting their sights on particular species, they picked up animals from wherever they could find them—usually in one of the colonial ports. Paris was not a primary port, and so it would have been cheaper to get the animals from cities such as Marseille, Le Havre, Bordeaux, or Amsterdam, where ships arrived daily from abroad and sailors hawked their exotic wares.[10] Martin Endric bought his monkey in Toulon sometime around 1710 from an animal merchant *(marchand ou traficant des animaux sur mer)* for 10 livres; a serious investment for him, but much less than it would have cost in Paris (see chap. 5). The owner of a large baboon from Ceylon purchased it in Marseille; a touring American buffalo was bought in Holland; and a ram supposedly from the Cape of Good Hope was acquired in Tunis. The proprietor of a fierce genet bought the animal in London but had no idea where it had originally come from.[11] In a few cases, the person who brought the animal to Europe also traveled around showing it.[12] Exhibitors may also have picked up cast-off exotic pets: sieur Soldi let it be known that "he buys all rare and curious animals that anyone would like to get rid of."[13]

Once bought, the animals had to be tamed and trained, fed, and carted around. Elaborate modes of transportation could even be part of the attraction. The publicity poster that circulated when the rhinoceros came to town in 1749 announced proudly, "It is necessary to transport this monster in a wagon and to cover the wagon. It is necessary, sometimes, that is when the roads are bad, to use up to twenty horses to pull it."[14] The proprietor of a seven-foot-long seal hauled it in a huge container, which he refilled with fresh water every time the animal defecated. At night he drained the water and slept next to the sleek beast.[15] Although we do not know how Endric transported his monkey Petit-Jean (were tame monkeys welcome in public carriages?), the two were constantly on the road: in a five-year period between buying Petit-Jean in Toulon and arriving in Paris, Endric toured southern France, visiting Marseille, Arles, Boquère, Bayonne, Bordeaux, and elsewhere. At first he kept his job playing in the orchestra of an itinerant acting troupe while he trained the monkey. He taught Petit-Jean to tumble and dance to the drum but, unable to excite the public with such routine tricks, he spent six months teaching the monkey a tightrope act and then struck out on his own.[16]

In Paris, lodging had to be found and permission of the police obtained. Animal owners probably often lived with or near their valuable charges; Endric had a room in the unappealing sounding Entrepôt de la Mercière (merchant's warehouse) just off the Saint Germain fairgrounds. Joseph Manfredi, Ogimbel Toscan, and their wives, who between them owned several animals, rented a boutique together on the quai Pelletier, which they furnished with wall coverings, chairs, and benches; they were to split the profits from showing Manfredi's lion and Toscan's "rare and strange bird." As far as I can tell, the two couples also lived there (probably on the second floor), but not very amicably: Toscan started badgering Manfredi to sell his lion, refused to pay his share of expenses, and even threatened the lion's life. Conditions sound atrocious enough, even before we hear what else was going on in the boutique: Manfredi tattled to the authorities that Toscan was conducting an unlicensed tooth-pulling business on the side and had not obtained permission for two monkeys he owned, which had already bitten several people because they were not confined—a precaution Manfredi said he always took with his own monkeys.[17]

Lions and monkeys may have caused trouble, but at least they were relatively tough. Investing in more delicate species must have been a risky business. Although many died, it is only in a few cases, such as that of Laurent Spinacuta's cassowary, that we have an account of their miserable last minutes. The bird was being shown at the Saint Ovide fair, but, because of a dispute, it had by police order been left temporarily in the care of a horse merchant, M. Duchemin. Duchemin made a little hut for the bird in a shed off the courtyard of his lodging and fed it what he had been told: some bread, apples, and pears. Half an hour later, however, the bird "released from its belly a large quantity of greenish water and then, after having lain down on the floor, trembled and died." Duchemin, worried that the death would be blamed on him and that he would not be able to collect the 12 sous per day that he was owed for its care, hurried to get the *commissaires* to his house to attest that he had cared adequately for it.[18]

A few traces remain to give us a sense for the emotional connections that the exhibitors formed with their charges, and vice versa. Unlike other business assets, animals could be companions, too, and could provoke affection as well as anger. The seal exhibited "an extreme tenderness" for its master and even seemed to talk to him, reported the duc de Croÿ.[19] The monkey Petit-Jean showed affection for Jean Vallet, who protected him from his master's beatings. Vallet, one of Endric's fellow fair performers, reported that he had snatched Petit-Jean away several times to keep him from Endric's blows, and, according to the police report, "the animal always recognized the witness [Vallet] as a dog would his master, because the witness had caressed him in the different sites where he had encountered the said Endric and he had several times given him something to eat."[20]

What were the animal displays themselves like? Imagine a couple deciding to spend an evening and a day's wages at the Saint Germain fair. They are drawn in by a crier announcing the arrival of a unique animal never before seen in Paris, with the head of a this, feet of a that, and tail of something else. They pay their money. Do they now go inside or stay outside? Is the animal on a stage or in a cage? Are they rubbing shoulders with the dukes and countesses there or, since they paid the lowest admission fee, are they on tiptoe looking over wigs and coiffures? Do they care if the marvel is real or fake?

A fairgoer with enough spare change to view only one or two of the animal displays might have trouble deciding among the alleged marvels. Exhibitors hyped their animals with epithets such as "remarkable," "very remarkable," "very rare," or "never before seen in Europe."[21] The entrepreneurs described unfamiliar species by comparing parts of the animals' bodies to those of creatures the audience would be familiar with. The poster describing the 1749 rhinoceros, for example, explained that it had skin like an elephant, ears like an ass, swam like a duck, and was as tame as a gentle dove.[22] Or the animal might lend itself to only a schematic but tantalizing description: sieur Soldi had on display in 1775 an animal "unknown to all naturalists," two feet long, fifteen inches high, with a lion's mane and a thirty-one-inch tail.[23] Handbills claimed that mythical animals could be seen in the flesh: the rhinoceros was "said to be apocryphal until now," and the pelican, renowned for tearing open her breast to feed her young with her own blood, "doesn't exist at all, according to history; yet you can see it at the Saint Germain fair," just brought to Paris from Turkey by sieur Chequer.[24] Biblical animals, too, came to life. The "Verus bubalus," suggested its proprietor, might well be the very behemoth from the book of Job, chapter 40.[25] In the strongest appeal to visual witnessing, a creature was simply described as indescribable.[26]

Exhibitors played up not only uniqueness but also the monstrous and bizarre: a beast with a beard trailing to the ground; an animal that moves so slowly it takes six months to climb a tree.[27] The amount of food consumed by large animals was a constant source of wonder: the rhinoceros was said to eat sixty livres of hay and twenty livres of bread per day, and to drink fourteen buckets of water and beer, and the "amphibious bear" from the Arctic required thirty-five livres of meat per day and, in summer, twenty livres of cooling ice.[28] A description of the *diable des bois du Perou*—the devil from the Peruvian forest—combines the bizarre, the unique, and the apocryphal: "Their faces are flame red, their ears are pointed like horns, their arms are two times as long as their bodies, their tail is of an extraordinary length and width and they move it like a snake, they have claws at the ends of their fingers; in a word, they are so misshapen that one would say that painters had used them as models to represent the devil."[29] Diable des bois was only one of the imaginative names that animal entrepreneurs invented or borrowed. The *belzebut, satyre, couxcousou,* and *subsilvania* vied for audiences that had no idea what they might be about to view. They knew they hadn't seen them anywhere near Paris, however, for they came from such places as the Indies, White Russia, the African island of Gorée, and "the Amazon mountains."[30]

From the few accounts that exist, it seems that large animals such as elephants and rhinoceroses appeared outside in pens where one would pay for a viewing spot, whereas other animals were kept inside, chained or caged in enclosures *(loges)* that were, as a contemporary periodical put it, "small, cramped, and badly lighted."[31] Enclosed booths or cages protected the animals from the weather and might double as wagons, and their shadowy interiors must have been a boon for charlatans passing off run-of-the-mill species as marvels.[32]

The only illustrations I have seen of the booths are two engravings that the animal painter Jean-Baptiste Oudry made to illustrate the fable "Le singe et le léopard" (The monkey and the leopard) in an eighteenth-century edition of La Fontaine (figs. 3.1, 3.2). The two fair animals in the fable compete to draw in customers: the flashy appeal of a leopard's fancy coat is contrasted with the skillful performance of a hoop-jumping monkey (moral: "Oh, that great nobles, like leopards, had more than just their dress as accomplishments!").[33] The illustration of the leopard's booth shows a banner waving outside the building, with a picture on it of a fierce beast, that would have lured both the literate and the illiterate. Narrow alleys and neighboring booths evoke the tight quarters of the fair environment, although Oudry does not depict the crowds. Artistic license may also explain why the leopard (which seems in fact to be a tiger) growls behind bars that face the street (why would you pay to enter if you could see the animal from outside?). The other engraving, of the agile monkey, shows an interior scene: the monkey performs on a crude stage; rough wooden benches seat an audience of eight people; and a small hole, high in the ceiling, supplies light. These two illustrations capture a dichotomy in the way stage animals were presented. Trained monkey and fierce leopard, the tame and the wild: exhibitors often presented their animals as being at one of these two poles, both of which celebrate human control. "Good" animals learn and obey; "bad" animals are kept in prison.[34]

LE SINGE ET LE LEOPARD . Fable CLXXII.

FIGURE 3.1. A wild "leopard" (which looks more like a tiger) in a fair booth. Jean-Baptiste Oudry, illustration for Jean de La Fontaine's fable "Le singe et le léopard" (The monkey and the leopard) in *Fables choisies mises en vers* (Paris, 1755–59), vol. 3. Bancroft Library #ff PQ1808.A1 1755 (vault). Courtesy of the Bancroft Library, University of California, Berkeley.

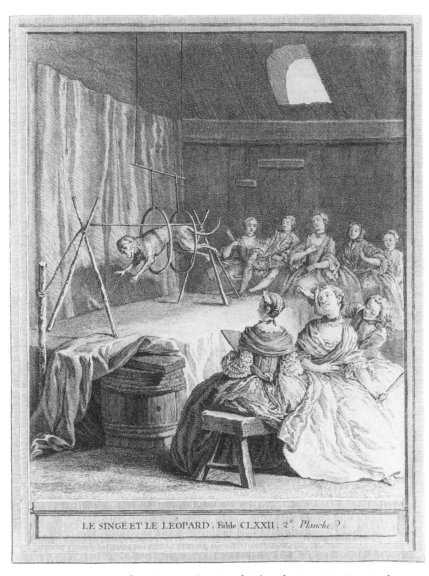

LE SINGE ET LE LÉOPARD . Fable CLXXII . 2ᵉ. *Planche* .

FIGURE 3.2. A performing monkey in a fair booth. Jean-Baptiste Oudry, illustration for La Fontaine's fable "Le singe et le léopard," in *Fables choisies mises en vers* (Paris, 1755–59), vol. 3. Bancroft Library #ff PQ1808.A1 1755 (vault). Courtesy of the Bancroft Library, University of California, Berkeley.

Tame or accomplished animals gave evidence of the human ability to control and teach. They might be caged, but only for convenience. Lions, tigers, and leopards were said to be as obedient as dogs, and the mysterious subsilvania was described as extremely gentle and tame. The 1758 Saint Germain fair featured a tiger that was "astonishingly" gentle; according to the poster advertising it, the animal could read, write, and do arithmetic, knew the playing cards and colors, and could point to them with its paw.[35] Fairgoers in 1775 marveled at a pair of cockatoos that raised their crests on command, answered questions (such as, How many francs are there in an écu?) by nodding their heads, and indicated the time of day and the number of people in the crowd.[36]

Trainers played with their tamed animals and often let the crowds touch or feed them. Twelve-foot-long snakes twisted around their mistress's neck and arms and kissed her on command, and "the spectators could also hold, touch, and stroke them."[37] A trained seal exhibited by the same woman offered its paws to the duc de Croÿ, just like a dog.[38] When the crusty rhinoceros licked people's hands, they were surprised to find that its tongue was as soft as velour.[39] The kinkajou, too, offered its soft tongue, but after a couple of months of pestering it began finishing off its licks with a bite.[40] The hyena's master constantly touched its back with a staff to make it raise its mane and demonstrated its gentleness by putting his hand in its mouth.[41] Although in a large cage with its head restrained by four ropes, the American buffalo let itself be stroked by its handlers.[42] Little boys could take rides on the back of a tame cougar.[43] Best of all was a chimpanzee (called an "orang-outang"), which led visitors by the hand, sat at a table, wiped its lips with a napkin, poured and drank a cup of tea—after stirring in sugar and letting it cool—and ate so many bonbons presented to it by the crowds that, a year later, in London, it died.[44]

At the opposite extreme, exhibitors exploited the image of some animals as fierce and even untamable, playing on the fantasy of danger by detailing the strength of the bars or chains that "for public safety" restrained the ravenous beasts.[45] A verse attached to the polar bear's cage read:

> I am always dressed in white
> My suit is all I own:

I please both small and great
But only when I'm caged.

And what a cage! The audience was told that the "terrible" animal was kept in an oak crate surrounded by fifteen hundred livres of steel.[46] Some animals could be counted on to provide a scary show. Buffon and Daubenton, in the *Histoire naturelle,* described two large monkeys that raged in their cages: the *ouanderou* (fig. 3.3) and a baboon, which grimaced, masturbated shamelessly, and shook the bars forcibly enough to frighten spectators.[47]

In the animal fights, spectators could watch animals display their bloodthirsty ferocity (provoked by trained dogs) in a less restrained but still safe environment (fig. 3.4). Like the Saint Germain fair, the combat d'animaux took place when other spectacles were closed: "ordinary" fights were on Sundays, "special" fights on festival days (e.g., Easter, All-Saints', Christmas) when the only other police-approved activity was a performance of classical music, the *concert spirituel*. Mrs. Cradock, an English visitor who found the mixture of religion and amusements on Sundays in Paris unsettling, described the curious pairing in her journal: "Here is the terminology used on the posters announcing this ceremony [Pentecost]," she wrote: "'High mass with music at Notre Dame, where Monseigneur the archbishop [of Paris] will officiate. In the evening, *concert spirituel* and bull fight.'"[48] A contemporary guidebook explained that "because the Concert Spirituel is neither to everyone's taste nor suitable for the faculties of all citizens, permission has been granted to what is called the Combat du Taureau to take place on the days when the Spectacles are closed."[49] Although this statement suggests a class division, with educated elites sedately listening to music while the rabble yelled at animals tearing each other apart, the audiences were not so cleanly divided: many observers noted that high-ranking people were enthusiastic spectators at the animal fights. One critic who attended a fight in 1781, for example, encountered "people of quality, of condition, of every age and sex, military men, women, abbots, in sum what is called good company, and even children."[50]

Fights were announced by criers and posters, and notices appeared in both the *Affiches de Paris* (from 1759) and the *Journal de Paris* (from 1780). A typical announcement reads: "The day of Pentecost at the barrière de Sève there will be a fight of a *large* female LION from the African islands

FIGURE 3.3. The *ouanderou,* or lion-tailed macaque (now an endangered species), was shown at the Saint Laurent fair. Although it looks mild-mannered here, Buffon wrote that it often shook the bars of its cage in fury and that ouanderous had been known to rape women. From Buffon et al., *Histoire naturelle,* vol. 14 (1766), plate 18. Reproduced by courtesy of the Department of Special Collections, General Library System, University of Wisconsin–Madison.

Par permission de M. le MAIRE, & sous la surveillance de la Police.

AUJOURD'HUI DIMANCHE 10 Novembre 1805.

GRAND COMBAT D'ANIMAUX,

Sous la direction de M. Rossi, Espagnol, élève de M. Padovani, ancien maitre de combats, connu en France, par les spectacles extraordinaires de ce genre qu'il y a donnés, à la satisfaction générale.

Il se propose de donner le combat général. On verra de plus Theodore, premier garçon de combat, qui se battra corps à corps avec un ours de Pologne, de la plus forte espèce, & qui par son adresse le terrassera.

Ce spectacle, dont cette ville n'a pas joui depuis passé 10 ans, au a lieu dans une enceinte construite à cet effet, et pour la solidité de laquelle on a pris toutes les précautions nécessaires.

Il commencera par plusieurs batteries de chiens de diverses espèces, qui combattront gueule à gueule, à force egale. — Suivra le combat du féroce loup-cervier contre des demi-dogues & chiens communs. — Viendra ensuite le combat d'un ours des Monts Pyrénées, de la plus grande force, et démuselé, contre des chiens de forte race. — A ce combat succedera celui d'un puissant & vigoureux taureau contre des dogues & boule-dogues. Il sera suivi du combat du ridicule *Peccata* & du Bacchanal mal-monté.

Le combat sera terminé par l'enlèvement d'un boule-dogue dans un parasol de feu. On l'élevera à trente pieds de hauteur et il se tiendra toujours ferme par la force de ses dents.

Ce spectacle aura lieu dans la cour du sieur Dominique, à la ville de Paris.

Prix d'entrée : Premières, 24 sous; secondes, 12 sous; & six sous pour les enfans.

Vu la longueur du combat, on commencera à trois heures 1/2. —— Le bureau sera ouvert à deux heur 1/2.

NOTA. Des amateurs qui ont des chiens vigoureux, pourront les amener au combat; mais ils sont invités à déclarer au directeur du spectacle, contre quel animal ils désirent que leurs chiens combattent. Il ne sera reçu aucun chien qui n'ait sa chaine & son collier.

FIGURE 3.4. Poster for the Paris combat d'animaux, final show of season, 1805. Note, upper left, the *enlèvement du bouledogue*—the dog being hoisted amid fireworks. A note at the bottom reads, "Amateurs who have vigorous dogs may bring them to the fight; but they are invited to tell the director against which animal they wish their dogs to fight." Bibliothèque historique de la ville de Paris.

against *dogs*. The fight will be preceded by the hunting of a *young boar,* an amusing *Hourvari,* fights of a *bear,* a *wolf,* and a *bull,* the dressage of a *little Prussian horse,* races of a *stag* and a *doe,* &c. The Spectacle will start at 5:00 sharp, and will end with *pretty fireworks,* during which a *bulldog* will be elevated."[51] The *enlèvement du bouledogue,* a feature of every fight, consisted of lifting into the air a bulldog that had grasped a baton in its jaw and held on even while all around it fireworks exploded.[52]

Announcements notified potential spectators of special combatants, which often made their debut at the Easter show: a bull that was to be dispatched because it was so vigorous that it had killed numerous expensive English bulldogs; one of the largest and fiercest bears from Poland, just arrived in Paris; a "beautiful TIGER"; a "large and ferocious LEOPARD from Africa," whose claws, teeth, and agility were guaranteed to make the 1778 Easter fight a bloody one; and a "man of the woods *[homme des bois],* or male Mandrill."[53] The mandrill, explained the announcement in the *Journal de Paris,* "is so cruel that the first dogs to attack are in danger of losing their lives." The 1777 Easter show featured the polar bear that had been on display at the Saint Germain fair.[54]

Two paintings of violent animal fights that Louis XV commissioned for his dining room at Choisy may well have used the Parisian combatants for models; they show in gruesome detail the slashing teeth and straining muscles of the fighting dogs that so excited spectators (fig. 3.5).[55] The short-lived combat du taureau, modeled on Spanish bullfights with human toreadors, may have failed because it lacked this excitement; the old bull brought out for the first and only fight had been so weakened by being bled beforehand (to reduce the danger) that it stumbled into the arena "almost inanimate."[56]

Despite a doubled admission price, crowds swelled for the mortal combats that took place only on special days—the Virgin's birthday, All Saints' Day, the day of purification of the Virgin (2 Feb.), Palm Sunday, Pentecost, Assumption Day.[57] A young Hungarian count who visited Paris in 1760–61 and spent much of his time there frequenting scientific and court circles, described such a fight in his journal:

> This afternoon I went to see a fight between different animals and dogs. Such
> diversions take place only on great festival days like today: Festival of the

FIGURE 3.5. Jean-Jacques Bachelier (1724–1806), *Un lion d'Afrique combattu par des dogues* (An African lion being attacked by dogs), 1757. The dogs in this painting, which was commissioned by Louis XV for his dining room at the chateau of Choisy, may have been modeled on those that fought at the Paris combat d'animaux. Courtesy of Musée de Picardie, Amiens, France.

Purification. First a wolf fought against two or three dogs, but he was soon defeated. Next it was the turn of a bear, who was more courageous. Then, a bull, trained for fighting; he was pitted against only one dog, which he teased. Then came a horse and a deer; they just ran around, which wasn't very entertaining. Finally they brought in a bull that was tormented to death by the dogs. It took about an hour; by the end, three dogs had expired. After that they hung a dog by a cord and set off fireworks all around it. The poor dog turned round and round until the end of the display. It's not a very interesting show, yet every foreign visitor is obliged to go at least once to see what it is. The best places cost 3 livres, the others less. The show takes place in the open, although there is a roof to keep off the rain.[58]

The association of animal displays with religious holidays (feast days and Lent) might be seen as a holdover from the traditions of popular religious rites, where festivals, especially Carnival, featured drunken revelry, animal fights, and people dressing up as devils or wild animals. For at least a century, however, the church had been discouraging activities that smacked of paganism, and the fact that city authorities permitted and condoned animal shows suggests that they were not spontaneous popular manifestations. It is possible that authorities encouraged such controlled wildness as a "safety valve" to permit regular and moderate release of tension by potentially dangerous masses.[59] If such was the thinking, it changed at the end of the eighteenth century, when some commentators began to decry the cruelty of animal fights and to warn that the shows would spark rather than dissipate violence (see chap. 8).

Audiences, Social and Scientific

People of all social groups heard about and went to see stage animals. Popular journals reported on doings at the fair, and one publication specialized in such events: the small, inexpensive *Almanach forain*, which covered the fairs, boulevards, and combats d'animaux, described events in detail and included funny anecdotes and brief histories. Posters, handbills, and public criers announced news of new creatures to the poor and illiterate. Everyone in Paris seems to have turned out for a few sensational attractions, especially the rhinoceros of 1749 and the elephant of 1771. Yet different audiences and individuals had different experiences and made sense of what they saw in different ways. Where the *beau monde* found material for social satire, the amateur naturalist saw a scientific curiosity. A disguised bear advertised as an exotic creature might infuriate the scientific observer but delight the entertainment seeker.[60]

Not all spectators had the same physical vantage point. Those who paid more got closer seats in booths and at the animal fights, and nobles might even at times have been accorded private showings. The *Almanach forain* announced, for example, a "Monster from the Amazon mountains" that was being shown on the boulevard every day except Wednesdays, "which are reserved for high nobility *[les Hautes-Puissances]*."[61]

For the fashionable set, announcements concerning exotic animals provided fodder for social satires, much as, today, widely recognized advertisements are take-off points for jokes and puns. Stage animals lent themselves easily to humor through the regularly traversed boundary between animal and human worlds. A conversation about muzzling husbands, for example, arose after the duc de Chartres and his friends dressed up as wild animals from a menagerie and crashed a ladies'-only dinner party.[62] A satirical journal of 1776 printed a joke page in the style of fair announcements in the *Affiches de Paris* that announced: "Sieur Baladini, just arrived in Paris, has on display four tamed wild animals: a lion, a tiger, a solicitor *[procureur]*, and a leopard. This spectacle can be seen at any time on the boulevard du Temple."[63] The year before, a more vicious satirical pamphlet verging on the pornographic had named some of the best-known society ladies in a list of "curiosities at the Saint Germain fair or that can be seen in Paris." Demoiselle Arnould appeared as proprietor of "a very naughty beast that throws itself indiscriminately at everyone, and can't be tamed; despite being old the animal hasn't lost its ferocity; luckily it has lost its teeth, which means that it poses a risk only to those touched by its venom or its odor." Mademoiselle d'Hervieux was said to own a beautiful monkey from the Indies, which she was very willing to show. "It is very lively and very engaging. It has very nice manners; although not wild, it doesn't acknowledge any master: today it's one, tomorrow it's another. It is very fond of brilliant stones, and by means of them one can gain its favor for a few moments. . . . A disease has made it unable to bear children, but its graceful little ways make up for that loss. One can't help but admire above all how skillfully it uses its little hands, etc." The satire was later presented as a play by the Ambigu-Comique on the boulevard.[64]

For other spectators, stage animals offered an opportunity for enlightenment in the area of natural history. Naturalists had long been in the habit of seeking out new species wherever they could find them. By the later eighteenth century, not only were naturalists continuing to get close to animal exhibits, but the exhibitors themselves began touting the natural history value of their displays. These come-ons were undoubtedly geared not only to serious naturalists but also to a wider audience. A similar trend occurred in experimental physics shows, where audiences laughed at magic tricks and jumped at electric shocks at the same time that they learned about new theories of gravity and electricity.[65] Spectators at animal shows must have been a mixed crowd, including serious naturalists, people attempting to participate in the fashion for natural history, and those who wanted a brief encounter with exotic creatures, either real or fake.

Entrepreneurs of animal shows competed to bring in spectators looking for natural history knowledge. Sometimes they described an animal with an enticing phrase such as "unknown to naturalists" or "so rare that its name is not found in dictionaries of natural history."[66] In a more direct appeal, sieur Padovani, who showed a menagerie of twenty-four animals in 1777, hoped, he declared, "*to attract the attention* of connoisseurs and amateurs of Natural History."[67] Similarly, sieur Ruggieri's publicity sheet for the 1772 Saint Germain fair is addressed to "the Public and especially . . . Amateurs of Natural History." The sheet included a description of Ruggieri's tapir from Buffon's *Histoire naturelle* article (since the animal had just died, however, it could only be seen stuffed).[68] Another exhibitor, a sieur Latour, also cited the great naturalist, identifying his captive seal as the animal that Buffon called the *tigre marin,* or aquatic tiger; the *Almanach forain* disputed this identification, however, along with Latour's claim that the animal would eventually grow to be fifty feet long.[69]

Naturalists frequented the fairs, the boulevards, and the combat d'animaux to get information available nowhere else, as did a wider public interested in natural history, especially in the second half of the eighteenth century. One avid observer, the duc de Croÿ, who had so appreciated the Versailles menagerie, often went to the fairs to see the *"animaux curieux."* There he compared his observations with those available in natural history texts. At the 1771 fair he took detailed notes on the elephant, the polar bear, and other species "for my own *Histoire naturelle,* especially because M. de Buffon was extremely ill."[70] Buffon recovered from his intestinal abscess, and the duke never published his work of natural history.

Buffon was himself attracted to the fairs and boulevards, although more often it was his collaborators and illustrator who visited, as well as other authors of natural history works on quadrupeds and birds. These specialists examined the animals and pumped their owners for information. Historians have underestimated the importance of such popular venues for naturalists, perhaps because of the unscientific-sounding hype that accompanied the displays.[71] Buffon did, indeed, criticize the inaccuracy and sensationalism of some of the exhibitors, for example the owner of the violent Ceylonese ouanderou, who advertised the large baboon-like monkey under an incorrect name and geographical origin: "This happens frequently, especially among the bear and monkey exhibitors, who, when they do not know the origin and the name of an animal, do not hesitate to give it a strange name that, whether true or false, is equally good for their purposes." To his surprise, however, he found that an animal described as "unknown to naturalists" turned out in fact to be new to him—a kinkajou. Not knowing that the animal was a South American species, he passed along the proprietor's claim that it was from Africa and had possibly been named after the country or island that it inhabited.[72] Other exhibits also provided important material. In addition to the ouanderou and the kinkajou, Buffon or his collaborators observed at the fair and on the boulevards a rhinoceros, an elk, a pair of ocelots, four kinds of monkeys and apes, a bison, two exotic rams, a marten from Cayenne, a hyena, a genet, an elephant, a seal, sieur Ruggieri's tapir, a tiger, a porcupine from Malacca, cockatoos, and Laurent Spinacuta's condor.[73] Naturalists not only looked, they also listened to the keepers' behavioral observations (presumably more reliable than information about names and origins): the snake woman showed the duc de Croÿ how she removed venom; Mauduyt learned from the cassowary's owner that it defended itself by kicking; and Buffon related stories gleaned from the exhibitors of the ocelots, the elephant, the seal, and the cockatoos.[74]

The combat d'animaux also proved a useful source for behavioral information and, sometimes, bodies to dissect. Buffon tapped sieur de Saint Martin, the proprietor, for notes on the lifespan of lions (Saint Martin said he had kept some lions for sixteen or seventeen years), and roaring behavior (they gave voice five or six times per day, but more often when rain was on the way). The lions, one African and one Asian, were apparently too valuable to be sacrificed for dissection, but Saint Martin allowed Daubenton to take measurements of one of the old warriors. Daubenton also recorded the color differences among the three bears kept at the combat; but for internal anatomy, he and Buffon bought and killed a bear in Burgundy. The naturalists were luckier with a female black wolf that a navy officer had brought back from Canada and had given to the combat d'animaux when it became too fierce to remain a house pet. It yelped and lunged at its chain when they approached its cage and they could not even measure it, but its dismal performance in battle soon brought it to the dissecting table; Buffon remarked that "[it had] ferocity without courage, which made it cowardly in combat, despite having been trained" (fig. 3.6).[75]

In acquiring what they considered to be data from entertaining animal displays, naturalists believed they were winnowing the wheat from the chaff—discarding exaggerations while retaining confirmed observations. Confirmation is in the mind of the judge, however, and often the reports that naturalists passed on about animals' behavior seem to us now to be frivolous or sensational. A typically titillating description of an ocelot pair, for instance, made its way with only minor changes from fairground to natural history tome. The description from the *Almanach forain*, probably drawn verbatim from the poster, reads, "The ocelot is naturally perfidious and ferocious. In 1764 two were on show at the Saint-Ovide fair. . . . Ungrateful and cruel, at the age of three months they tore off the nipples and sucked the last drop of blood from the dog that had nursed them. The male is so brutal and savage that he has no regard for [the female]; she waits patiently until he has satisfied his voracious appetite."[76] Buffon's article on the ocelot reported these facts, slightly toned down (the cubs simply "killed and devoured" their dog nurse).[77] Valmont de Bomare, Duchesne and Macquer, and the abbé Sauri all passed on more or less gruesome versions in their natural history texts, and Sauri concluded his description with a little moral commentary: after describing how the male wouldn't let the female eat until he was done, Sauri remarked, "It is not rare to find men who, in that regard, have the morals of the ocelot."[78]

FIGURE 3.6. The black wolf *(loup noir)* that was ignominiously defeated at the combat d'animaux and subsequently dissected by Daubenton. From Buffon et al., *Histoire naturelle*, vol. 9 (1761), plate 41. Reproduced by courtesy of the Department of Special Collections, General Library System, University of Wisconsin–Madison.

Although naturalists found the fairs and the combat d'animaux useful resources, they and the editors of several periodicals often complained that exhibitors exaggerated or exotified names or manufactured fake animals. One of the earliest critiques came from an English physician, Martin Lister, who visited the Saint Germain fair in 1698:

> I was surprised at the Impudence of a Booth, which put out the Pictures of some Indian Beasts with hard Names; and of four that were Painted, I found but two, and those very ordinary ones, *viz.* a Leopard, and a Racoun. I ask'd the Fellow, why he deceived the People, and whether he did not fear Cudgelling in the end: He answered with a singular Confidence, that it was the Painter's fault; that he had given the Racoun to Paint to two masters, but both had mistaken the Beast; but howeer, (he said) tho' the Pictures were not well design'd, they did nevertheless serve to Grace the Booth and bring him Custom.[79]

As the eighteenth century progressed and demands for education and utility grew, so did complaints about charlatans from those trying to gather solid information or who were contemptuous of the overly credible. The naturalist Le Vaillant ridiculed gullible Parisians for wasting money to see a *Gangan* (actually just a camel) instead of trying to educate themselves.[80] A visiting Swiss naturalist wrote to the *Journal de Paris* in 1781 exposing the sham *grand Tarlala de Tartarie*, an animal being shown at the Saint Germain fair as rare, unique, and unknown. He complained that he had spent a week's wages to see what turned out to be a hairless bear wearing clothes and performing "disgusting caresses." He hoped that his exposé would keep others from being duped. Using the tarlala as his example, the social critic Louis-Sébastien Mercier exclaimed that years could go by before anything worthy of the naturalist's eye appeared at the fair.[81]

The *Almanach forain*, the *Avant-coureur*, and the *Affiches de Province* exposed obvious fakes, derided the silly names given to the animals, and, for legitimate curiosities, supplied the "real" names along with quotations from or references to writings by naturalists. A long article in the *Avant-coureur* noted the presence of many exotic *(étranger)* animals at the 1773 Saint Germain fair that could be of interest to amateurs of natural history. It complained, however, that the small, badly lit booths impeded observation of the animals, as did the tight leashes that restricted their movement—but one could at least observe their stature, form, and external appearance, which was better than nothing. The article described a dozen species and corrected mislabeling, especially of the cougar, noting "we have described this animal at some length so that it will be easy to recognize; because those who display it announce it as a kind of monster, and do such a good job in their ridiculous description that one would not be able to find it among the quadrupeds that naturalists have spoken about."[82]

These criticisms were part of a larger trend among contemporary commentators to ridicule the credulity of the masses. As historian Arlette Farge has pointed out, however, this rhetoric tells us more about the attitude of the elite toward popular culture than it does about gullibility, a slippery concept that could be applied to many beliefs of the elite as well—belief in bloodsucking ocelots, for instance.[83] And indeed it seems that rather than feeling duped, many spectators enjoyed a battle of the wits as part of the fun of fairs. Several anecdotes suggest that the crowd often responded with admiration rather than anger when they—or better yet, others—were cleverly tricked. People were said to be impressed, for example, when they found out that a fierce animal growling from the depths of a dark cage was actually a skillful human mimic.[84] A story that highlighted a charlatan's quick thinking told of a young educated man at the Saint Ovide fair who contemplated a monstrous animal for a minute and then yelled out, "Hey, that's nothing but a pig!" "Monsieur," responded the master, "since you know each other so well, you must have been a pig-keeper in the past."[85] Even so, the proponents of enlightenment were probably right to suspect that for many exhibitors and spectators, amusement took precedence over education.

Critics who accused fair exhibitors of sensationalism and spreading misinformation also faulted the combat d'animaux for exalting cruelty and inciting base passions, and some proposed transforming it into a more edifying spectacle. A typical commentary comes from Mercier, who described the combat d'animaux as a show where animals were made to rip each other to shreds so that people could enjoy their torments: "Eh! It is on the most holy days that the church allows the celebration of such horrors! What shameful barbarism, for public order [la police], for the Parisian nation, supposedly so humane!"[86] Guidebooks referred to it as "a spectacle worthy of butchers" and a site of "horror and ferocity."[87] A 1780 edition of Valmont de Bomare's dictionary of natural history remarked that animal displays are "more enjoyable for people of a gentle and humane nature when they take place without bloodshed."[88] The *Almanach forain* called the fights a "bloodbath," but asked, "Wouldn't it be possible to make a more rational amusement, by imitating those Nations that make animals fight for only a minute, and only to study their forms of self-defense?"[89] A letter to the *Journal de Paris* in 1781 also recommended a more instructive study of animal behavior:

> Man can use animals for diversion and amusement because his skill gives him absolute dominion over them, but not in making them tear each other to pieces. In large cities I've seen monkeys strike a light and light a candle, and a lion put it out with its paw; an elephant uncork a bottle of wine, grasp it in its trunk, empty it in its mouth, etc. If a man wants to make use of public curiosity and educate the most intelligent animals, let him set up a studio, a circus, well supplied with tricks and singeries; the people will come running, and honest men can amuse themselves there for a quarter of an hour.[90]

Such criticisms parallel a widespread trend in Europe away from public displays of the suffering and death of both animals and humans, one of the results of which was the animal-protection movement of the nineteenth century. These criticisms of brutality were often directed at the lower classes, and they had political as well as humanitarian motives; "Love for the brute creation was frequently combined with distaste for the habits of the lower orders," remarks Keith Thomas.[91]

The Rhinoceros and the Elephant

Two genuinely novel animal visitors to Paris that attracted both the low and the high, the ignorant and the educated, were the rhinoceros of 1749 and the elephant of 1771. The contrast between the reception of the two animals shows us how much changed in a space of only twenty-two years. In 1749, when the rhinoceros arrived, the natural history fad was just beginning to catch on. The animal was represented as a rarity and repository of curious facts, mostly culled from the ancients. The elephant, too, stirred up classical tales, but was greeted explicitly as an item of interest to natural historians. In addition, it carried more political symbolism, in accord with the growing tendency to see exotic animals in terms of metaphors of tyranny.

Both animals, unlike many other exotic imports, were already familiar from classical as well as more recent history. Everyone knew of the legendary enmity between the two titans. The rhinoceros supposedly used its horn to gore the elephant's underbelly, while the elephant retaliated with its mass, falling on the rhinoceros and crushing it.[92] In 1517 the king of Portugal had reportedly attempted to provoke such a battle of the titans, with anticlimactic results (the animals snorted and postured, but then the elephant ran away).[93] A rhinoceros had visited London as recently as 1739, but had never been seen in France. Elephants had lumbered into France several times—appropriately, during the celebrated reigns of Charlemagne, Louis IX, and Louis XIV—but the last one had died in 1681.

Douwe Mout, the Dutch ship captain who escorted the young female rhinoceros around Europe, headed first for Versailles, expecting an audience with the king. When he found himself put off, he set up at an inn in town; a few days later members of the court paid a visit, and, finally, the king. After having saluted the royalty, Mout and his rhinoceros went into Paris for the Saint Germain fair, preceded by rumors that the king would have bought the animal for the menagerie had the captain not quoted a tremendous price.[94]

The poster announcing the rhinoceros's arrival in Paris began with the equivalent of a drumroll: "In the name of the King and of Monsieur the Lieutenant General of police, Messieurs and dames, you are notified that there has recently arrived in this city an animal called Rhinoceros, which was believed until the present to be apocryphal." Its attributes, as outlined in the poster, practically guaranteed that it would be a sensation: it was exotic (from Assam), bizarre (ears of an ass, skin like seashells, swims like a duck), fabulous (sworn enemy of the elephant), enormous (five thousand livres in weight and twelve feet wide), naturally fierce (gores elephants' bellies), but raised to be tame and gentle as a dove (at the age of two it ran around in the house like a dog).[95]

The chronicler Barbier predicted that the Dutch captain would make lots of money because few people failed to satisfy their curiosity to see the animal. Indeed, it seems to have become the dominant attraction of the day: a dramatist whose play was being performed at the same time complained that the monster was stealing away his audience. Tastemakers capitalized on the novelty, generating a "rhinomania" of poems, coiffures à la rhinoceros, engravings, and fancy clocks with rhinoceros bases (fig. 3.7).[96]

Among the curious onlookers were naturalists and scholars who had never had an opportunity to see a live rhinoceros before. The anatomist Petrus Camper made drawings and a clay model from the animal when it was in Leiden, and its Parisian visitors included Daubenton and the animal painter Jean-Baptiste Oudry. Buffon, who was busy seeing to the publication of the first three volumes of his *Histoire naturelle*, does not seem to have joined the rhinoceros-watching crowds. Volume 11 of the *Histoire naturelle*, published fifteen years later, included the animal's portrait, engraved from a drawing made from Oudry's life-sized oil painting, and an anatomical description by Daubenton, also based on the 1749 animal. Most of the material in Buffon's account, however, came from observations of the 1739 London animal by the naturalist James Parsons.[97]

Amazingly, the first scholarly account of the animal appeared within a month after the opening of the fair, published in pamphlet form and offered for sale at the entrance to the booth. This anonymous, thirty-page pamphlet, written in the form of a letter to a friend at the Royal Society of London, filled a gap for those who wanted to read up on the rhinoceros; the era of dictionaries, encyclopedias, and manuals of natural history was still a decade away, and there really was nowhere to turn for information.[98] The author, later identified as the abbé Ladvocat, librarian at the Sorbonne, used his own observations to assess the credibility both of Douwe Mout's claims and of rhinoceros lore from travel accounts and sacred and classical texts. Although he accepted some of the claims of the animal's keepers—concerning its gentle character, its ability to swim like a duck, and its love of having tobacco smoke blown in its nose and mouth—he questioned or corrected others. His own measurements proved it to be smaller than reported, and he thought it sounded more like a wheezing cow than a squealing calf. He dismissed as a fable the claim that other animals stand aside with respect when the rhinoceros drinks from a river, but found the famous enmity between the elephant and rhinoceros plausible, based on reports that Jesuits in Abyssinia had often seen elephants that had been gored to death by rhinoceroses. However, he could not confirm Pliny's assertion that the rhinoceros sharpens its horn on rocks to prepare for these fights. The Paris rhinoceros

FIGURE 3.7. Bronze and ormolu rhinoceros clock (Paris, ca. 1749). Although produced at the time of the rhinoceros's visit to Paris, this clock is modeled on Dürer's 1515 woodcut. Sotheby's, New York.

often rubbed its horn on a board, but this must be simply a "natural motion," he concluded, for it obviously was not preparing for a battle against an elephant. And finally, the rhinoceros was much too big to be crushed by an elephant after goring it.[99] Buffon made no reference to Ladvocat's pamphlet, but another leading natural history encyclopedia, by the popular lecturer Valmont de Bomare, copied the *Lettre sur le rhinocéros* almost word for word.[100]

The next big hit was a young male elephant (the one with which this book opened), the first to be seen in Paris since 1668, shown at the Saint Germain fair in 1771 and 1772 and on the boulevard. As it had on the street, the elephant demonstrated its agile trunk at the fair by grasping, uncorking, and downing bottles of beer. Another elephant, a young female, was on display in 1773, but the *Avant-coureur* noted that she would probably receive less attention—presumably the novelty had already worn off.[101]

In 1771, in contrast to the situation in 1749 when the rhinoceros was in town, one could now educate oneself before observing the beast: Buffon's long and complimentary history of the elephant had appeared as the lead article in volume 11 of the *Histoire naturelle* (1764). The *Affiches de Province* advised its readers to read the description in the portable edition of Buffon before going to see the animal; dandies who might otherwise consider its ugliness unfit for their regard were particularly urged to do so. The *Avant-coureur* recommended either Buffon or Valmont de Bomare's *Dictionnaire d'histoire naturelle* for further reading.[102]

Buffon himself visited the male elephant in 1771 and the female in 1773, and he used his observations for a supplemental article published in 1776. He conveyed the information he got from their keepers in a description that confirmed behavioral differences between the sexes as well as the elephant's notorious sense of justice:

> In elephants, as in all species, the female is gentler than the male; this one was even affectionate toward people she didn't know; on the contrary, the male is often fierce [*redoubtable*]. The one we saw in 1771 was prouder, more aloof [*indifférent*], and much less tractable than the female. . . . He even tried to grab with his trunk people who came too close, and he often snatched away the pocket books and caps of onlookers. His keepers even had to take precautions with him, whereas the female readily obeyed. The only time she

became upset was when she was being loaded into her traveling wagon. She refused to go in, and the only way to force her was by poking her rear with an awl. . . . Annoyed by the bad treatment she had just experienced, not being able to turn around in her tight prison, she took the only form of vengeance she could; she filled her trunk and sprayed a bucketful of water on the face and body of the one who had harassed her the most.[103]

The elephant did not pay a state visit, as the rhinoceros had, but a published print carried deferential verses portraying it as a loyal subject:

> I leave Asia with no sorrow
> And renounce my country with no regret
> To live forever under the laws
> Of the most powerful, the most loved of Kings.[104]

This anonymous verse could well have been propagated through official channels. Louis XV would undoubtedly have enjoyed receiving an expression of tribute from India, which the British had routed the French from not many years before.

The elephant, like the rhinoceros, generated a free-standing publication, *Mémoires de l'éléphant,* but the two works differ in style and tone. Unlike the *Lettre sur le rhinocéros,* the elephant's memoir is written in first person, following a very popular genre that took off from Montesquieu's *Persian Letters* (1721) and included works written by an Iroquois, a monkey, and many others, in which outsiders commented on French society. In keeping with this genre, the content has a critical edge, as well as a more sentimental tone. Although the memoir summarizes natural history information and elephant stories from the classics, it has an element missing from the rhinoceros work; the elephant is presented as a noble savage who has provocative things to say about French society. While in Paris, the elephant narrator notes, he will be observing and learning about people: "I'll be able to brag that they paid to see me, while I saw them for free." He claims that elephants bend their knees before kings but not tyrants, and that they use reason all the time, unlike some men. He hopes that in writing his memoir he will convince men that simple, modest, and natural customs are preferable to artificial ones. Elephants themselves do not do well away from their natural environment. They mate only in the deep forest. When in captivity, "the individual alone remains a slave, while the species stays independent and refuses to reproduce for the profit of the tyrant who has taken away his liberty" (apparently the elephant had read Buffon, from whom this passage is taken).[105] Meshing duly documented facts and observations about elephants with political and social commentary, this tract shares the tone of many popular natural history works from the late eighteenth century, as we will see in chapter 6. The critical tone, however, earned a slap from the *Avant-coureur,* which objected to the sarcasm against "respectable nations."[106]

In their enclaves at the fairgrounds and on the boulevard, zoological expatriates offered Parisians an encounter with the wild and exotic. They also acted as convenient vehicles for moral lessons, social satire, and political criticism. But their owners offered animals for observation only; a different group, the oiseleurs, offered a chance to buy.

CHAPTER 4

The Oiseleurs' Guild

AS HE WALKED by the corner of the rue de Sèvres and the rue de la Jouial-
lerie on a Sunday morning in June 1749, Jean Baptiste Gautreau, a forty-
two-year-old master box maker, noticed a big crowd gathered outside the
Cabaret des Trois Bouteilles (Tavern of the Three Bottles). He stopped to ask
what was going on, and someone told him that the *oiseleurs*—members of
the bird-sellers' guild—were trying to keep some people from selling birds.
When he got closer he saw that a thin, tall man had knocked down one of
the oiseleurs he knew, François Bertin, and was yelling to his son to join in.
Several people piled on top of Bertin, who "would have perished" if fellow
oiseleur Gabriel Adam hadn't come to his aid. Adam himself got beaten
up—Gautreau said he saw him bleeding from a wound on his neck—but
more reinforcements arrived on the oiseleurs' side, and after about fifteen
minutes the fight was over.[1]

Bertin and Adam brought charges of battery against their assailant, a
middle-aged cobbler named Pierre Bernard Thuillier. During his interro-

gation, Thuillier gave a different version of the tussle. He told the king's counselor that he and his son had gone down to the tavern that morning with a cage containing five "large-beaked sparrows," which they were intending to trade with a Mlle Boucher for five canaries. They had some drinks with their friend Carabin, a maker of ornamented buttons, and a few other people. Adam, Bertin, and several other oiseleurs were also in the tavern. Thuillier's son accused Adam of having called his father a jackass (*Jean foutre*) and of having said other insulting things about him; Adam responded with his fists, and the fight was on. Thuillier contested Adam's and Bertin's claims that they had been injured, declaring that the oiseleurs had hardly gotten a scratch, while he had had his "head broken." His friend Carabin, who had stayed out of the fight, guarding the birds, agreed that the only blood that had been spilled was Thuillier's, who had fallen and hit his head during the scuffle inside the tavern and then was knocked down into a gutter in the street.[2]

The questions that the interrogators asked of Thuillier and Carabin in trying to decide how to settle the case revolved around not only who hit whom but who was at fault in a more fundamental way. The counselor asked Thuillier, who admitted to sometimes selling birds on Sundays and feast days, "if he had the right to engage in bird selling." Thuillier claimed that he had twice paid off the jurors (*jurés*—elected guild officials) in return for letting him continue his "inconsequential" business.[3] Indeed, after the fight, he had gone to the shop of another oiseleur he knew, named Chateau, to sell him the five sparrows. Carabin, who also sold birds on the side, was asked whether he was aware that there was a guild in Paris that had the sole right to engage in such commerce. Carabin replied that he wasn't really in the business, but simply bought birds for pleasure and then sold them, often to the oiseleurs themselves, and that he had on occasion bought parakeets from Gabriel Adam.

This altercation was typical of the problems that beset the oiseleurs, one of the smaller and least known of Parisian guilds, or artisans' and merchants' associations (fig. 4.1).[4] The oiseleurs held the privilege on the sale of birds and small animals of pleasure: Parisians wanting a goldfinch, canary, parrot, or monkey, or simply interested in seeing exotic animals, knew to head for the shops of merchants like Adam and Bertin along the quai de la Mégisserie, just across the Seine from the Palais de Justice on the île de la Cité.

FIGURE 4.1. A bird merchant selling canaries, 1774. The illustration may show a traveling merchant rather than a member of the oiseleurs' guild (oiseleurs were forbidden to sell birds on the street). Detail from anonymous illustration, "Marchande d'huîtres et Marchand d'oiseaux, en 1774." Musée Carnavalet. © Photothèque des Musées de la Ville de Paris / photo: Pierrain.

Like all guilds, it was officially connected to the monarchy through the network of privileges and responsibilities that constituted its founding statutes. Guild members supplied the king's menagerie and aviaries and also provided a ceremonial function by releasing hundreds of birds from cages during certain royal spectacles.

Political and economic shifts during the eighteenth century disrupted the guild's monopoly and its privileged status. Some of these changes were connected with the movement for reform of ancien-régime institutions that affected all of the guilds; others related to the commercial revolution and, particularly, the growing availability of exotic animals. Oiseleurs took advantage of this influx by stocking macaws, African firefinches, and capuchin monkeys along with the traditional native species and by pitching their wares to amateur naturalists. But at the same time, the oiseleurs found themselves competing more and more with private entrepreneurs like Thuillier and Carabin. Although Gabriel Adam and other feisty oiseleurs found plenty of reasons for legal suits and fistfights, unauthorized sales became one of the central points of contention in the later eighteenth century. The history of the king's menagerie and of street fairs and displays revealed the fate of spectacular and rare exotic fauna. Through the history of the guild and its members, it is possible to trace the growing traffic in small exotic birds and animals—those brought back in large quantities by seamen—and to witness in microcosm the clash between ancien-régime structures and commercial growth.

Guild Privileges and Responsibilities

As members of an official guild, the oiseleurs were governed by a set of statutes and regulations authorized by the king and approved by the Parlement of Paris. These statutes were the basis of their status as a *communauté*. Such communities were *corporations*, institutions in ancien-régime France that were accorded privileges by the king in exchange for certain responsibilities. Groups as divergent as universities, the *bourgeois de Paris*, and water carriers enjoyed corporate status. Although a few trades were free, and "privileged" workers in some areas of Paris were not required to join corporations, most skilled workers belonged to one of the approximately 120 Parisian guilds. In theory, the guild structure guaranteed to the public the availability and quality of the product sold or fabricated, and it offered to its members, the masters, financial, social, and legal services.[5]

Bird selling is quite a different sort of trade than bread baking or barrel making; the fact that its privileges dated back to an ordinance of 1402 indicates the long tradition in France of keeping caged birds as pets, as well as the important ceremonial role performed by the oiseleurs. The guild fell under the jurisdiction of the branch of royal government called Eaux et Forêts (streams and forests), which oversaw activities such as lumbering, hunting, and fishing, and its statutes, which were periodically updated, covered the internal functioning of the guild, buying and selling practices, and duties to the Crown, as well as rules applying to bird merchants from foreign lands.[6] As in other guilds, the profession tended to run in families because both sons and sons-in-law could become masters without serving apprenticeships; others were required to serve a three-year apprenticeship and pay a larger fee. Widows of masters could (and did) take over their husbands' businesses. The guild was administered by three jurors *(jurés)*, who were elected for two-year terms and required to visit each master two times a year to make sure the regulations were being followed.

Only master oiseleurs were allowed to sell birds and other small animals of pleasure in Paris and its vicinity and to make birdcages and traps; but this commerce was subject to various restrictions. Trapping was forbidden on royal lands and during nesting season (from mid-March to mid-August), street-peddling was not allowed, and female birds could not be sold as male birds (the latter were more expensive because more gaudy and vocal). In addition to maintaining shops, oiseleurs could sell on the pont au Change (Money changers' bridge) and the adjoining vallée de Misère (Valley of Misery) on Sundays and holidays from 10 A.M. to 1:30 P.M., except on solemn feast days such as Christmas and Easter, the day of the feast of Saint John the Baptist, and days of jubilees and processions.[7]

The oiseleurs' regulations also covered the activities of the *marchands forains*, or traveling merchants, who were supposed to follow a strict protocol upon arrival in Paris. First, they had to display their wares from 10 A.M. to noon at the Table de Marbre (tribunal of Eaux et Forêts) during the entry of the Paris Parlement, at which time the maître des Eaux et Forêts determined how much tax they had to pay and what price they could charge for their birds; these rates had to be affixed to the cages. Before any could be

FIGURE 4.2. Oiseleurs performing the duty of their guild by releasing birds
for Charles VII's entry into Paris. Miniature from *Chroniques
d'Enguerrand de Monstrelet*, BnF, fonds Lavallière, no. 20361, costumes
du seizième siècle.

sold, the governors of His Majesty's aviaries (who were to be notified by
the jurors) picked out the choicest specimens. Then each juror received one
bird from each hutch. After that, the merchants sold first to people bearing
the designation bourgeois de Paris, then to the oiseleurs, and finally to the
general public.[8] Like the oiseleurs, the merchants were forbidden to sell
female birds as male birds. They also were required to keep sick birds in
separate cages, so that buyers would not be tricked into buying unhealthy
birds. Oiseleurs were forbidden to intercept these merchants on their way
into Paris in order to buy birds at cheaper prices.

In exchange for their privileges, the guild had responsibilities to perform.
The primary one required the three jurors to go to Reims when a new king
was crowned to release four hundred birds in the cathedral during the ceremony (the Sacre du Roi), "as a sign of joy and liberty." Birds were also to
be released when new queens entered Paris (Entrées des Reines), on the day
of Corpus Christi (Fête-Dieu), and on other important occasions (fig. 4.2).

The physician and ornithologist Pierre-Jean-Claude Mauduyt de la Varenne explained the symbolism of this ritual in his entry for the term *oiselier* in the *Encyclopédie méthodique* (1784): the release of the birds during the coronation, he wrote, "seems to indicate that our kings are regarded as our liberators; it is they in effect who have freed us from the barbarity of feudal government, and the prince under whom we live proved soon after having witnessed this custom performed in his honor that it could and should be applied to him, in the moral sense, as legitimately as to any of our kings who have been most zealous for the happiness and liberty of their subjects."[9]

In February 1779 the oiseleurs fulfilled their obligation to the monarchy when they obeyed an order to release four hundred birds in Notre Dame Cathedral to honor Marie Antoinette's delivery of a child. Although the ceremony featured traditional symbolism, the queen decided to include an innovative event that would capture the spirit of reform. As part of the festivities, she arranged for a mass marriage ceremony of one hundred "poor and virtuous girls" with one hundred "honest artisans" in Notre Dame.[10] The event did not escape the sarcasm of social commentators; one wag was inspired by the ironic juxtaposition of marriage ties and liberated birds to compose a poem that circulated around Paris:

> In Notre Dame de Paris
> One hundred birds leave bondage;
> At the same time one hundred girls and one hundred boys
> Fall into the trap of marriage.[11]

In addition to ceremony, monarchs wanted their own birds—rare birds, rarer birds than anyone else had. Thus, not only did the king have first pick from the marchands forains, he also designated one of the oiseleurs as the *oiseleur du roi,* or the king's bird seller. At the end of the seventeenth century the oiseleur du roi lived on rue Saint Antoine, with a gold-lettered sign outside his shop reading "governor, preceptor, and regent of the birds, parrots, [and] monkeys of His Majesty."[12]

In 1762, Ange-Auguste Chateau and his son Ange became the oiseleurs du roi and thus acquired privileges and duties beyond those of regular guild members. The royal warrant *(brevet)* stated that Chateau and his son had been chosen because of "the knowledge that [they] had acquired about different kinds of birds and animals from foreign countries and about how to feed and conserve them." The title allowed them "to go freely into all the ports, towns, regions, and maritime coasts of the kingdom in order to buy by mutual agreement birds and animals of foreign countries, even to have the preference, above all others, of those that are suitable for the menageries and aviaries of His Majesty." The usually cumbersome customs regulations were also relaxed:

> His Majesty commands all governors of provinces, towns, and regions under his dominion, all *intendants* and *commissaires* in the provinces and *généralités* [administrative subdivisions], and all judges, mayors, municipal magistrates, town consuls, and all other officers to let the said Chateau father and son freely pass, with the birds, animals, and grains for their food, that they might have bought in order to supply the aviaries and menageries of His Majesty, without being molested or disturbed; he commands, on the contrary, that they be given help to further this commerce.

In 1782, following Ange-Auguste Chateau's death, his son-in-law Gérard Auguste Bastriès acquired the title oiseleur du roi, and four years later, Chateau's son Ange renewed his warrant.[13]

The elaborate rules and regulations that anchored the guild to ancien-régime institutions suddenly gave way in 1776. In February of that year, the finance minister, Turgot, convinced by arguments in favor of the unrestricted circulation of goods, abolished all of the guilds, cutting them loose to float in the free market. Barraged with complaints, the king reinstated approximately half the guilds six months later, but the others, including the oiseleurs, remained "free" professions. Now anyone could become a master oiseleur simply by registering with the maître particulier des Eaux et Forêts. Limited regulations forbade taking quail, partridge, and pheasant eggs and restricted collection on land owned by the king or nobles, but there was no penalty for failing to register.

Despite having been freed, however, the oiseleurs were still constrained to provide winged liberty symbols by continuing the "ancient custom" of releasing no fewer than four hundred birds for each coronation and for other royal ceremonies.[14] The oiseleurs complained about this new system in a petition addressed to Eaux et Forêts in 1779—just six months after they had supplied Marie Antoinette's birds—in which they detailed the "infinite" abuses that had occurred since the reforms. The oiseleurs requested that anyone desiring to exercise the profession take an oath at the head-quarters of Eaux et Forêts or else pay a fine of 100 livres and have their merchandise confiscated.[15] The petition begins by praising the king for being so concerned with the happiness of his people that he wants everyone to be able to exercise their talents without having to buy the right to do so, but it points out that the profession of oiseleur must be distinguished from the others that were liberalized; surely the king does not want people to collect eggs during prohibited times or from the king's lands without permission. In addition, if the profession remains free, it will be impossible to compel the members to pay the expenses required to release birds for various festive occasions, and the syndics and jurors cannot afford it themselves. In other words, only in a regulated society can people be forced to furnish symbols of freedom.

The Business of Selling Birds

With a broad picture of the guild's structure and transformations in mind, let us now take a closer look at the master oiseleurs, their goods, their customers, and their competitors. A reconstruction of the daily lives and business practices of the oiseleurs is limited by the available sources: legal records for the most part, supplemented by notices and advertisements in periodicals, and references in natural history texts. Court documents (from the tribunal of Eaux et Forêts) reveal some basic characteristics of the guild and what sorts of conflicts agitated it. Many of the disputes were similar to those that arose in other guilds—members neglecting to pay their dues, arguments over elections of masters and jurors, maintenance of boundaries with related guilds (such as the turners, whose fabrication of birdcages encroached on the oiseleurs' territory), and general bad behavior and rowdiness (for instance, when Gabriel Adam burst into a dinner he hadn't been invited to and smashed all the dishes).[16] There were also, however, a number of cases like the one with which this chapter opened, which tell something about what kinds of birds the oiseleurs sold, how their stocks changed, what sort of competition they faced from uncertified bird sellers, and a little bit about the oiseleurs themselves. Notices in periodicals and comments in natural history texts provide additional information about the oiseleurs' wares, their business practices, and their customers.

The oiseleurs' guild was a small one: the number of masters remained under 50 during the entire eighteenth century (as compared with 580 bakers and 2,500 mercers in the mid-eighteenth century).[17] A 1746 roll listed 38 masters and 3 widows, and meetings generally mustered between 25 and 30 masters and widows.[18] Several families and individuals dominated the guild: the Bourrienne family, for example, was represented by Jean in the early eighteenth century, his son René through midcentury, and grandson Jean René toward the end of the century. New family names continued to appear, however, suggesting continued outside recruitment (although some might be sons-in-law). Because most shops were family affairs, the oiseleurs' guild did not experience the disputes between apprentices and masters that frequently broke out in the larger guilds.

The majority of oiseleurs for whom I have found addresses lived on or near the quai de la Mégisserie, a bustling, congested area on the right bank of the Seine, not far from Notre Dame Cathedral. Pet shops are still found there, and on Sunday mornings tourists and city pigeons gather around to gawk at the cages on the sidewalk. Unlike today, however, in the eighteenth century most oiseleurs probably rented the small building containing their shop and lived in rooms above it.[19]

Indirect evidence on income and educational level suggests that oiseleurs were on the low side on both counts, compared with other guilds. From signatures on documents, one can estimate that about one-half to two-thirds of masters who attended meetings were literate (though of these, some spelled out their names in the shaky block letters of someone who rarely uses a pen); this figure places them near the low end of guild members at the time.[20] An inventory of the possessions of Antoine Vallin (or Vaslin), one of the more prominent masters, after his death in 1785, shows that although he did not live in poverty, neither did his standard of living approach that of the top artisans in the bakers' and other large guilds. Vallin did own a mirror in a gilt frame and a watch in a gold box, but so did growing numbers of lower-class people by the end of the eighteenth century.[21] This inventory, six pages that look as though they were scrawled speedily by a busy notary, affords the rare opportunity to take a tour of a oiseleur's house and shop.[22]

Vallin became a master in the 1760s, after some controversy. Several oiseleurs filed a suit in 1763 opposing his reception because he was illegally operating his own shop before having completed the three-year apprentice-ship. Vallin, identified as a "bourgeois de Paris," apparently said he would donate 300 livres to be admitted, and some members felt obliged to accept his offer because of a recent theft of guild funds.[23] By 1768 Vallin was a juror and active in guild affairs. When he died, in 1785, he was identified as "oiseleur de Monsieur"—that is, he had some sort of arrangement with the comte de Provence (Louis XVI's brother, known as Monsieur), probably an agreement to supply birds in return for use of the title. He and his wife, Genevieve Vingue, had three children: a daughter who married the oiseleur Jacques Duval, a son who was a student at an academy of painting, and a second daughter who had married a metal caster but had since died.

When the notary entered the dwelling on the quai de la Mégisserie, he noted that it consisted of three rooms, as well as an attic, a cellar, and some storerooms. Home and shop intermingled, with cages stacked everywhere. In the bedroom were typical furnishings—two nice beds and one rudimen-tary one, the gilt-framed mirror, a couple of chairs, and blue-and-white cur-tains of a cotton and silk fabric *(siamoise)*—along with the trappings of Vallin's trade: nine canary hutches, several parrot stands, a bird trap, and more than a dozen cages. The room behind the shop, where the Vallin fam-ily cooked and ate by the fireplace, contained a table and seven old straw chairs, an armoire and two buffets, dishes, linens, brass candlesticks, the watch in its gold case, a pair of silver buckles, and fifteen bottles of red wine. Also, in what must have been a very crowded room, were a commode filled with hardware for cages, five aviaries, three dozen other cages, boxes full of birdseed, red partridge eggs, and a little monkey in a crate *(boete)*. It is im-possible to tell if the monkey was merchandise being kept warm by the fire, a family pet, or both; but surely the family must have interacted with it in some way. A storeroom at the top of the building housed specialized cages for parrots, canaries, squirrels, and pigeons, along with several for decoys *(chanterelles* and *appellants)*, suggesting that Vallin must have trapped birds himself. The boutique, which opened off the street, would have resounded with squawking, quacking, singing, and cooing: there, the notary counted ten parrots, three parakeets, one cock, five pigeons, one guineafowl, one Chinese pheasant, two quails, a large aviary filled with ducks and hens, canary cages (with canaries inside? the inventory does not say), forty-three cages of different sizes, four aviaries, and several other hutches and cages.

The quantity and variety of exotic species in Vallin's shop reflect the changes in the kinds of birds and small animals sold by oiseleurs and in the ways they acquired them. Although I do not have an inventory from earlier in the eighteenth century to use for comparison, it is likely that it would have had a much larger proportion of native species. Traditionally, many oiseleurs (although probably not all) had trapped native birds themselves. During the course of the eighteenth century, a greater proportion bought birds from others. At the same time, oiseleurs began to stock more and more exotic species—first canaries, and later birds and miscellaneous mammals from West Africa, the Americas, and the East Indies.

The guild's trapping skills were stretched during coronations, when they had to collect and release four hundred birds in the cathedral at Reims. For ceremonies that took place in Paris, each master and widow supplied a certain number of birds, which they might well have bought rather than caught (masters were regularly fined for not supplying their quota).[24] But birds could neither be taken from Paris to Reims (they would not survive) nor bought in Reims (there were no supplies), so the oiseleurs had to catch them locally in the days before the ceremony. The contrast between the two eighteenth-century coronations, in 1722 and 1775, suggests that trapping was a diminishing skill among Parisian masters. Before both coronations the entire guild met to decide who would go to Reims, along with the three jurors, to do the hunting. In 1722, the oiseleurs elected Guillaume la Cour, Pierre le Comte, and Jacques Landiers to be the hunters. Two of the three were relatively prominent members of the community: Guillaume la Cour and Pierre le Comte had both been jurors in 1698. In 1775, the choice was easy: the guild picked two obscure masters, Pierre Baveux and Antoine Chaumont, who were described as the "only hunters in the community."[25]

Rather than do their own trapping, oiseleurs might go out to the country to buy birds. When they returned, they were supposed to divide the birds up among all the masters: a guild was a community, not a collection of individual merchants trying to accumulate goods for themselves. In the mid-seventeenth century, Noel Panchardy received permission to travel twenty lieux (about fifty miles) from Paris to buy siskins, goldfinches, linnets, chaffinches, and buntings; after supplying the needs of the king's aviaries, he was required to divide up his purchases among the other masters if he brought back more than seven dozen.[26] In the eighteenth century, we find masters traveling as far as Picardie and Soissons, in northeastern France.[27] The *lotisse-ment,* or doling out, took place at the guild office, and did not always go smoothly. In 1697, Marie Racine, a master's widow, was fined for "violence" and lack of respect to the jurors, and in 1720 a fight broke out between the Briere and Rossignol families over whether Briere or his son should be sent to claim their share of starlings.[28] On occasion, the communal spirit lagged and masters tried to keep the birds for themselves. One who got caught was Jacques Laubet, a former juror, who was fined 20 livres after fellow oise-leurs found thirteen to fourteen dozen goldfinches that he had illegally bought in the country during nesting season and had spirited into his boutique at night.[29]

A more serious event occurred in 1721, when the jurors complained to the maître des Eaux et Forêts that three or four members of the community were engaging in the practice, contrary to the guild's statutes, of going *sur les lieux* (i.e., out into the countryside) to buy birds from marchands forains before the merchants could bring them into Paris; these individuals then sold the birds at high prices to the other oiseleurs, rather than bringing them to the guild office for inspection and allotment. Most of the masters weren't able to do this, they complained, and it wasn't fair that these few should profit by their abuse. The situation arose from a general lack of birds after the devastating winter of 1709. The jurors also pointed out that this prac-tice was keeping the bird population from recovering (one purpose of the guild was to protect the resource) and that the birds were of low quality because they were not being checked by the jurors (another purpose of guilds was to guarantee high quality). Warning that this massive abuse was "of great importance, [and] capable of ruining the entire guild," the jurors asked for an order forbidding anyone to buy birds out of town for ten years, with a fifty-livre fine for any breach.[30]

The Growing Trade in Exotic Species

As colonial commerce grew in the early modern period, Parisian merchants began to sell imported goods such as coffee, sugar, tea, and Indian fabrics, and the oiseleurs began to sell exotic birds and small animals that they had bought in port cities or from merchants who came to Paris. The records reveal some aspects of this commerce, although they are mute on topics such as what it was like to travel for several days over rough roads with cages full of noisy birds or active monkeys.

Proclamations regulating commerce in exotic species date back to the earliest days of foreign trade; regulations from 1618 specified who could import exotic birds, how the birds were to be distributed once they arrived in Paris, and how much could be charged for them. In all of these decrees a more or less explicit concern is the king's desire to skim off the best specimens for himself. A 1618 judgment, for example, gave Pierre Cosson, "bird seller in Paris and keeper of the king's parrots," permission to go to Rouen, Saint Malo, Havre de Grace, and Dieppe to buy canaries, on condition that he first supply the king's aviaries "so that they will not get the leftovers of other cities, as in the past."[31]

Although there are records of parrots, parakeets, and other African and American birds being imported in the late seventeenth and early eighteenth centuries, the most common exotic import at that time was the canary. Acquiring rare varieties became a fad among well-to-do Parisians, who also did some of their own breeding and hybridizing, and prices soared. The inflated prices, along with the fashion, diminished in the latter part of the eighteenth century, perhaps because of greater availability and because other exotic birds began to arrive in larger numbers.[32] During what might be called the "canary bubble," the oiseleurs had a difficult time; they were competing both against Swiss merchants who set up shop in Paris twice a year and against Parisian fanciers.

Early in this period, canaries were actually exported to France from the Canary Islands. In 1618, a Portuguese merchant landed at Saint Malo with twelve hundred canaries; he sold three-quarters of them in Nantes and Rennes and took the rest to Paris.[33] But some time in the seventeenth century, bird merchants from the Tyrol and Switzerland began raising and breeding canaries and transporting them all over Europe. Twice a year, in fall and spring, they took several hundred canaries of different plumage types to Paris, where they set up at an inn, the *Boule Blanche* (White ball), on the rue du faubourg Saint Antoine. They renovated one wing of the inn with a special floor and Swiss-style oven.

The arrival of the bird merchants was such an event that a near-riot among canary fanciers broke out in 1696. Three Swiss merchants explained in a petition that on their way to Paris they had crossed paths with Joseph and Jean Steur, homeward-bound compatriots who warned them about the trouble they had just experienced there. The Steurs reported that after they had gone to the city authorities to obtain permission for their commerce, a group of people had followed them to the inn and made vehement demands—to buy several dozen birds at once, or only female birds, or that the merchants bring their wares into the center of the city. The next day these people had returned and forced their way in, "swearing and threatening to kill all the birds." Some took birds without paying; others cheated by pulling white feathers from special grey-and-white birds and then trying to pay the lower price for the common grey bird (at this time common canaries were grey, not yellow). Because of the harassment that the Steurs had been subjected to and the rigor of the cold weather, the merchants asked that the prices of the canaries be set higher than in the past, and they warned that if such disorder continued, they would take their birds to countries where they could find more protection and better facilities. Eaux et Forêts officials set prices (from 50 sous for a grey female to 22 livres for a white male) that were lower than what the merchants had requested, but they established rules of order: a guard had to be present during the sale, no more than two people could be in the room at once, and each person could buy only eight birds at most. They reconfirmed that the birds had to be exposed at the Table de Marbre and that the king got first choice, followed by the bourgeois de Paris, and then the bird sellers.[34]

Not everyone viewed the Swiss merchants as mistreated victims. The author of a treatise on canaries published in 1705 depicted them as greedy and sly: if you show them your money, he wrote, they are courteous and humble, but if you ask questions without buying, they treat you very rudely—thus confirming the proverb "No money, no Swiss." He accused them of not caring if the birds died after they were sold, but remarked that they had been unwise to sell so many pairs, for Parisians had begun breeding canaries themselves and would no longer have to wait to buy them from the foreigners.[35]

With the bourgeois getting their pick before the oiseleurs, it is not surprising that conflict arose between the two. In 1692 the bourgeois brought a sort of class-action suit against the oiseleurs because of incidents such as one in which the oiseleurs seized eighty canaries from Jean Micolet, a tailor, and all but three or four of the birds died while being guarded. The bourgeois declared their right to own, feed, raise, and breed "as many canaries and other birds as they liked" and requested that the bird sellers be prohibited from entering their houses and bothering them. The oiseleurs were fined 100 livres. In a later suit, the oiseleurs declared themselves willing to go along with the bourgeois' right to buy birds for twenty-four hours before the oiseleurs and to raise and sell them, but they continued to request that bourgeois be prohibited from selling them from their homes, or buying them in order to sell them, because "that would constitute engaging in the practice of bird selling."[36] Maintaining their privileges was a constant effort.

To acquire exotic birds other than canaries, oiseleurs sometimes traveled to ports to purchase them there, for which they required special permission. In 1720, for instance, one master obtained a certificate allowing him to travel to Le Havre:

> Today, the 18th of November seventeen hundred and twenty, Claude Cagny, master oiseleur in Paris, residing on the quai of the Valley of Misery, parish of St. Germain Auxerrois, appeared at the clerk's office [and] declared that he is leaving this city to travel to Havre de Grace to buy parrots and parakeets and other foreign *[estrangers]* birds for his bird-selling commerce. . . .
>
> [signed] cani[37]

Merchants also brought exotic birds other than canaries to Paris. In the 1680s, a Dutch merchant twice obtained permission to sell parrots after he had "satisfied" the king's aviaries and been visited by the jurors.[38] Only a few sporadic documents, mostly from before the surge in exotic animal imports, record these exchanges; I discovered little about sources of birds in the later eighteenth century. Some generalizations can be made, however, about the variety of species acquired, how the market changed, and how the influx of exotic animals led to competition not from other privileged groups, like the bourgeois de Paris, but from individual entrepreneurs like Thuillier.

The example of sales of cardinals over a hundred-year period shows just how much more prevalent and affordable exotic species became during the course of the eighteenth century. In 1682, the prince de Condé, who headed the most powerful family in France after the royal family, bought a cardinal and a monkey from a oiseleur after prolonged negotiations. The abbé de la Victoire acted as intermediary, writing from Paris to the prince at his estate in Chantilly that a oiseleur had shown him an extraordinary bird, a "coq de Virginie," just arrived from the Indies. It was the color of fire, crested and red-beaked, gay, lively, and well proportioned. The oiseleur assured him that it sang like a nightingale—though he had heard only its pleasant whistle—and that it would live a long time. The oiseleur was also selling a gentle young female monkey that the abbé thought might enliven the prince's elderly male monkey; and wouldn't it be pleasant if the two produced offspring? In his next letter, the abbé detailed the bargaining process. The oiseleur initially asked 25 louis for the two (600 livres, or about two years' wages for an artisan), but the abbé talked him down to 10 louis for the bird and 6 louis for the monkey. Although the price was still high (10 louis could have bought five writing desks), the abbé agreed, because he did not want to risk losing out to an unexpected competitor—a M. de Crequy, who muttered that he would pay 12 louis for the bird and would take it the next day to Versailles (presumably for the king). The abbé again described the charms of the bird and its ease of care ("no more trouble than a caged canary"), and added that 6 louis wasn't much for the monkey, especially since she did whatever you wanted and didn't bite.[39] During the late seventeenth century, oiseleurs probably supplied expensive and rare pets primarily to people, like the prince, in the upper reaches of the aristocracy.

A century later, cardinals were no longer rare luxuries. They often appeared in oiseleurs' shops—except during wars, when transatlantic traffic slowed down—and their availability was widely advertised in the *Affiches*.[40] After the liberalization of the guilds, they even escaped from the hands of the oiseleurs. A *limonadier* (café owner) offered through the pages of the *Affiches de Paris* in 1788, "10 CARDINALS with scarlet plumage, having a 1-inch crest on the head, and the beak crowned with black; they have a very pleasant song and quickly learn airs from a bird-organ."[41] Two years later, the *oiselier du prix fixe* (fixed-price bird seller) announced a cardinal for sale for 72 livres (3 louis).[42]

Cardinals were only one of many exotic species sold by oiseleurs. Knowing that exoticism appealed to consumers, oiseleurs frequently mentioned the foreign origins of their goods in their advertisements.[43] Reflecting trade routes and hardiness as well as customers' preferences, the most common species stocked were parrots, parakeets, and small African finches. In the *Encyclopédie méthodique,* Mauduyt noted having seen in oiseleurs' boutiques fifteen to twenty different types of parrots and parakeets from Africa, South America, and the Caribbean (seven of which he described as "common"), as well as waxbills, grenadiers, and whydahs from Africa, indigo and painted buntings from North America, and Java sparrows from the East Indies.[44]

The oiseleurs particularly sought out species that were pretty, talked well, and were not mean or too noisy. The *jaco*, or grey parrot, from West Africa, was the most popular, being a quick mimic and an affectionate companion. Oiseleurs also valued distinctive or rare animals for which they could charge high prices, like parrots of the kind that Bougainville had attempted to take home from Montevideo, with multicolored feathers created by injections at the feather roots.[45] The most prominent oiseleur in Paris, Ange-Auguste Chateau, the oiseleur du roi, offered, along with the usual parrots and finches, an extraordinary variety of species, from cockatoos to whydahs to blue jays.[46] He allowed an African crowned crane, or Royal Bird (appropriate for the king's bird seller), the run of his boutique, where it play-attacked customers when they came in the door.[47] Chateau's privileged position must have helped him put together this impressive collection. In the 1780s, his son expanded the business, keeping the shop on the quai de le Mégisserie but also opening a menagerie on the rue des Postes. In addition to pet cage birds, Chateau and later both his son and his son-in-law supplied ducks, swans, and peacocks, and eggs of pheasants, partridges, and guineafowl, both native and exotic, for purposes of hunting and ornamentation on large estates.[48]

Oiseleurs also sold small animals: advertisements in the *Affiches* listed a loris (a type of lemur), gentle capuchin monkeys, a green monkey, and monkeys identified as *Mongous* (from Madagascar—lemurs?), *la nonne*, *Magabe*, and *Indiens* (a pair).[49] Buffon saw a rare kinkajou at the boutique of a oiseleur on the rue de Richelieu.[50] Chateau sold squirrels and a gentle and odorless "English mouse," his son sold little American goats, and his son-in-law advertised a very interesting sounding animal, "the Insect called the *Curculio imperialis* from Brazil" (probably a large weevil).[51]

This variety of species attracted naturalists to the oiseleurs' boutiques. Oiseleurs do not figure in histories of natural history or ornithology, but they probably should. Mauduyt frequented oiseleurs' shops, and Buffon and his collaborator Guéneau de Montbeillard also visited them, especially that of Chateau. Sometimes the only live individual a naturalist ever observed of a species was *"chez un oiseleur."*[52] Mauduyt asked Chateau what he fed different species and how long they lived. Although he referred to him as a distinguished bird seller *("oiselier distingué dans son état")*, he did on occasion disagree with his recommendations; if Chateau had tried feeding his Java sparrows with rice rather than birdseed, he suggested, they might have lived longer.[53] Buffon found out from "very intelligent oiseleurs" that it was not a good practice to replace canary eggs with ivory ones so that they would all hatch at once.[54] When Guéneau de Montbeillard was trying to differentiate the siskin from the redpoll, he showed a picture of the latter to a "oiseleur with much experience but little learning," who recognized it and explained that the females of the two species resembled each other. Chateau gave him information about the indigo bunting and the cordon bleu. "We owe everything we know about [the indigo bunting's] history" to Chateau, he remarked, and he credited Chateau for information about how cordons bleus are caught in Senegal, their molting, and their lifespan.[55]

The oiseleurs knew that amateur naturalists as well as prominent men like Mauduyt were interested in unusual species. Chateau capitalized on the natural history fad by stressing in his publicity the scientific value of his merchandise. The preamble to the list of species in his boutique, which appeared in the *Avant-coureur* in 1773, touted the rarity, beauty, and singing ability of the birds, but it also addressed itself to those with more intellectual interests: "Each family of these birds . . . has different characters that the Naturalist will take pleasure in remarking. . . . We will give a short notice of these birds with the prices that sieur Chateau has proposed for amateurs of this part of Natural History."[56] A similar list that had appeared five years earlier referred to Chateau's collection as an "animated and interesting spectacle" that would give pleasure equally to the amateur *("le Curieux")* and the scientist *("le Physicien")*, who would enjoy contemplating the fascinating differences between these foreign birds and those of "our continent." The article, which listed twenty-two species, referred readers to Valmont de Bomare's *Dictionnaire d'histoire naturelle* for a description of one of Chateau's animals, a flying squirrel.[57]

The oiseleurs may have contributed to public interest in exotic species; they certainly made use of that interest in their sales pitches. In this way, the influx of exotic species probably benefited their business. At the same time, however, this increased availability made it easier for people outside the guild to acquire and market exotic animals. During the second half of the eighteenth century, the oiseleurs were constantly on the lookout for people trying to bypass their monopoly, although, as the story of Thuillier and Carabin showed, they winked at transgressions when it suited them. They often seized unauthorized goods, and they requested again and again that their privileged status be reconfirmed. In petitions requesting permission to seize merchandise, the oiseleurs complained about the effects of illicit commerce on their business. A plea to the maître des Eaux et Forêts in 1750 painted a dire situation: they claimed that their commerce was weakening every day due to the "infinite number" of uncertified *(sans qualité)* individuals hawking and selling openly, either in so-called privileged places, in public fairs, in the streets, or, most often, right in front of the oiseleurs' shops, where they even set up their own businesses. These uncertified merchants had an advantage, the oiseleurs noted, because, unlike guild members, they did not have to pay dues and other fees. Not only was the oiseleurs' business being damaged, but the public good, too, was suffering, because these people sold low-quality merchandise and the public had no recourse against them.[58]

The officials of Eaux et Forêts were generally willing to give the oiseleurs permission to seize illegal goods from people out on the street who were just trying to make a little money, but they were clearly afraid to step on the toes of more reputable people. In 1767, when the oiseleurs kept asking for a blanket order that would allow them to seize merchandise from uncertified merchants, the officers of Eaux et Forêts declared that they first had to supply the names and addresses of the offenders.[59] With or without permission, the oiseleurs scoured the streets for interlopers. One day they confiscated three Caribbean green parrots and a cage from Joseph Stoll of Equignon, another time they took an Amazon parrot from a "man in a red jacket" who refused to give his name or address, and they fomented brawls such as the one involving Thuillier and Carabin.[60]

Animal exhibitors constituted an additional source of trouble for oiseleurs, for they could all too easily slip from showing to selling. In 1769, Jean-René Bourrienne and Antoine Vallin heard that Jean-Marie Gernovich, an animal exhibitor on the boulevard, was illegally selling birds in his boutique. When they went to investigate, they saw that he had a pair of rare parrots and was trying to sell one to a woman for 14 louis (an astronomical price); they returned the next day "in force" (that is, with several more oiseleurs) and tried to seize the parrots, but Gernovich resisted. When the guards were called and he was taken before the authorities, he stated that he had simply been exhibiting the birds.[61]

No matter how tough, a gang of oiseleurs could do little about large-scale economic and social transformations, and their monopoly eroded with the liberalization of the guilds in 1776 and the increase in colonial traffic and exotic goods. They did try valiantly to hold on, though: in 1779, for example, a bird merchant accused a group of master oiseleurs of trying to force him and other new merchants out of business through forging and antedating sham documents authorizing seizure of their merchandise.[62] The extent of increased commerce in small birds and animals is apparent in advertisements in the *Affiches de Paris*. At first, beginning in 1772, Ange-Auguste Chateau was the only oiseleur to place notices in the *Affiches*. After his death in 1776, his son, his son-in-law, and the oiseleur Louis le Goubey began to advertise, but so did a sieur Chavanet, who placed more notices than any of the oiseleurs. He identified himself at first as a *fripier* (old-clothes dealer) and later as a *tapissier* (upholsterer). Twice he called himself a oiseleur; but perhaps he was discouraged from claiming the title.[63] Chavanet specialized in parrots and parakeets—he once offered a gentle, unique, articulate parrot for 2,400 livres—but he also listed a mishmash of other objects for sale, making the birds seem like just another fancy consumer item: "18 parrots of all species, which speak well; 2 macaws, one of which has a 26-inch tail; dresses for all seasons; bed-clothes and beautiful counterpane from Persia."[64] After 1782, when the American Revolution ended and the slave and Caribbean trade boomed, birds must have accompanied many a shipment of goods; at least they proliferated in the pages of the *Affiches*, where we find among others a wig maker offering a large collection of live birds from Senegal and Galam and a hatter advertising thirty birds from the Gold Coast. Ads listing individual birds for sale were also common.[65]

Along with a few exhibitors at fairs, the oiseleurs had the most direct ties of any eighteenth-century Parisians with exotic animals: they lived with them, depended on them for their income, fought over them, and capitalized on their attractiveness both as pets and as scientific specimens. In their shops, birds and small animals took on a dual identity of commercial objects and living companions. As the eighteenth century progressed, they became less and less the products of the countryside and more and more the products of foreign trade. Via the hands of the oiseleurs and, increasingly, those of other enterprising merchants, exotic animals made the transition from shop to drawing room.

CHAPTER 5

Pampered Parrots

CONSIDERING how many exotic animals passed through the shops of the oiseleurs and other merchants, as well as the number that accompanied travelers returning from abroad, it is not surprising to discover that they were popular pets in eighteenth-century households in France (as well as in other countries engaged in foreign commerce). Historians have inventoried and analyzed aspects of daily life in prerevolutionary Paris such as clothes, furniture, and food, but pets, both exotic and native, have so far avoided scrutiny.[1] Yet when you look, you find them everywhere. There were so many canaries that lost birds were sometimes identified by the song they could whistle: "Ah! vous dirai-je, Maman" was the specialty of one that escaped from a Monsieur Deperey on a June morning in 1783.[2] Parrots babbled away in chateaus and artisans' homes; monkeys jumped onto princesses' dinner tables and slept on little girls' beds. They invaded texts as well, from satires to naturalists' tomes. Exotic pets often provoked different sorts of musings than did long-domesticated pets such as dogs or animals at fairs and in menageries. Their status as expensive consumer items and their association with women provoked critical comments that tied in with contemporary concerns about changing social and gender roles. The mimicking abilities of monkeys and parrots provided fodder for satirists but also stimulated savants to think about the boundary between the human and animal realms. And finally, their place by the hearth and in their owners' hearts raised questions about domestication and attachment: of animals to people and people to animals.

To find out something about the daily life of exotic pets and their owners, I looked at books for pet owners, letters, memoirs, paintings, engravings, and periodicals; especially informative are the for-sale and lost-and-found columns of the *Affiches de Paris*, which from 1745 through December 1778 appeared biweekly, and thereafter daily.[3] Most of the for-sale ads were for horses and carriages, but beginning in the 1770s the ad columns usually included several pets per month. A typical listing reads: "Young and beautiful amazon PARROT, very tame and beginning to speak well: 96 liv. Inquire of the porter of M. le chevalier *du Chaylar*, rue des petits Augustins."[4] Lost-and-found notices, which first appeared in the 1750s, became a regular feature in 1777. The column grew so much that in October 1788 the editors started charging 3 livres per notice (they had previously been printed free of charge, as had for-sale notices), because they threatened to take over the whole paper.[5] The most common lost items were portable consumer goods that were becoming popular at the time—watches, muffs, umbrellas, snuff-boxes—but escaped parrots and canaries, and the occasional monkey, appeared as well. The *Affiches* were widely read, and so well known that a play at the Saint Laurent fair in 1780 portrayed an editor besieged by people requesting publicity in the pages of his newssheet—including a little girl who had lost her bird—and in the same year an entire fake edition was printed, with the following satire of a for-sale ad: "Very beautiful green parrot, who can say only *Come on up sir, pay, kiss me, and then go;* one hopes that it will learn more in the future. Price 1 louis [24 livres]. Chez Mlle Felix, rue St. Julien le pauvre, at the sign of the Babbler."[6]

Notices in the *Affiches* reveal what types of animals were available as pets, who owned them (the names, addresses, and professions of owners are often listed), what monetary value they had, and what characteristics were prized. Parisians incorporated these animals into their culture in interesting ways;

but before exploring these topics, let us take a look into people's homes for a glimpse of daily life with animals from abroad.

Animals from Abroad

Canaries

Canaries existed in a state of limbo between the native and the exotic. Appropriately, their name derives from another domestic animal, the dog— the birds' native home, the Canary Islands, was named by European explorers after the distinctive breed of dogs *(canis)* found there. Canaries became domesticated by the sixteenth century in Europe; by the seventeenth century, breeders in both Italy and Germany were producing different plumage varieties; and by the eighteenth century, they were being bred in France.[7] Many commentators assumed that other wild species would eventually follow the same path.

Canaries are pretty, easy to take care of, affectionate, and able to learn melodies; these characteristics, along with their affordability, made them popular cagebirds. As I described in the preceding chapter, the fad among the bourgeois de Paris for breeding fancy varieties led to high prices in the early eighteenth century, but prices came down later in the century, and ordinary birds could be had relatively cheaply. Canary owners had a convenient handbook in the *Nouveau traité des serins de Canarie* by J.-C. Hervieux de Chanteloup, which was reprinted several times after its first publication in 1705. Hervieux, the Paris timber inspector *(commissaire ou inspecteur des bois à bâtir)*, was also "governor of the canaries of Mme la princesse de Condé." The book lists the variety of plumage types available in Paris and their cost: from 3 livres 10 sous for a common grey canary to 25 livres for the coveted lemon-yellow canary with a black crown. Rare hybrids could go for much more. As well as giving advice on prices, Hervieux apparently hoped to reform some bad canary-keeping habits. He found inflated prices distasteful, reporting with scorn that a family (parents and three offspring) of a goldfinch-canary cross had sold for 500 livres. He also discouraged greedy acquisition; he warned that large numbers of birds were difficult to care for properly and suggested keeping no more than five or six pairs.[8] For-sale ads confirm that people often did own large numbers of canaries; several dozen were frequently offered for sale at one time, sometimes in large aviaries also containing goldfinches and linnets.[9]

Hervieux presented a list of rules for canary care: do not feed them rich food or too poor food; do not keep too many in one cage; change water every day; do not teach more than two songs (one prelude and one air) or they get confused; and do not give them lessons more than five times a day. A solicitous owner should also pay attention to a bird's temperament, he counseled: canaries exhibit four different temperaments, each of which requires different care. Melancholy birds, for example, tend to get dirty feet and then are unable to perch well; their feet should be cleaned with spit rather than water, which, if too cold, might cause the bird's death. Indeed, if one happened to acquire a melancholy canary, it was best to get rid of it.[10]

Although canaries' talent for singing constituted one of their major attractions, their ability to mimic words was also prized. A M. Collet lost a bird that could whistle an air and frequently repeated the phrase "Collet, my friend." An apparent prodigy in this realm was offered for sale in 1782 by a hairdresser, who claimed that the bird "speaks very distinctly and pronounces, among others, the names of Louis XVI, the queen, and the dauphin."[11]

Numerous sentimental poems lauded the charms of gentle canaries, and Buffon gave them top billing in volume 4 of the *Histoire naturelle des oiseaux* (1778). The canary is preferable to the nightingale, Buffon explained, because it can speak as well as sing, it is not as vain, and it learns what we teach it rather than singing whatever it pleases. The canary "even contributes to our happiness; because it gives enjoyment to all young people and delight to recluses; it charms the boredom of the cloister, and brings gaiety to innocent and captive souls; and its love affairs, which we can observe up close when breeding them, have evoked tenderness in devoted hearts thousands of times. It does as much good as vultures do evil."[12]

Parrots

Parrots, like canaries, were valued for both their looks and their voices, but some species have a much greater talent for mimicking human speech and other sounds. The higher prices commanded for parrots derived not just from this ability, but from their continued rarity and exotic origin. Despite a few successful attempts, parrots were not bred in Europe in significant numbers until the nineteenth century; thus each bird had been born overseas and imported, from Africa, South America, or some other distant place (fig. 5.1).[13]

Judging from the number of parrot anecdotes and "lost parrot" or "parrot for sale" notices appearing in the *Affiches,* parrots had become familiar pets by the end of the eighteenth century. One suggestive, if difficult to interpret, piece of quantitative evidence is that at least 150 notices referring to lost parrots and parakeets appeared in the *Affiches* between 1778 and 1790. Considering that many people who lost parrots must not have listed them, and that most owners of parrots must not have lost them, that would seem to add up to a lot of parrots in Paris.[14] Qualitative evidence is more clear-cut. Writers ranging from natural historians to satirists assumed that readers were familiar with parrots; a notice in the 1768 *Avant-coureur* of the birds that the oiseleur Chateau had for sale, for example, listed two macaws as "parrots of the largest species, too well known for it to be necessary to describe them."[15]

A huge variety of parrots and parakeets appeared in for-sale ads, but several species predominated: most common was the African grey parrot, also know as the Jaco for its favorite enunciation, followed by Amazon parrots, macaws, and green parakeets.[16] Cockatoos were occasionally listed. If not identified by species, birds were usually described either by their place of origin (Senegal, the Indies, the island of Margache) or by their plumage—generally, the gaudier the costlier. In the 1760s, one author noted that the flashy scarlet macaw was most popular among nobles.[17]

Asking prices listed in for-sale notices (which are probably higher than the prices the birds sold for after bargaining) ranged from 48 livres for a "young Amazon parrot just arrived from Le Havre and starting to talk" to 960 livres for a "pretty parakeet that speaks well" (table 5.1). The highest price I have seen was in a notice placed by the dealer Chavanet, who asked 2,400 livres for a "very gentle parrot, one-of-a-kind, speaking very well and with a crowned head and feet like a pigeon."[18] The median price was about 96 livres for parrots and 200 for parakeets. The adjectives most often used to describe them were *parlant bien, joli, beau, jeune, doux,* and *rare* (speaks well, pretty, beautiful, young, gentle, rare). Sometimes the notices specified what language the parrot spoke or its particular mimicking

FIGURE 5.1. Louis XV's youngest daughter posing with a showy parrot. Adelaide Labille-Guiard, *Portrait posthume de Louise Elisabeth de France*, 1788. Chateaux de Versailles et de Trianon, Versailles, France. Réunion des Musées Nationaux / Art Resource, N.Y.

TABLE 5.1. Representative Asking Prices for Birds and Monkeys,
from *Affiches de Paris*

Price in livres	Description	Date
48	Young Amazon *parrot* just arrived from Le Havre, starting to talk	27 Dec. 1788
72	Beautiful *parrot*, speaks very well, with cage	13 Mar. 1790
72	Beautiful *parakeet* with pink and black collar	24 Nov. 1784
96	Pretty grey *parrot*, beginning to speak	10 Mar. 1780
96	Young grey *parrot*, speaks and whistles well, imitates dogs and cats	6 Dec. 1789
96	Beautiful Senegal *parakeet*, very tame, collar nicely shaded	19 July 1788
96	Young and pretty *capuchin monkey*	20 Dec. 1787
96	Flying *squirrel*, with aviary	30 July 1785
100	Beautiful blue and yellow *cockatoo*, speaks well	9 Oct. 1787
120	Young amazon *parrot*, speaks well	2 Oct. 1787
144	Beautiful green *parrot*, streaked with yellow and red, speaks and sings very well	8 Dec. 1787
192	Small *monkey*, very young and very gentle	7 Feb. 1789
240	Beautiful *parrot*, speaks and sings well	26 Nov. 1785
240	Pretty 15-mo.-old *parakeet*, with the brightest and most varied colors	7 Jan. 1786
360	Pretty green *parakeet* from the Indies, with black collar, 6 years old, very gentle and speaks well	6 Mar. 1784
720	Pretty green *sapajou*, 3 years old, with black cowl (or to trade for beautiful clock or beautiful ring)	5 June 1788
720	*Parakeet* that sings and speaks well	20 Dec. 1778
960	Pretty *parakeet* that speaks well	29 Dec. 1781

abilities: appearing in the *Affiches* were bilingual parrots that spoke French and Portuguese, French and Spanish, and French and Breton; one that imitated dogs, cats, and chickens; and another that imitated the song of a canary.[19]

Parrots added affection, humor, and flair to a household. Part of the fun, of course, was teaching them to talk. A manual on pets explained how to instruct them: lessons should take place in the evenings, always at the same time, preceded by a snack of bread soaked in wine. Place a cloth over the cage, dim the light, and repeat the word or phrase you want it to learn; then, in the light, talk while holding a mirror in front of the parrot, so that it will think another parrot is talking.[20]

Mauduyt described the liveliness a talking cockatoo brought to a house he lived in. The cockatoo, which enjoyed playing with the pet cat, would call the cat's name when it was ready for a romp; if the cat was sleeping, it would pull on the cat's tail or ears. When the two chased each other or wrestled over a toy, the cockatoo would sound cries of "joy and gaiety."[21] Buffon, too, tried to explain why a talking parrot is so appealing:

> It entertains, it distracts, it amuses; in solitude it is company; in conversation it is an interlocutor, it responds, it calls, it welcomes, it emits peals of laughter, it expresses a tone of affection, it plays with the seriousness of a sentence; its little words tumble out at random, amusing us by their disparity, or sometimes surprising us by their justice. This play of language without ideas has a *je ne sais quoi* of bizarreness and grotesqueness, and without being emptier than much other talk, it is always more amusing. In imitating our words, the parrot seems to take on something of our inclinations and habits: it loves and it hates; it has attachments, jealousies, preferences, caprices; it admires, applauds, and encourages itself; it becomes cheerful or mournful; it seems to be moved and touched by caresses; it gives affectionate kisses; in a house of mourning it learns to moan; and, accustomed to repeating the dear name of a deceased person, it reminds sensitive hearts of their pleasures and sorrows.[22]

The "disparity" of parrot language provoked hearty laughter among the duchesse de la Vallière's acquaintances; she had inherited from a refined friend a surprisingly foul-mouthed parrot, which had learned its bad language from the military men who had frequented the household.[23]

Life with parrots was not all laughter and tenderness, however; many of them bit, tore apart furniture, shrieked loudly, or refused to talk. In the *Encyclopédie méthodique* volumes on birds, Mauduyt noted the negative characteristics of several kinds of parrots. He recommended displaying macaws in vestibules, where their beauty could be admired in passing but their raucous voices were kept at a distance. Their destructive tendencies, and those of cockatoos, could be curbed by giving them pieces of wood, which they would spend the day happily shredding. Some birds could never become satisfying pets. Mauduyt owned a blue-headed parrot from Guyana that was dissatisfying because of its extreme dullness. Although beautiful and rare, the bird always appeared sad, never spoke or uttered any sound, and could have been mistaken for a stuffed bird except when it moved two or three times a day to get its food.[24] Many owners must have regretted their purchases or tried to sell off unwanted animals, and some no doubt thought of their parrots more as ornaments than as household companions.

Monkeys and Other Animals

Monkeys were never as widespread as birds, but they, too, became more available in the latter part of the eighteenth century. A book full of monkey anecdotes from 1752 stated that they had recently become quite common and that one encountered them in Paris "at the homes of a relatively large number of people."[25] Many different species were available, and they are hard to identify now because they were often described vaguely. Animals offered for sale in the *Affiches* were often listed simply as "monkey" (*singe* or *guénon*) or were identified by their place of origin (Senegal, Guinea, Madagascar). Of those that can be identified, the most common were New World capuchin monkeys—sometimes known as the organ-grinder monkey—and African green monkeys. Like parrots, monkeys were imported rather than bred locally. Natural history texts and journals reported a few cases of monkeys reproducing in captivity, however, and some authors believed that with the proper care they would in time become domesticated.[26]

Prices for monkeys were high, but variable, ranging from 96 to 720 livres in the *Affiches*. Actual selling prices were probably lower than asking prices, for bargaining was de rigueur. The teenaged Geneviève de Malboissière described such a process while she and her family were enjoying an afternoon walk on the boulevard. They stopped before a "parade of monkeys" to look at a particularly charming one. The merchant asked 480 livres; her mother's companion offered 240. "It would have been crazy," she wrote to her friend, "but it's such a pretty animal!"[27]

Judging from the adjectives used in for-sale ads, buyers looked primarily for monkeys that were *très-doux* (very gentle) or *très-privé* (very tame), no doubt because they had a reputation for bad behavior. *Jeune* and *joli* were also favored, the former presumably because it meant the animal would be more easily tamed. Sellers occasionally stressed their animals' resemblance to humans or their abilities to perform human services. A bailiff offered a "monkey of a very small species, aged around 10 months, brown fur, with the face and hands of a Negro," and another ad listed an "Arabic monkey, large-sized, very gentle, and serving as a domestic."[28]

Many monkey owners cherished their pets for their affection and funny antics; some people, though, found them dirty and malicious. When the baronne d'Oberkirch visited Paris, she was not too pleased to encounter one on the dinner table. She had been invited to a party at the home of the princesse de Chimay, the queen's lady of honor, after the opera. While they were at the theater, a monkey, a favorite of the princess, escaped from the room it shared with a small dog, next to the princess's bedroom. After refilling the dog's water dish, the monkey went to his mistress's makeup table. He applied perfume, rolled in the powder, then applied rouge and beauty spots, "just as he had seen his mistress do; only he put the rouge on his nose, and the beauty spot in the middle of his forehead . . . suddenly, . . . during the middle of dinner, he came into the dining room, jumped on the table in that get-up, and ran to his mistress." The baroness recorded in her memoirs that although the animal did look very funny, she did not approve of keeping monkeys indoors.[29]

Although canaries, parrots, and monkeys were the most popular exotic pets, and the only ones to become regular actors on the cultural stage of ancien-régime Paris, people kept or tried to keep many other kinds of animals as pets. Some of the more popular species were small seed-eating birds such as Java sparrows and African finches.[30] Aristocrats, through connections to traveling friends or outlays of large sums, accumulated an impressive array of rare and unusual species. The duc de Penthièvre acquired an

East Indian bird *(veuve des Indes)* with stunning plumage: a black head and back, auburn-and-yellow breast, and very long, blue-black tail feathers. He placed one of the first lost-and-found notices in the *Affiches* when it flew away, stating that "this bird is very rare" and offering a fair reward to the finder.[31] The marquis de Montmirail owned hamsters (not yet regularly available as pets), and visitors to his private menagerie could marvel at a sloth hanging upside down with its long nails curled around a tree branch.[32] Some of these animals, which must have been kept primarily as curiosities, were really caged wild animals rather than pets. The comtesse de Marsan, for example, kept a "rat" from Madagascar for three years, but it bit, never became tame, and ventured from its cage only at night.[33]

Cages and Caretakers

A worker who found a parrot in October 1778 let it be known that he would gladly accept a reward, and that he also expected reimbursement for the cage he had had to buy.[34] Cages and chains were necessary to confine semiwild animals, but the size and the style depended on the type of animal and the means of the owner. Small birds could be confined in modest cages or kept in groups of several dozen in larger structures. Some people had aviaries fitted into windows, and expansive outdoor aviaries were popular among nobles; some of these were described in travelers' guides to Paris.[35] Parrot cages were big: one advertisement for a macaw cage gave its dimensions as five feet by three and three-quarter feet by two and one-half feet, or about the size of a modern household refrigerator. Instead of being caged, parrots were sometimes chained to parrot stands, and monkeys were restrained with ball-and-chain contraptions (fig. 5.2). Some cages were incredibly elaborate: songbird cages might feature fountains, statuettes, containers for fresh flowers, or a kennel underneath for the dog. One advertisement described a "pretty aviary suitable for an apartment or a garden, representing a water tower with a colonnade, four rivers, several waterfalls and fountains, and forty figures forming a battalion of cavalry that parades in two lines at the sound of a bird-organ." Monkeys could also be housed in style: a confectioner's marmoset went on sale along with its Indian-style cabin and silver chain. Prices ranged from a low of 72 livres for a wheeled aviary with

FIGURE 5.2. A pet parrot participating in a conversation from its stand. "L'occupation," engraving by Lingée after Freudenberg, BnF Département d'Estampes, 0a80 rés plate 4. Photo: Bibliothèque nationale de France, Paris.

canaries, goldfinches, and other small birds, to a high of 4,000 livres for a seven-foot-high parrot cage with gold-plated brass ornamentation.[36]

Some animals let their dislike of cages be known. Macaws and cockatoos shrieked or pulled out their feathers when unwillingly restrained, and pets frequently flew out of open cage doors or broke their chains and made mischief (fig. 5.3).[37] Well-behaved animals could be given free rein, however: Valmont de Bomare's grey parrot ate at the table with him, and the one in Buffon's household flew free indoors during the day. Buffon's mona monkey frequently got off its long chain and ran away, but it always returned readily when apprehended.[38]

Just as wild animals responded differently to being in captivity, so did the sight of their cages and chains arouse conflicting responses in observers. These contrasting views reflect the tensions of a culture that was based on chains of authority, but in which freedom was becoming a popular refrain. In the positive portrayals of confinement, the cage was seen as providing a refuge, protecting the creature from a harsh world. These representations often made explicit connections between the caged creature and a vulnerable human, usually a young girl or woman. In a poem entitled "The Dangers of Liberty: Moral Tale Addressed to Young Girls," a fifteen-year-old girl feels imprisoned because her mother neither allows her to go out in society nor admits young men to the house. One morning, young Julie's canary cries and struggles to escape from its cage. "Poor dear! I'm holding you prisoner," she exclaims, comparing its situation to hers, and she releases it. The bird takes off over the rooftops, but not to a happy life of freedom: it immediately becomes a tasty morsel for the neighborhood cat.[39]

Balanced against these positive images were numerous criticisms of cages, both real and metaphorical. A popular natural history text described the parrot as a beautiful bird that ornaments our homes but noted that "it is appropriate, nevertheless, to observe that it only really ornaments them when outside [in a large aviary]. All house birds *[oiseau de chambre]* or birds in small aviaries are unhappy, and thus not enjoyable for anyone with a touch of humanity." Louis-Sébastien Mercier spoke contemptuously of artisans who kept pets in cages, "as though to make them share the tedium of their own slavery."[40] I explore the symbolism of animals and slavery more fully in chapter 7.

In the late eighteenth century, many of the most expensive and unusual

FIGURE 5.3. A white-faced capuchin monkey *(saï à gorge blanche)* showing its broken chain and the results of its mischief—a shattered pot and unwound ball of yarn. From Buffon et al., *Histoire naturelle*, vol. 15 (1767), plate 9. Reproduced by courtesy of the Department of Special Collections, General Library System, University of Wisconsin–Madison.

animals were still owned by the wealthy, but exotic pets had become much more accessible to those lower on the economic ladder, even to the artisans that Mercier referred to. The range of owners is nicely illustrated in two "lost parrot" notices that appeared in the *Affiches* a few days apart. One was placed by a carpenter, who described his pet simply as "a grey parrot that limps on its left foot." The second notice announced the escape from a *conseiller honoraire au Parlement* of a "pretty apple-green parrot with a ring around the neck, red on the head and black under the throat, a red beak, the tail green and long."[41] Historians have documented the "trickling down" of luxury items in eighteenth-century France, and luxury animals are no exception.[42] A list of the professions of people who lost parrots and parakeets ranges from the mighty (the duc d'Orléan's chancellor) to the ordinary (including a saddler, a baker, and four café owners) (table 5.2). Bakers and café owners could indeed be quite well-to-do, but even so, this list shows that the exotic pet had spread well beyond the realm of the nobility.

Men, women, and children all owned exotic pets and sometimes devoted considerable thought and time to them. In a letter printed in the *Journal de Paris,* a man requested advice from other *pithécophiles* (monkey lovers) on how to cure his beloved monkey from biting its tail. Two weeks later, the editors reported that readers had suggested a variety of remedies, including returning it to its native country, thrashing it (a technique said to work for hunting dogs), switching it to a vegetarian diet, and feeding it large quantities of snails.[43] Another letter to the *Journal* told the story of a woman who had spent two hours delivering mouth-to-beak resuscitation to her parrot after it had been caught in a flood.[44] The young daughter of a wig maker had a pet monkey that slept on the foot of her bed every night.[45]

A close attachment between two girls and their pets is evident in the letters that Geneviève de Malboissière wrote to her friend Adélaïde Méliand. Geneviève had canaries, a goldfinch, and a capuchin monkey, and Adélaïde, too, had a monkey, plus canaries and a parakeet. When she took care of Adélaïde's birds, Geneviève fed them tidbits, talked to the parakeet, and kept up the canary's singing lessons: "I teach *[serine]* him every day, and it's not his fault if he doesn't progress; he tries very hard," she wrote. She asked after Adélaïde's monkey Brunet every few days when he became fatally ill, closing one letter, "I kiss you, you and Brunet."[46] Parents and educators may have tried to discourage their children from coveting exotic creatures, how-

TABLE 5.2. Professions of People Listing Lost Parrots or Parakeets, from *Affiches de Paris*

Profession/Title	Date and page number of advertisement
Actor *(comédien ordinaire du roi)*	12 Nov. 1788, 1628
Baker *(boulanger)*	2 Aug. 1782, 1784
Boardinghouse keeper *(maître de pension)*	16 Apr. 1783, 920
Box maker *(layetier)*	6 May 1788, 1301
Businessman *(négociant)*	19 July 1788, 2061
Button maker *(maître de boutons)*	19 Aug. 1788, 2347
Cabinetmaker *(ebéniste)*	12 Sept. 1787, 2531
Café owner (3 *limonadier*, 1 *limonadière*)	10 June 1786, 1540; 31 July 1786, 2035; 3 June 1787, 1579; 3 Aug. 1787, 2164–65
Carpenter *(menuisier)*	15 July 1782, 1636–37
Caterer *(traiteur)*	23 May 1790, 1428
Chancellor of the duc d'Orléans *(chancelier de mgr. le duc d'Orléans)*	8 Sept. 1789, 2576
Clerk of buildings *(greffier des bâtimens)*	4 Oct. 1787, 2732
Clock maker *(horloger du roi/horloger)*	28 May 1789, 1592; 26 Oct. 1784, 2819
Commissioner *(commissaire)*	19 July 1788, 2061
Courier of the Royal Post *(courier de l'hôtel royal des Postes)*	9 June 1786, 1540
Director of the Bureau of Water Conveyances *(directeur du bureau général des diligences par eau)*	17 June 1779, 1340
Draper *(drapier)*	31 July 1788, 2116
Goldsmith *(orfèvre)*	19 Oct. 1783, 2512; 6 June 1785, 1528; 19 June 1786, 1725
Grocer *(épicier)*	7 June 1788, 1628
Hatmaker *(chapelier)*	20 Sept. 1785, 2524
Honorary consul to Parlement *(conseiller honoraire du parlement)*	19 July 1782, 1669

(continued)

TABLE 5.2. *(continued)*

Profession/Title	Date and page number of advertisement
Lawyer *(procureur)*	3 Jan. 1780, 21; 12 Oct. 1787, 2803
Notary *(notaire)*	24 Feb. 1788, 548; 18 July 1788, 2053
Perfumer *(parfumeur)*	20 Jan. 1783, 164
Pharmacist *(maître en pharmacie)*	14 Oct. 1787, 2828
Receiver-general of finances *(receveur général des finances)*	1 Jan. 1779, 75
Saddler *(sellier)*	21 June 1781, 1441; 19 Dec. 1781, 2909
Stationer *(papetier)*	13 Jan. 1781, 99
Surgeon *(maître en chirurgie / chirurgien-herniaire)*	12 Sept. 1781, 2119; 7 Oct. 1786, 2652
Tile merchant *(marchand de tuiles)*	17 Feb. 1780, 381
Vinegar maker *(vinaigrière)*	15 July 1783, 1723–24
Wallpaper maker *(fabricant de papiers peints)*	5 Sept. 1784, 2339
Wheat merchant *(marchand de bled)*	18 Jan. 1778, 925

ever; the young protagonist of one children's book tells his teacher, in words obviously written by an adult, "I would love parrots madly if they weren't so dirty, so bad-tempered, and so noisy."[47]

Taking care of pets was an onerous job. It involved preparing food (usually a bread paste—mixed with fresh meat or worms for insect-eating species), cleaning cages and food dishes, assuring that the animal was not too hot or too cold (which often meant moving it from room to room), and preparing and administering remedies when it got sick. In wealthy households, these tasks must often have fallen to servants. Relying on servants had its drawbacks, however, and careless employees received the blame for disasters. The resuscitated parrot, for example, had supposedly been caught in a flood because a servant had left it out in the garden for air.[48]

A letter to the *Journal de Paris* in 1780 warned that some servants were not just careless, but malicious. The writer, who signed himself Darée, recounted an incident that had occurred when he had stepped into a bird seller's shop in the fauxbourg Saint Germain to get out of the rain. While he was there, a servant came in with a cage containing a dead canary and said to the merchant, "This bird won't sing any more. Give me another one." Darée was certain that the servant was trying to replace the bird, a male that looked as though it had been killed by a cat, with a similar-looking one in order to keep his mistress in the dark about the fate of her bird. The servant examined only female canaries rather than the higher-priced, vocal males, however. Darée asked the servant how he would explain to his mistress why the bird didn't sing any more. "I'll say it's molting," he replied, and left to look for a better match in another shop. When the shocked Darée commented to the merchant on the roguery he had just witnessed, she told him that it was nothing—this kind of thing happened every day. Servants also traded well-trained canaries for cheaper, untrained ones, she revealed: they would pocket the difference in price, 3 to 4 livres, and then twist the neck of the new bird (presumably to pass off as a naturally caused death). Darée explained in his letter that this incident would not only be enlightening to people who, like him, enjoyed keeping birds, but that he believed it was important to alert "one half of Paris against the other; that is, masters toward their servants, or honest people against knaves."[49] Servants might be left with the dirty work of pet care, but some of them found ways to make their masters pay.

All was not lost, however, when a pet died. Owners could diminish their losses by engaging in a practice that was common in the eighteenth century and through much of the nineteenth: having their pet stuffed. Many taxidermists placed advertisements in periodicals. Most were aimed at people who wanted to preserve animals to put in their natural history cabinets, but some also offered pet preservation. A Mlle Baudouin, who seems to have run a thriving taxidermy and natural history specimen business, with preparations guaranteed to last more than fifty years, advertised frequently in two journals during the 1760s. One of her advertisements began with this appeal to bereaved pet owners: "The loss of certain domestic animals distresses us; we wish we could keep, at least, their form *[figure]*."[50] In the 1780s, M. Fromageot de Verrax, from Strasbourg, described in a letter to the *Journal de Paris* a new preservation method he had invented that would, he claimed, preserve birds for several centuries. Prices ranged from 36 to 150 livres. He reported later that among the orders he had since received was a request from a lady to preserve her dear deceased parakeet. Unfortunately he could not prepare a proper tableau because the bird she had sent was missing some important feathers, and he could find no fake parakeet eyes in Strasbourg. He requested anyone sending him parrots, parakeets, or other exotic *(étranger)* birds to make sure the birds had all of their essential feathers.[51]

Luxury, Gender, and Mimicry: Portrayals of Pet Owners

When a woman sends her dead parrot to Strasbourg to be transformed into an incorruptible tableau, does her action indicate a touching bond between human and animal or, rather, a crass attempt to extend the life of a valuable and showy possession? In *Man and the Natural World*, Keith Thomas claimed that sympathy toward animals became widespread in England in the late eighteenth century, after humans felt secure in their domination of nature, and wilderness was safely distant. Hester Hastings also documented a shift toward humane attitudes—or rather, verbal expressions of humane attitudes—after 1750, in France.[52] I have encountered many expressions of strong affection toward pets that seem to support these generalizations about attitudes toward animals in the later eighteenth century. But everyone knows that sentimental language and even actions can conceal baser motives, and in the late eighteenth century critics began to question pet owners' sincerity. In letters to journals and in anecdotes, poems, and satires, critics accused people—especially women—of hypocrisy and oversentimentality. This was part of a more general critique of ostentation and luxury by the philosophes and others against the financial excesses of aristocrats and the monarchy, including the menagerie (see chap. 2). When directed at women, these accusations melded with broader criticisms that castigated women for being superficial, frivolous, and flighty.

At this time, wealth was beginning to replace family background in determining social status, and it is easy to see how pets could become items of conspicuous consumption. They could be seen as luxury possessions, like art works or fine clothes or furniture, desired for the message they conveyed that their owner was a person of means. As portraits of the time attest, the rich flaunted not only fine fabric, furniture, and coiffures, but also fancy breeds of dogs, along with pet parrots and monkeys. Exotic pets had the benefit of value associated both with rarity and with the world of global trade, sharing the cachet attached to foreign goods like coffee, which appealed, according to Daniel Roche, because "it is exotic, and thus represents modernity and commerce."[53]

In the first chapter of his *Tableau de Paris,* Louis-Sébastien Mercier wrote of the proliferation of exotic commodities in late-eighteenth-century Paris, where dining in a well-appointed household would take your imagination on a promenade around the world. As you sit on a chair covered with Indian textiles, drink Asian tea from Chinese and Japanese porcelain, use silver wrested from Peruvian mines to spoon sugar grown in America by unhappy Africans, and read in a newssheet about wars around the world, "everything, up to the monkey and the parrot of the house, reminds you of the miracles of navigation and of man's fervent activity *[ardente industrie].*"[54] For a growing number of those who, like Mercier, questioned the social and economic consequences of growing consumption, expensive pets were a symptom of the excessive spending that was draining Europe of precious metals, inciting wars, and, if anything, increasing poverty.[55] Exotic creatures were popularly associated with the heedless rich who advanced on the backs of their fellows. A letter printed in the *Journal de Paris* in 1787, for instance, told of a pedestrian who had been run over by a carriage driven by two fops, one dressed like a macaw and the other like a zebra.[56]

Though rare pets made convenient vehicles for ridiculing the rich, they could also be used to knock the supposed greediness of the lower orders. A widely circulated story, which was also performed as a play, combined satirical jibes against both high and low. A woman on the way to the opera with her noble lover sees a parrot perched in a shop, chatting adorably with all the passersby. The woman exclaims that she absolutely has to have it, and her lover promises that he will get it for her, no matter the cost. The next day he returns and offers the proprietor of the shop a huge sum for the parrot. The man apologizes, saying that the parrot belongs to his wife, that she is very attached to it, and so he doesn't dare sell it. Later his wife comes home and he tells her what happened. She roundly chews him out for being such an idiot, exclaiming that she would sell both him and the parrot for half that price. He in turn is so incensed by her lack of gratitude that he wrings the parrot's neck and throws it in the soup pot.[57]

This story, like many others, centered on the female pet owner. Women received by far the most criticism attacking their relations with their animal companions, and indeed from stories, illustrations, and even some natural history books, one would think that women and children were almost exclusively the pet keepers of the family. Pet-care books were oriented toward them, and the vast majority of stories, anecdotes, and poems featured women as pet owners; one even told of a woman who believed she had metamorphosed into a parrot.[58] It is impossible to know what proportion of pet owners were in fact women, but the portrayal of female ownership was certainly out of synch with reality. A number of male pet owners are mentioned in memoirs, letters to journals, natural history texts, and other sources, and only about one-third of the lost parrot and parakeet notices listed a woman as the owner or contact person.[59]

One reason writers associated women and pets was because of the romantic and erotic connotations aroused by the close physical and emotional bonds between owner and animal. Male writers and artists tended to portray women's relations to their pets either lasciviously or petulantly, depending on whether the pet was a metaphor for the male lover or was an all too material rival. Poems described the kisses showered on pets and the sweetness of canary courtship; the lover imagined himself as a canary brought to a beautiful woman's lips or as a parrot awakened by love. One poem, for instance, told of a parrot that remained mute and unmoved with the old woman who bought it and caressed it tenderly; but when it saw her beautiful young daughter it immediately began to talk, gallantly expressing its love.[60] Artists also exploited the theme: a typical example is Jean-Baptiste Greuze's popular *Young Girl Crying over Her Dead Bird*, which depicts a budding young beauty with voluptuous lips and invitingly undone clothing. She rests her head sadly on her hand, her eyes downcast; in the foreground, the lifeless bird lies on its back on top of its cage. Diderot, in a lusty analysis of the painting, interpreted the girl's tears as lovesickness: "This child is crying about something else, I tell you. . . . Such pain! At her age! And for a bird!" he leered, "I wouldn't be too displeased to have been the cause of her pain."[61]

The fact that sometimes the tears and kisses were indeed for the bird and not the lover provoked poems such as one that accused women of ignoring their ardent lovers in favor of parrots, music, astronomy, and other caprices.[62] A humorous anecdote took this idea to an extreme. A fashionable lady owned a parrot that she said she loved more than her life, and for which she would have given all of her lovers. When the bird flew away, she lamented, pulled her hair, and cried out loud, "Ah! my poor parrot, I would give anything to see you again. I'll sleep with whoever brings you back to me." A robust water carrier appeared the next day bearing her bird and saying that he had heard her promise when he was in the kitchen, and that he had come to claim his reward. Embarrassed to have a water carrier defile the bed that had received dukes, bishops, and the president of Parlement, she offered him money instead. He demanded that she keep her word, so she gave in with a sigh, saying, well, he's a man like all the others; then she ran to be with her parrot and to forget the price she had paid.[63]

Many commentators suggested that people—especially women—kept pets more for show or frivolous entertainment than because of any genuine affection, and they linked pet keeping with other aspects of a superficial lifestyle. The satirist Caraccioli pointed a finger of rebuke at the lady's toilette: there, he wrote, she puts on makeup in a ritual that must have been learned from monkeys, she spends hours having her hair done, dressing, gossiping, and strategizing, and her parrots, canaries, and dogs enter to admire and kiss her. The same author's *Critical Dictionary* continued the ridicule of pets, describing capuchin monkeys as "little monkeys that people have for show, or because they resemble them," and parrots as the resource of idlers. Young dandies amuse themselves by teaching old countesses' parrots to gossip, he continued, remarking that scholars never keep either monkeys or parrots.[64] A humorous print mocked both monkey ownership and fantastic hair arrangements: it shows a well-dressed lady's tall wig catching on fire as a monkey climbs the curtains and knocks over a candle (fig. 5.4). Mercier portrayed pet monkeys as temporary fashion accoutrements, noting, in the 1780s, that society ladies were casting aside the monkeys they had formerly been so crazy about in favor of young black servants *(petits Nègres).*[65]

Numerous popular anecdotes and plays showed women treating pets as commodities. One story begins with a young woman earnestly asking a judge to grant a request; when he finally agrees, it turns out that she wants

FIGURE 5.4. "Incendie imprévue arrivée à la coëffure moderne de Mme Monte-auciel" (Unexpected fire in the modern hairdo of Mme Rise-to-the-sky). The devil-like monkey who started the fire sits on top of the screen, holding a candle. BnF Département d'Estampes, 0a20, vol. 12. Photo: Bibliothèque nationale de France, Paris.

some feathers from his parrot to add to a fancy headdress, since they are just the right color to go with her gown (the judge responds that he will have to ask his wife, who owns the parrot).[66] Another pet story that was produced as a play told of a woman whose two lovers were constantly trying to fulfill her demands. She convinces lover A to buy her a parakeet that she saw in a cloth merchant's shop and demands that the other buy the famous monkey owned by the performer Astley. Unfortunately, lover B cannot convince Astley to sell his monkey, so he acquires an animal that looks identical from a poultry shop, where it works plucking chickens. One day the happy mistress leaves the feather-plucking monkey and the parakeet in the same room. . . . Lover A is convinced that the denuding of his parakeet was an intentional trick played by lover B, and a bloody duel ensues.[67]

To those inspired by Rousseau's praise of nature over modern civilization, the boudoir was more than a place for wasteful consumption; it could even be harmful to the wild creatures kept there. The naturalist Le Vaillant, for instance, in an account of his travels in southern Africa, compared a troop of wild monkeys that he encountered there to pampered Parisian "bastardized" ones that "languish in Europe, in slavery, fear, and boredom, or perish suffocated by the caresses of our women, or even poisoned by their candies."[68] All pets were criticized as symptoms of luxury, but parrots and monkeys differed from native animals not only because they were exotic and were not bred locally, but because they had no possible usefulness. Dogs as a group were often praised as guards, hunters, and herders, and cats as vermin chasers, even if those inhabiting boudoirs might not perform these functions.

A number of critics suggested that the time and effort spent on pets would be better used on more important objects or pursuits. To some of these commentators, pets appeared to be a wasteful outlet for affection that should be directed toward humans. A popular natural history book warned that women should not keep just a single canary or dog "the size of a nut," because they risked becoming more attached to the pet than to their friends. A satirical dictionary defined canary as "a little yellowish-white bird that comes from the Canary Islands and that serves as the amusement for I don't know how many women and nuns, who would often rather see a person perish than their bird."[69] Other commentators recommended study as a more valuable pursuit. A letter to the Journal de Paris described a foolproof method for focusing children's minds on their studies: the author claimed to have scared a little girl into learning to read by torturing and killing her pet parrot, and concluded that it is worth having one less parrot in the world in order to have one more virtuous woman. In the introduction to his series of textbooks, the abbé Fromageot recommended natural history as eminently suitable for girls to study; not only would it provide an antidote to boredom and idleness, but once girls became educated, they would no longer prefer to converse with a dog, a monkey, or a bird.[70]

All of this vituperation no doubt tells us more about prevailing attitudes toward women than it does about the actual relationships of women and their pets. Historians who have studied the debate on women's roles as consumers have revealed a complicated picture that reflects conflicting opinions about commerce itself. Some commentators contended that consumption driven by fashion and the desire to be up-to-date provided a beneficial boost to production; others identified women's insatiable demands as a corrupting influence that distracted men from serious business and encouraged immoderate spending.[71] A highly publicized scandal that seemed to support the corruption view was the Diamond Necklace Affair, in which a young woman, by impersonating Marie Antoinette, bilked an expensive necklace out of a prominent nobleman who had a crush on the queen.[72] The representation of women pet owners in the popular press suggests that this critical view was widespread.

Pets, especially parrots and monkeys, became associated with a different kind of criticism as well, which, instead of focusing on the behavior of pet owners, used the mimicking behavior of these popular pets as metaphors for the behavior of imitative people. These types of metaphors exist in many languages and cultures—for instance in the English verbs *to ape* and *to parrot*—and were popular themes in the baroque artistic genre of *singerie*, but in eighteenth-century France they had particularly strong meaning.[73] Anxiety about eloquent imposters was intense in some circles during this period—that fear combined with the prevalence of parrots and monkeys as pets among the fashionable made for convenient analogies between mimicking animals and *petits-maîtres* or *petits-maîtresses* (which could be roughly translated as social climbers) (fig. 5.5). As consumer items and the vocabulary of the educated became available to more and more people, how was

FIGURE 5.5. "Le singe à la mode: Dedié aux petits Maistre francois" (The modish monkey: Dedicated to French dandies). The monkey displays the most fashionable accessories of the time: umbrella, pocket watch, walking stick, and muff. BnF Département d'Estampes, Collection Hennin, vol. 109, #9571, Qb 1775. Photo: Bibliothèque nationale de France, Paris.

one to distinguish the wellborn from the nouveau riche, or the classically educated from someone who skimmed through specialized dictionaries?

Representations of mimicry in the later eighteenth century often referred back to Jean-Baptiste-Louis Gresset's best-selling poem *Ver-vert* (1734). The poem told the story of some nuns who taught their pet parrot to pray. He became very famous, and another convent down the river asked if the parrot could be sent to them. The nuns put Ver-vert on a boat to Nantes. During the voyage, he found that his saintly language was laughed at, and in order to fit in he learned to swear just like the boatmen. When he got to the new convent all the nuns gathered round expectantly as the mother superior addressed him. He opened his beak and squawked, "By Jove! Aren't nuns crazy!" *(Par la corbleu! que les Nones sont folles!)*, along with a few more choice phrases he had learned. The nuns immediately returned him to his home convent, where he was confined to his cage with no treats or garden walks. After a time he regained his devout patter, however, and the nuns honored his release with a big party. Poor liberated fellow, he overdosed on candies and promptly died. The poem was reprinted many times and was later produced as a play.[74]

Parrot anecdotes and poems throughout the century often played on this theme of language as a deceptive veneer covering ignorance. Where Gresset used Ver-vert's meaningless Ave marias to make fun of people who mechanically mouth religious vows, other anecdotes mocked mindless society prattle, as in this "Epitaph for a Parrot":

> Here lies Jacquot, dead of old age
> And tenderly loved by his gentle mistress:
> He never spoke but after others;
> How many people have died and will die like him![75]

Mimicry stories could also function as humorous ways to cut through facades to reveal normally hidden social or political truths. Gold diggers were mocked in a poem that told the adventures of a petit-maître's monkey who stole away his master's lovers simply by dressing well and pretending to be rich. On the political side, a joke went around Paris that the corrupt finance minister Terray had a parrot that knew how to say only "Au voleur!" (Stop, thief!). A more extended satire appeared in an anonymous pamphlet published during the period of upheaval in 1771 when Louis XV exiled the Paris Parlement. The story featured Astley, the popular Parisian performer. In the tale, Astley dresses his trained monkey, Général Jacquot, as a magistrate and introduces him to people. At first the monkey is received well, but when he begins to declaim with more wisdom than the real magistrates, he is exiled.[76]

Pets and the Savant

Social critics were not the only ones writing about house pets; philosophes and naturalists also observed and described the animals and their behavior. Pets were a more legitimate and important source for eighteenth-century naturalists than they are today for two main reasons. First, sometimes the pet was the only living individual of a particular species to be seen in their local area, or possibly in all of France or Europe. Second, naturalists were more interested in an animal's capacity for domestication and in how it behaved in human society than they are now.[77] The natural history of pets also gave naturalists a way to personalize their texts. By mentioning the names of people whose animals they had observed, they widened their audience and gave pet owners the pleasures of participating in a natural history project, as well as seeing their names in print and their pets' charms publicized.

Buffon, Daubenton, and Mauduyt paid close attention to pet animals and used observations made by themselves, their colleagues and friends, and other contacts. We have already seen how much information Mauduyt obtained from the oiseleurs. He also owned many birds himself, which he kept in his house and garden and tried to breed.[78] Buffon, too, had a large personal menagerie of mammals and birds, both native and exotic, and his connections with other aristocrats brought him access to several other unusual species.[79] He observed a gentle mongoose that de Robien, president of the Parlement of Brittany, carried everywhere in his hat, and two charming black squirrel monkeys, one belonging to the marquis de Montmirail and the other to the duc de Bouillon. The duke's monkey became a display in the Cabinet du roi after its death. Daubenton watched Montmirail's sloth clutch the edges of a gaming table and turn slowly beneath it, unable to clamber on top. Montmirail sent Buffon two live hamsters, the duc de Bouillon gave him several capybaras, Mme de Pons gave him a hornbill from Pondicherry, and Mme de Pompadour willed him her epileptic scarlet macaw, her monkey, and her dog.[80]

Buffon praised Montmirail in print for doing all he could to advance natural history and also acknowledged the duc de Bouillon's contributions, writing that he is "interested in exotic animals; he has sometimes done me the honor of calling me to come see them, and because of his devotion to a good cause *[par amour pour le bien]* he has given us several."[81] Other amateur naturalists wrote letters or made notes about their pets, from which Buffon often quoted at length. His sister supplied several pages of observations on her grey parrot, and his friend the local priest père Ignace Bougot also offered notes on his "yellow-headed parrot."[82] The article on the cockatoo included a gallant compliment in a note contributed by the naturalist Sonnini de Manoncourt describing the gentle nature and endearing habits of a bird "presently in Nancy, where it is the delight of a beautiful and kind lady."[83]

Readers of early volumes of the *Histoire naturelle* sometimes sent in observations that were published in later volumes of the decades-long series. The marquis de Courtivron, for example, a member of the Académie des sciences, challenged Buffon's assertion that otters are not educable by reporting of a tame one he had often played with at a convent in Autun. Another reader, owner of a tame kinkajou, apologized for sending only a letter: "Several times I resolved to give this live animal to you, to submit to your observations, but each time it came up to caress me so gently and to play around me with such gaiety that, seduced by its endearing behavior *[gentillesses]* I never had the courage to part with it."[84] Thus the *Histoire naturelle* inspired some readers, at least, to think of their pets in terms of their ability to advance knowledge of natural history, and the project ended as something of a dialogue between Buffon and his audience.

Pet behavior provided a perfect forum for naturalists to meet the often-stated goal in eighteenth-century texts, to both "amuse and instruct." Buffon obviously relished amusing and moving his readers by describing the antics of pets with which they might be familiar (for example, the canary and parrot descriptions quoted above). When writers succumbed to a good pet story, however, they sometimes felt the need to remind readers of their serious intent. Following his long and touching description of a friendship between a cockatoo and a cat, Mauduyt explained that he had recounted their playful interactions only "because it seems to prove that the name of the cat had become by habit for the *cockatoo* a sound representing the object itself and consequently that intelligence in this *parrot* extended to combining two ideas and attaching that of a form to the consonance of a sound."[85]

Two areas that interested naturalists and other thinkers most about the behavior of pet animals, especially exotic ones, were attachment and mimicry. The subject of the emotional bond between human and animal had a utilitarian aspect, since it was an important element of domestication, but it also raised the question of whether the attachment between animals and humans was equivalent to friendships between humans.

A number of naturalists were curious about why pets became particularly attached to certain people. Mauduyt observed that parrots seem to like some people and dislike others, and Buffon described the preferences of his household's green macaw, of his mona monkey, and of a suricate, which invariably bit certain people it disliked. The macaw, he wrote, jealously attacked anyone its mistress paid attention to, because it "does not want to share . . . the least caress or the slightest attention of those it loves."[86] Attachment is not a mysterious trait, however, claimed Mauduyt: animals like people who treat them well and dislike those who mistreat them. Buffon's sister also found a practical explanation for her grey parrot's passionate love for the kitchen girl, whom it followed everywhere. She concluded that this love had more to do with the girl's access to food than to her personality, because when a new girl was hired the parrot immediately transferred its affections to her. The naturalist Salerne proposed that parrots like women and children better than men not only because of the caresses they receive from them, but because their voices are more similarly pitched.[87]

Attachment might look like friendship, but Buffon claimed that it was really an inferior sentiment, because friendship required a spiritual element that animals lacked. If animals could not experience true friendship, however, humans could form animal-like attachments to things, when they became attracted to them through blind sentiment that involved no reflection. Thus, he explained, the love of a woman for her canary or a child for its toy was the same sort of sentiment as that of a dog for its master. The dog's sentiment, however, was more natural, since "it is founded on need, whereas the object of the other [the child or the woman] is nothing but an insipid amusement in which the soul has no part. These infantile customs exist only because of idleness, and owe their strength to empty heads."[88] Sentimental pet owners, in Buffon's eyes, then, descended to the level of the pets themselves in failing to exercise their higher human capabilities.

Like attachment, mimicking behavior raised questions about the distinction between the animal and human realms. It is an odd coincidence that the two most common exotic pets in the eighteenth century, parrots and monkeys, are groups that imitate human gestures and sounds (or is it a coincidence?). Both appeared as subjects in numerous works on the nature of human nature. In his *Essay Concerning Human Understanding*, John Locke reported a story of a parrot that had conversed with Prince Maurice in Brazil, and he struggled to explain why a rational parrot should not be classed with humans, the rational animal.[89] La Mettrie, citing Locke and other evidence, simply did away with the distinction in *L'homme machine*, declaring that language was not a barrier between realms, and that it was likely that monkeys could be trained to talk.[90]

Many people resisted this argument, which raised disturbing ideas on several fronts, from the nature of the eternal soul (supposedly a property only of humans) to the legitimacy of human domination of animals. One of the most forceful was Buffon, who, despite his praise of parrots, frequently pointed out the vast gulf between animals and humans. Another was Mme d'Épinay, who, in *Conversations d'Émilie*, used the example of mimicking animals to encourage children to understand and use the capabilities that only humans possess. The book opens with Émilie looking out the window and saying "Maman, I have studied my catechism well; can I come work next to you? Ah! Mama, come, come, I hear the drum. The monkeys are going by." After Émilie watches the parade, her mother asks her a series of questions to make sure she understands that although monkeys look like humans, they are different because they do not reason. Later she compares children to mimicking monkeys and tells Émilie to be like a little monkey with a sound mind *(esprit juste)*, not a false one, by imitating only what is good and reasonable rather than just any extravagant nonsense. She refers to parrots, too, in a lesson about not imitating words without understanding what they mean. She stresses, however, that although humans and animals differ in the intellectual realm, monkeys and all other animals are able to suffer, and so should be treated with kindness.[91] As in so many discussions of animals, especially tame ones, the author digs a moat separating the intellectual capacities of humans and animals while constructing a bridge linking their faculties of sensation.

Mauduyt and Buffon both placed emphasis on this moat between human and animal in their discussions of mimicking. After describing parrots' capabilities, Mauduyt added that they were, however, devoid of real intelligence, because they were unable to understand the relation between word, gesture, and thing. He referred the reader to Buffon for a more complete discussion.[92] Although Buffon addressed the distinction between humans and animals in a number of places throughout the *Histoire naturelle*, two of the most extensive discussions arose in reference to monkeys and parrots. The article on the parrot begins with the generalization that humans most admire animals which "participate in his nature": the monkey, "by its resemblance in external form," and the parrot, "by its imitation of speech." Because of these resemblances, people commonly believe that these animals are in some way intermediate between man and beast, but they are mistaken; these imitative species are not even the most admirable animals. It is lucky that nature did not unite imitation of word and gesture in one animal, mused Buffon, or the philosopher would never be able to convince people that such a creature was only an animal. What differentiates animals, even mimicking ones, from humans is that they learn as individuals, not as a species. Civilized human societies are able to progress constantly through the cultural accumulation of knowledge. Imitation is not learning: thus, "the monkey who gesticulates, the parrot who repeats our words, are no closer to increasing in intelligence and perfecting their species." Indeed, the parrot is not superior to other birds, nor is the monkey the closest of the animals to man.[93]

If parrots and monkeys, contrary to popular belief, are not the best animals, which ones are? Buffon gave high marks to the dog, the elephant, and the amazing South American trumpeter (fig. 5.6). In the case of dogs and elephants, unlike that of indocile and repugnant monkeys, "it suffices to treat them well to communicate to them the gentle and even delicate sentiments of faithful attachment, voluntary obedience, unpaid service, and unlimited devotion." These superior species, although they will never progress as a species in the same way as humans, can be ennobled through their association with and instruction by humans.[94] Both Buffon and Mauduyt lauded the

FIGURE 5.6. The trumpeter *(agami)*, an easily tamed South American bird.
Some naturalists suggested that these friendly birds would be superior to
dogs as pets and guard animals because they did not transmit rabies.
Buffon et al., *Histoire naturelle des oiseaux*, vol. 4 (1778), plate 23.
Reproduced by courtesy of the Department of Special Collections,
General Library System, University of Wisconsin–Madison.

trumpeter, an extremely tame chicken-like bird from South America whose attributes had been described by de La Borde, one of the king's physicians in Guyana. De La Borde had contended that tame trumpeters were as faithful as dogs; they obey and follow their masters and exhibit joy in seeing them, but chase away ugly or bad-smelling people and bite "negroes" or other servants who come too close to their masters. They come when called, love to be caressed, and can learn to guard flocks of sheep. On the basis of the trumpeter's instinct for human society, Buffon concluded that it was as superior to other birds as dogs are to other quadrupeds. He and Mauduyt both recommended that the bird be domesticated in France.[95]

Like schemes to import and breed zebras, the idea of replacing dogs with trumpeters never panned out. It made a nice dream for the late Enlightenment, though—a good-looking bird that had never been domesticated in its native habitat, coming to France to breed a race of faithful, affectionate, and useful companions.

CHAPTER 6

Animals in Print

PEOPLE CAME into contact with exotic animals not only by observing or owning them but also by reading about them. As more species of animals arrived in France, as voyage literature proliferated, and as interest in the exotic and in the sciences of nature spread, people began to buy and read factual books about the natural world. They also began to prefer fables and other fictional works that were based on actual, not invented, behavior of the animal protagonists. Yet, while views of nature were becoming increasingly scientific (in the modern sense of the word), the most popular works consistently related animal facts to human concerns, whether as aids for creating collections, as metaphors, or as religious or moral lessons. Animal stories could be about much more than animals; and fables did not disappear, but rather changed form in Enlightenment France.

The growth of interest in natural history during the eighteenth century coincided with rapid growth in the publication of books and periodicals, both legal and illegal (i.e., without permission of the royal censor), especially after 1750. Literacy rates also increased, reaching, by the end of the century in Paris, 50 to 80 percent for people in the lower classes. Although rates of book ownership increased, one could participate in the reading public without purchasing a book or journal by joining a reading club, perusing a newspaper in a café, or listening to someone reading aloud—a popular activity, especially in salons.[1]

Reading was considered an emotional activity, more akin to moviegoing today, and readers prized authors whose words elicited tears, laughter, or sexual arousal. Some physicians even warned that reading novels could endanger a young woman's health.[2] A few readers recorded their responses in letters or journals—Suzanne Necker, for example, thoughtfully analyzed her responses to Buffon—but because most readers left behind no trace, a reader-centered approach has to rely on other sources.[3] Book reviews in widely circulated journals are useful sources, since they express the opinions of editors addressing a broad public. Similarly, abridgments and compilations, which were very common in this precopyright era, reflect the compilers' and publishers' judgments of which sections of longer works the public would most want to read. Prefaces to books do not tell us who read them or how they responded, but they do indicate what audience the author hoped to reach. I should point out that I use the term *popular* for books oriented toward an audience beyond that of naturalists and scholars; this audience would have been largely urban rather than rural, and from the middle rather than the lower classes. Because distribution figures are so hard to come by, it is not always possible to claim that a work was also popular in the sense of being widely available. Where possible, I use the existence of new editions and printings or statements of contemporary judgments as indications of a work's popularity.

Literature about animals catered to a variety of tastes and can be divided roughly but conveniently into several (overlapping) categories based on audience and genre. One popular type of literature presented animals in fictional settings. People turned to poems, fables, utopian novels, and satires both for entertainment and for social or political morals. Later in the eighteenth century, however, fictional works began to be based more and more on verifiable characteristics of their animal protagonists. A second type of literature was oriented toward those who wished to read about animals primarily to think about religious or philosophical issues. Books in the natural theology tradition, many intended for children, explained God's design of the natural world and usually focused on insects. Works on philosophical topics dealt with related issues but often took a decidedly antireligious stance. Animals lent themselves to philosophical discussions about what distinguished them from humans; many works discussed the nature of animal souls and the relative happiness of animals and humans. A third type of literature about animals catered to utilitarian tastes. Dictionaries and manuals were published for the use of people who had natural history cabinets and needed guides; others provided veterinary or agricultural advice. In these works, plants and minerals, because of their more practical uses, featured more prominently than animals. Finally, many readers wanted encyclopedic works that presented an overview of the entire natural world. Following in the tradition of Aristotle, Pliny, Gesner, and Aldrovandi, such works tended to highlight large animals and birds, to include exotics, to stay away from explicitly theological themes, and to emphasize literary style. The second half of the eighteenth century was dominated by the most prominent of these, Buffon and Daubenton's *Histoire naturelle*, along with a host of abridgments and take-offs.

Animal Fiction

Fictional animals thrived in eighteenth-century literature.[4] These works drew on earlier traditions such as Aesop's fables, biblical parables, fairy stories, bestiaries, and emblem books, all of which linked animal tales with moral lessons. Emblem books paired pictures of animals with stories about their behavior that related to a specifically Christian meaning. The pelican, for instance, was said to feed her young by tearing open her own breast and spilling forth her lifeblood, in the same way that Christ sacrificed his life for others. Folk stories and fables had long circulated in cheap, popular chapbooks, but in the late seventeenth century they found an enthusiastic audience in elite circles, when La Fontaine and Charles Perrault produced sophisticated new versions of old tales. Perrault, whose brother Claude had coordinated the Académie des sciences's *Mémoires pour servir à l'histoire naturelle des animaux*, brought wolves and barnyard animals to a wide audience in his Mother Goose Tales. La Fontaine created a menagerie of beasts that are still widely familiar today. These tales, like their predecessors, conveyed morals in an entertaining fashion.[5]

Fables were one of the most popular types of animal fiction. Although commentators disagreed about the precise definition of a fable, the term was generally used for a brief poem or story bearing a moral message and often featuring animals as characters.[6] All eighteenth-century fabulists wrote under the giant shadow of La Fontaine, whose eleven masterful books of fables had appeared in 1668 and 1678–79. Although he wrote in a variety of genres, his fables, which reveal a perceptive, worldly wisdom, received the greatest acclaim. The fox flattering the crow until it drops a hunk of cheese from its bill, the cicada blithely singing while the dutiful ant stores provisions for the winter, the wolf in sheep's clothing: most of the fables feature animal characters but convey insights about human behavior. So did the fables of Aesop, Phaedrus, and Pilpay that La Fontaine used as sources; so, too, did the medieval bestiaries and Renaissance emblem books that also served as models. No one expects to see a fox talking to a crow, and so it might be surprising to find eighteenth-century commentators discussing the behavior of La Fontaine's animals as animals, not as human stand-ins. Part of the charm of the fables, however, for audiences today as well as for those of three centuries ago, is in the careful depiction of the animals, which act simultaneously as humans in animals' clothing and animals in their own clothing.

La Fontaine himself brought up the issue of verisimilitude in "Le souris et le chat-huant" (The mouse and the owl). The narrator of the fable tells of finding an owl's nest inside a felled tree that contains a pile of fat mice with no feet: the prescient owl had maimed his prey to keep them from running away before he could eat them. A playfully ambiguous footnote to the fable states, "This is not a fable *[ceci n'est point une fable]*; the event, although incredible and almost unbelievable, really happened. I may have taken the foresight of the owl a little too far, since I do not mean to establish in animals a reasoning process like this; but these exaggerations are allowed in poetry, especially in the kind of writing that I do."[7] According to critic Patrick Dandrey, passages like this undermined the strictly allegorical nature of the traditional Aesopian style of fable: in La Fontaine, animals are simultaneously real and allegorical. The "singing" of a cicada, for example, refers both to the literal sound of the insect and to the figurative human action of wasting time. But, like cicadas, humans also literally sing, that is, produce melodic sounds: thus the word connects humans and insects through this common activity. This stress on the similarity in nature between humans and animals is new, argues Dandrey, or at least stronger, in La Fontaine's fables, especially the later ones, than in those of his predecessors. La Fontaine reinvigorated the tradition of the animal allegory by incorporating elements of naturalism associated with the new science.[8] Both different and similar, the human and animal worlds simultaneously split and merge in a dance that gives the poems their playful vitality. La Fontaine's successors continued to dance on the boundary of these two worlds, as did the most popular naturalists of the eighteenth century.

More than one hundred editions of La Fontaine's fables appeared in the eighteenth century, but other poets were not deterred: scores of them continued to compose new books of fables. Some authors relied on the same sources that La Fontaine had used; some drew on other sources; some composed their verses from scratch. Discourses on the fable also proliferated, and reviews in periodicals judged new productions based on several criteria, including their educational function, because fables had become standard fare for young children's instruction. According to Harriet Ritvo, after 1750 children's animal stories also proliferated in England, where they conveyed many of the same moral lessons as the French stories.[9]

Although fables were classified as literature, not science, much of the eighteenth-century debate about them centered on the verisimilitude of the animals. How closely should the behavior of fable animals match that of the real thing? Is it bad for children to believe that animals can talk? Does nature provide any guarantee for the truth value of the moral? In the developing discussion on these questions during the century, we see a continuation of the naturalism that had crept into La Fontaine's fables, growing in step with the general interest in natural history. Fabulists could have used animals simply as props for morals; the fact that they did not reflects, I think, a new concern for grounding moral truths in nature.

One indication of this increased naturalism is the inclusion of new species not found in earlier fables. The range of species in Aesop and Phaedrus was broad, including exotic animals like lions and monkeys, but La Fontaine introduced others, and his eighteenth-century successors continued to expand the fabulous menagerie. One fable features a parrot and a parakeet engaged in a plumage competition; another, a *sarigue* (opossum) who provides a model of maternal devotion; and a third, a sloth and a monkey who illustrate the virtues of a quiet versus an active life. Footnotes keyed to the titles of the latter two fables defined the unfamiliar animals. The *sarigue* was identified as "a kind of fox from Peru (Buffon, *Hist. Nat.*, vol. IV)," and the sloth was described in Buffon's words (without acknowledgment) as ugly, slow, cowardly, and stupid.[10] But new species were not the rule. Most fables stuck with the usual stock of familiar creatures; the new approach emerged in the way they were described.

The question of realism was raised in theoretical tracts (such as the discourses that opened many books of *Fables nouvelles*), in reviews of fables in periodicals, and in treatises about education. Houdar de La Motte, whose 1719 discourse on the fable was influential throughout the century, stressed that fables should be founded on nature. He followed his own precept, composing one that used—and justified the use of—a newly discovered fact about the regeneration of crayfish legs.[11] One of his successors noted that "the more appropriate and natural the image, the more perfect the fable. The image is natural when it is realistic, and when nothing is contrary to the instinct of the animals."[12] The *Avant-coureur* shared this opinion, which it expressed in a critical review of an anonymous volume of fables published in 1762. The reviewer stated that fabulists should give to each animal its genuine *(propre)* character and criticized a fable about a swallow family because "it is false that the swallow teaches her young to construct a nest." The *Journal de Paris* disliked a fable featuring an eagle that read Montesquieu: "The fable has its type of credibility; animals have their habits; it is necessary to observe them carefully."[13]

Educators began to wonder whether fables taught children lies about nature. The harshest attack came from Jean-Jacques Rousseau, in his novel *Émile*. He critiqued every line of La Fontaine's "Le corbeau et le renard" (The crow and the fox) in an entertaining dialogue that showed that the moral about vanity and flattery was too sophisticated for young children, that the vocabulary was obsolete and inappropriate, and that the tale conveyed false ideas about animal behavior. Crows and foxes can't speak, Rousseau pointed out, and even if they did, they wouldn't understand each other.[14] According to historian Robert Granderoute, sentiment against using fables for pedagogy began to be voiced around 1750; for instance, in a 1756 essay on education, H.-C. Picardet proposed substituting Buffon's *Histoire naturelle* for fables, which he claimed were "full of marvels and frivolity."[15] In an announcement soliciting subscriptions for a new four-volume edition of La Fontaine's fables with colored illustrations, the *Affiches de Province* praised the work for its educational merit:

> What determined the author to present this work by subscription is its util-
> ity for young people who receive an elite *[distinguée]* education. There has
> been a special attempt to make the animals match the originals painted from
> life by the best masters in this genre. The result of the authenticity found in
> the multitude of animals represented in this work is that the young people to
> whom it is shown will not acquire false ideas about the true colors of these
> animals; and this preliminary knowledge could inspire in them the taste for
> natural history and for painting.[16]

Natural history could be taught through the suspect medium of fables not only by including accurate illustrations, but also by supplementing them with verbal descriptions. Such was the goal of the abbé Grozelier, who composed his fables specifically for teaching. At the beginning of his second volume of fables, he announced his intention to include information about the featured animals so as to give young people some knowledge of natural history. A note accompanying "The Punished Parrot" tells the reader what parrots look like, that they eat bread dipped in wine called "parrot soup," and that there are now known to be more than a hundred different kinds in the New World.[17]

Other fictional works also combined commentaries about human society with the newest information about exotic species. The *Mémoires de l'éléphant,* for example, combined political criticism with scientific accounts of the elephant that was starring at the Saint Germain fair (see chap. 3). A similar work, *Lettre d'un singe (Letter from a Monkey),* appeared in 1781. This short tract, by the prolific author Restif de la Bretonne, purported to be the transcript of a letter written by César de Malacca, a hybrid monkey-human who had been adopted by a Parisian countess, to his fellow monkeys in Africa about the injustices reigning in the world of human beings. Melding features of natural history with fiction and piquant anecdotes, along with social criticism, the work included capsule descriptions of twenty-four species of monkeys, taken mostly from travel accounts.[18]

Animals in Natural Theology and Philosophy

A very different sort of book about the natural world inspired the duc de Croÿ to write an entry in his journal on 22 May 1779, after he returned from a vineyard he had just finished constructing: "It was a lovely spot, with a delicious view, and although I've done good jobs before, I've never made such a pleasant spot. It cost, in all, 1,700 livres. I took with me M. Pluche's book, and while reading, at the end of the third volume, the use one should make of the spectacle of Nature, I was exalted, and I felt, with the gratitude one should feel, the obligation we should have for the gifts of the Creator and the talents he has given us. What a delicious experience." To commemorate the moment, he decided to install a plaque to the abbé Pluche.[19]

The book that made such an impact on the duke was the *Spectacle de la nature,* first published in 1732–50. The eight-volume work related in an accessible and amusing style a conversation among a young chevalier, a count and countess, and an abbé (modeled on Pluche himself). Often reprinted and translated into several languages, it remained extremely popular throughout the century and introduced many readers to the delights of nature in a different context from the more secular genre of the fable. Pluche covered all three realms of nature, along with subjects such as music and politics, with the goal of demonstrating God's presence and benevolence. In the world of the *Spectacle,* everything had a reason, though it might be unfathomable. Animals provided lessons for the young nobles whom Pluche regarded as his audience: he held up the social insects as models of diligence, and the ass as a useful servant whose uncomplaining labor should, like that of the peasant, be gratefully appreciated.[20]

Other works combined a greater attention to the facts of natural history with an equally theological approach. One of the most important was Réaumur's *Mémoires pour servir à l'histoire des insectes* (1734–42), a more scholarly book based on direct experiment and observation.[21] This work reached a wide audience through Gilles-Augustin Bazin's popular abridgments, *Histoire naturelle des abeilles* (1744) and *Abrégé de l'histoire des insectes* (1747–51).

The royal censor approved of the "excellent morality" of the latter work and its potential to benefit the hearts and minds of young people.[22] Insects' impossibly intricate and tiny features, first clearly visible after the invention of the microscope, had been proffered as important evidence of God's handiwork since the seventeenth century, and they continued to be the featured actors in theologically oriented books through the eighteenth century.[23]

Many children's books presented the natural world as evidence of God's handiwork. Fillassier's *Ami de la jeunesse,* for example, used the catechism format preferred for children's literature in an ambitious work that covered religion, commerce, physics, natural history, mythology, chronology, geography, and French history, all in two volumes, with the stated aim of forming civil and moral virtue in young people of all orders.[24] In the section on animals, Eraste, the teacher, explains to his pupil Eugene the marvels of God's design. He describes condors, for instance, as the biggest, meanest birds on earth, which can eviscerate cows with their beaks and are known to devour young children! Wise Providence, however, has made sure that they are very rare.[25] Louis Cotte, author of a children's book focused specifically on natural history, informed readers that his goal was

> to incline their [children's] young hearts at an early age to the most tender feeling of gratitude toward their divine Author, to always look with interest at the admirable works of the Creator, and to pique their curiosity, in order to inspire in them the desire to cultivate in the future a science so suitable both for satisfying the yearning for knowledge that is natural to young people and for calming the passions with useful distractions by means of promenades that will be more pleasurable when all of the objects they see will interest them.[26]

The curious child asks questions about the nature and purpose of natural history, its connection with religion, man's relation to the rest of creation, and why insects are the most important class of animals to study. Insects, the wise teacher explains, are the most prevalent, the easiest to study, and they provide the greatest evidence of God's power in his ability to outfit such small creatures with all of the necessary structures and organs.[27]

This natural theological tradition self-consciously set itself against the more materialist, worldly approaches to studying nature. Cotte, for example, had his child protagonist ask why natural history is often more suitable for humble people *(les simples)* than for savants, to which the teacher replied, "It is because humble people usually have only God in mind when they study nature, and savants too often mix in human concerns, the desire to shine and to parade their knowledge, [and] too great a curiosity to uncover what God has hidden from view." Also taking sides in this skirmish between worldly admirers of flashy beasts (with Buffon often the unspoken target) versus humble observers of modest bugs, La Chesnaye-Desbois, in his *Dictionnaire raisonné et universel des animaux*, challenged those who so deftly described the elephant, tiger, or leopard to try their hands at depicting the subtle iridescence of insects.[28]

Contemplating nature could elicit comfortable admiration for God's creation, but it could also raise disconcerting thoughts about animal and human nature. In the wake of Descartes' claim that animals are like machines, seventeenth- and eighteenth-century thinkers chewed over the questions of whether animals had souls, whether they could reason, whether they could feel pain, and whether they were happier or more virtuous than people. Some of the letters printed in the *Journal de Paris* about animal exploits suggest that people who had read about these issues looked at animals through this particular lens, noticing and circulating incidents of animal behavior that related to instinct, virtue, or intelligence.[29] These issues also provided the backdrop for compilations of animal stories drawn from ancient sources or from travel literature. A representative example is the light-hearted *Histoire des singes, et autres animaux curieux* (1752), which praised animals' amazing reasoning abilities. The author strung together tale after tale of feats by monkeys, elephants, beavers, dogs, porpoises, ants, and birds, gathered from travel literature. He related, for example, the story of a smart monkey-servant owned by an Arab. The monkey had been left alone in the house, where he was supposed to guard a pot of cooking meat. While the monkey wasn't paying attention, a falcon swooped down the chimney and stole the meat. The quick-thinking monkey climbed into the pot and exposed his bottom. When the falcon returned for this tasty morsel, the monkey grabbed the bird, cut off its head, and threw it in the pot. Later he explained to his master what had happened by acting out the events in pantomime.[30]

Where fables used animals to convey insight into human social behavior and moral guidance, natural theological and philosophical works presented them in the light of more abstract issues: the wonder of creation or the nature of animal souls. All three genres took advantage of the growing knowledge of exotic species, but described the animals in ways that fit their own goals. Another genre, that of reference works, presented animals in a different guise, one that catered to readers with more practical goals.

Utilitarian and Reference Works

Eighteenth-century Parisians needed useful works to turn to when they bought a curiosity for their natural history cabinet and wanted to identify it, when they acquired a colorful parrot and hoped to find out something about its native land, or when they wished to look up the name of an animal that they had seen at the fair. Many reference books were produced to meet this demand.[31]

At the more scholarly end of the spectrum we find works like Duhamel de Monceau's *Traité général des pêches* and Brisson's *Ornithologie,* which updated the earlier works of zoologists like Belon, Rondelet, Ray, and Willughby. Brisson's book listed in numbing detail the external characteristics of fifteen hundred species of birds (mostly from Réaumur's collection), page after page for six volumes, and was meant primarily as a reference for other savants or for collectors of natural history specimens. The author himself declared in the preface that "this uniformity of description produces a monotony that would be unbearable in a work meant to be read all at once."[32] Other manuals appealed to a broader audience. Duchesne and Macquer oriented their one-volume, alphabetically organized *Manuel du naturaliste* (1771) toward people who visited natural history cabinets or wanted to assemble their own. Hoping to amuse while instructing (the standard formula for such publications), they chose to emphasize "the most piquant and the most interesting" facts of natural history. The manual, which sported its dedication to Buffon on the title page, was reprinted and updated several times.[33]

One book stood above the rest in popularity and scope: Valmont de Bomare's *Dictionnaire raisonné universel d'histoire naturelle,* first published in 1764 and translated, revised, and reprinted (both with and without the author's permission) numerous times. In his survey of private libraries, Daniel Mornet found it to be the third most popular science book after Buffon's *Histoire naturelle* and Pluche's *Spectacle de la nature.* Despite his fame at the time,

Valmont de Bomare is now hardly mentioned by historians.[34] Jacques Christophe Valmont de Bomare (1731–1807) was born in Rouen. Defying the wishes of his father, a lawyer, he pursued an interest in natural history developed during his education at a Jesuit school, where he studied the classics, pharmacy, and chemistry. He went to Paris in 1750 and set up shop as an apothecary, but managed to obtain a royal commission as a traveling naturalist, which permitted him to travel throughout Europe for several years. He visited natural history cabinets, met other naturalists, and collected specimens, especially mineralogical ones. In 1756, after his return to Paris, he began teaching what became an extraordinarily popular and long-lived lecture series on natural history, which attracted a number of high-ranking auditors.[35]

The course, offered yearly from 1756 to 1788, took place in Valmont de Bomare's cabinet in the rue de la Verrerie. In the 1750s he offered the course on Thursday afternoons and gave free public lectures at 3 P.M. on Sundays and holidays. From the early 1760s on, the four-month course met for three two-hour sessions a week, usually beginning in November or December. Some years a separate *cours particulier* was also offered: this may have been a smaller course for a more elite crowd. In 1763, Geneviève de Malboissière, a teenager from a wealthy family (whom we encountered as a pet lover in chap. 5), wrote to her friend Adélaïde to tell her of a plan to take what sounds like a made-to-order course with companions of their own choosing. "We want to do it [the course] with *gens de connoissance*," she writes: "We will find a time convenient for everyone to go see that man who lives on rue de la Verrerie and who is very celebrated (I don't remember his name) and it will be very pleasant to spend two hours together three times a week. Can you propose that arrangement to Madame your mother? If she consents, we will find out when the course opens, what time, and how much the instructor charges. We will tell him the names of the people we have chosen, and it will be charming."[36]

The courses covered the entire natural world: a description published in the *Journal de Paris* in 1777 listed the subject matter as including, in the mineral kingdom, theories of the formation and evolution of the earth, waters, soils, rocks, salts, metals, volcanos, petrifactions, and so forth; in the vegetable kingdom, agriculture and rural economy, roots, bark, stems, flowers, fruits, seeds, resins, gums, spices, and drugs; and in the animal kingdom, animalcules, polyps, worms, shells, insects, fish, serpents, lizards, amphibians, birds, quadrupeds, and man. The following year, Valmont de Bomare announced that the course would include, beginning in March, outdoor walks.[37] Animals, especially big ones, were a minor element in this scheme, which favored the tangible and the practical. Minerals led and humans came at the end, in contrast to the humans-first order of Buffon's natural history.

The duc de Croÿ appreciated the way Valmont de Bomare laid out on the table each item that he talked about: "everything in nature passes by your eyes, through your hands, and into your mind, by means of the explanations he gives, which are full of vigor, order, and eloquence."[38] The course also attracted other serious scholars, including some visiting Spanish naturalists, one of whom complained, however, that the course was more "curious" than instructive, because it focused more on individual items than on scientific ideas. Some attendees paid more attention to the other students than to the petrifactions. The author of an underground newsletter, after attending his first session in 1780, described the features of the "elegant ladies" who filled the front row of the auditorium, attentively following the lecture-demonstration. Critical as he was of these women's pretensions to be *savans,* it is not surprising that he found little to praise other than the professor's good memory. He described the lecture as displeasing and uninteresting, and he criticized Valmont de Bomare's mannered coldness, his puerile pretensions and cunning manner, and his guttural, honeyed voice.[39]

Valmont de Bomare's five-volume *Dictionnaire,* first published in 1764, offered complete coverage of the entire natural world and served as a textbook for his course. The dictionary received universal acclaim and was expanded and republished in a six-volume second edition in 1767–68, a nine-volume third edition in 1776, and a fifteen-volume fourth edition in 1791.[40] Valmont de Bomare explained that he chose to use an alphabetical arrangement not only because it would be most useful but also because it allowed him to imitate the "sublime disorder" of nature. However, readers could create their own more methodical approach by following cross-references and choosing a sequence of articles running from the general to the specific: for example, reading "natural history," "animal," "bee," "wax," "honey," and so on would enable the reader to put together a complete treatise on the bee.[41] Based in part on his own observations and cabinet, the book also freely drew from other books, journals, and Academy memoirs.

Although Valmont de Bomare stressed the practical and useful, he also wanted to attract pleasure-loving *gens du monde*. According to the preface, he tried to be "useful, instructive, and interesting" and hoped to appeal to a diverse audience: "The savant will find . . . the result of his knowledge and studies; the *homme du monde*, for whom sometimes everything in Nature is new, will look there for useful and instructive amusement; and I even flatter myself that this book will be incorporated into the educational programs of well-born people of both sexes."[42] Thus, instructive articles often included colorful anecdotes. In the article on ostriches, for instance, Valmont de Bomare told his readers not only about what ostriches looked like, what they ate (including scraps of metal), how they were hunted, and what they tasted like, but that the infamous Roman emperor Heliogabalus once had the heads of six hundred ostriches served up at his table so that he could eat their brains.[43] Valmont de Bomare became a well-known figure who was credited with spreading an interest in natural history among the gens du monde.[44]

How did gens du monde actually use the dictionary? One function was as a field guide at rural retreats: a reviewer in 1768 suggested that the book was principally for use in the countryside, where one turned toward nature, and a review of the 1791 edition predicted that it would become the "guide and the friend" for people whom the Revolution had driven out of the city. But it was frequently used in the city, as well. The duc de Croÿ looked in it for descriptions of the coati, the polar bear, and other animals he had seen at the fair. Some readers turned to it as a guide for creating a natural history cabinet and identifying specimens. One person who used the dictionary in this way was Geneviève de Malboissière, the young woman who also took Valmont de Bomare's course. After labeling her own specimens, she lent the book to her friend to do the same.[45] The dictionary helped to make a pleasure out of acquiring knowledge—just the thing for people with time on their hands and money to spare.

Buffon and the Encyclopedia Tradition

Despite the availability of so many animal books of so many different types, Buffon's *Histoire naturelle* (1749–89) outstripped all the rest in popularity. It gave readers something they could not find in other texts: reliable, up-to-date facts from all around the world embedded in a coherent, secular, interpretive structure, interspersed with observations and lessons applicable to their own lives, and conveyed in beautiful, even poetic, language. It combined the literary style of fiction with the moral element of natural theology (minus the overt religiosity), the speculative audacity of philosophy, the scope and utility of reference works, and the exoticism of travel accounts. It did not completely replace other works, however; Valmont de Bomare's dictionary, for example, continued to be essential, especially since the *Histoire naturelle*'s volumes on birds were not published until the 1770s and 1780s, and it did not cover reptiles, amphibians, or insects.

Although the *Histoire naturelle* stood alone, it grew from the roots of the classical encyclopedia tradition. Aristotle usually received credit as the first representative of this genre, having put together a comprehensive natural history combined with a philosophy of nature, but most naturalists took Pliny the Elder's *Natural History* as their model and touchstone. Pliny was still frequently cited in the eighteenth century, and a French translation and critical edition was published in the 1770s.[46] In the sixteenth century, the Renaissance naturalists Gesner and Aldrovandi produced updated encyclopedic treatments of nature that incorporated some of the species recently described from the New World. The Paris Académie des sciences sponsored a natural history of animals in the late seventeenth and early eighteenth centuries, the three-volume *Mémoires pour servir à l'histoire naturelle des animaux*, but it lacked the charm that attracted readers to authors like Pliny.[47] Not until the mid-eighteenth century did the classical and Renaissance encyclopedic natural histories meet their match, in Buffon's *Histoire naturelle*. Buffon's own claim to be following in the tradition of Pliny was generally accepted by the public, and he came to be known simply as "le Pline français."[48]

The public greeted the *Histoire naturelle* with unrestrained enthusiasm from 1749, when the first three volumes appeared in bookshops, to 1789, when the last supplement was published. It engendered scores of spin-offs and imitations that also reached wide audiences. Sniping by critics such as Grimm, Condillac, Condorcet, and Réaumur, who faulted Buffon for leaving the solid ground of fact to fly in the realm of hypothesis and conjecture, had little impact on general readers, who made Buffon a celebrity (fig. 6.1).[49] Daubenton, who wrote all of the anatomical descriptions in the volumes on

FIGURE 6.1. The natural historian Buffon portrayed as Moses, master of the earth, animals, minerals, and shells. Engraving by N. Ponce, after C. P. Marillier. BnF Département d'Estampes, N2. Courtesy of Bibliothèque nationale de France, Paris.

quadrupeds, and Guéneau de Montbeillard and the abbé Bexon, contributors to the volumes on birds, were mostly forgotten. Buffon's popular renown was so immense by 1772 that the *Affiches de Province* found it impossible to review a new volume of the *Histoire naturelle* because it had nothing new to tell its readers: "For a long time, the works of M. de Buffon have needed only to be indicated; the simplest mention sufficed. . . . Everyone has read, reads, and rereads the *Histoire naturelle:* what praise is worth more than that fact? What could one add to the idea of a work that surely will have readers as long as our language is understood, a work the reading of which, by people of all abilities, creates so many, constantly renewed, living witnesses?"[50] This extraordinary homage to public opinion very nicely illustrates the faddish side of Buffon ("everyone has read, reads, and rereads it"), but gives little insight into what attracted such devotion. For that we need to turn to the testimonies of other readers.

Some read Buffon for instruction about the natural world. Periodicals cited Buffon as an authority or as a source for further information about animals shown at the fair or on the boulevard, and many educators recommended using Buffon in schools for teaching natural history.[51] Geneviève de Malboissière included him in her daily educational schedule, as she told her friend Adélaïde: "Today I don't have any masters [coming to the house]; I am going to work like a little cherub until it's time for the Comédie [française]. This morning I have to read lots of Greek and lots of Buffon (that rascal cost me 8 louis), and write to Mlle Cust and to my brothers."[52] The *Histoire naturelle* could also serve as a reference work and ornament for natural history cabinets: a portrait of Joseph Bonnier de la Mosson, who had one of the largest natural history cabinets in Paris, showed him posing in front of his collection, holding a volume of the *Histoire naturelle* open on his lap.[53]

To many readers, the aspects of Buffon that appealed the most were the ambitious scope of the project, the beautiful style, and the human-interest angle of the content, particularly in the descriptions, or *histoires*. Numerous testimonials expressed awe at Buffon's audacity in taking on no less than a description and interpretation of the entire natural world (it was often forgotten that he had treated only quadrupeds, birds, and minerals, leaving aside reptiles, amphibians, fish, insects, and plants). By extension, his readers, too, could survey the globe. A statue erected of him in the Jardin du roi originally bore the inscription *Naturam amplectitur omnem* (He embraces all of nature), and poets liked to describe him in god-like terms: "The vast skies are open to him; / For him the sea is bottomless," wrote Fanny de Beauharnais; "You climb into the heavens, you reach down into the seas," apostrophized Nogaret. Only a great man could take on such a great subject, his admirers believed; by contrast, his critics were, in de Beauharnais's words, "cold pedants" and "impotent insects"—the latter a not-so-veiled gibe at Réaumur and his insect studies.[54]

In attempting to show Buffon's broad reach, some of his eulogizers had no qualms about extolling his attention to the humble species that he had ignored, as in this excerpt from a wretched poem by the chevalier de Cubières:

> He paints for us the lion, superb, audacious,
> With a style as noble as the king of the animals.
> How well he descends without baseness from this king
> To the timid insect, and how with grandeur
> He then attains the height of the elephant!
> How well he follows winged creatures in the air!
> How well he burrows into the reptile's lair!
> How well he unfolds and counts their convolutions!
> He even swims with the fish in the oceans.[55]

In addition to his scope, Buffon's style garnered fulsome praise. When elected to the Académie française in 1753, he delivered a "Discourse on Style" that laid out his views on composition and quickly became a classic text.[56] Even a critic of Buffon's theories such as Melchior Grimm admitted that "the nobility of his images and the grandeur of his pen make him a pleasure to read," and Rousseau dubbed him the greatest writer of the century.[57] He became the "painter" of nature; a poem by the marquis de Villette (one among many expressing similar sentiments) raved, "Seduced by the happy magic / The brilliance of his palette / Twenty times I reread his works."[58] Buffon's readers appreciated his attention to harmony, which he honed by testing the rhythm of each sentence, substituting longer or shorter words as necessary.[59] Mme de Genlis, for example, recommended reading aloud the celebrated description of the screamer *(kamichi)*, a South American bird: do so, she wrote, and you will hear slow music, "vague, melancholy, and somber," that will transport you to the vast swamps inhabited by the bird.[60]

Vivid language might not have enticed people to read about sloths and swans if Buffon had not incorporated another important element—human interest—through which he touched the *sensibilité* of his readers. Mme Necker (Suzanne Curchod), a salon leader who became Buffon's intimate friend in his later years, explained that when Buffon found his eloquence lagging while he was describing a place or an animal, he would bring in a reference to man, a sensory word, or a moral reflection. She concisely summed up this aspect of the *Histoire naturelle:* "Animals seemed to be very distant from us, and the art of M. de Buffon has been to continually bring them closer." "Closer," she meant, to human affairs: "M. de Buffon never speaks of animals without bringing us back to humans. Linking ideas is thus necessary for our interest, by reminding us of ourselves, the center of all our interests."[61]

When Mme Necker read her favorite author, she was no doubt reading from an elegant, leather-bound, folio volume, the format in which most modern scholars encounter him today. But the Buffon who became a literary sensation in late-eighteenth-century France appeared before his reading public in much more varied dress. Although members of the nobility, like Geneviève de Malboissière, had the money to buy and the leisure time to study original editions, many others did not. Publishers and writers spied this opportunity and responded by producing a variety of natural history texts, from shorter or cheaper versions of the original to compilations that lumped together chunks of prose from Buffon, Valmont de Bomare, travel accounts, and the ever favorite classical sources. Original works could take on different meanings in the hands of compilers, who often did not cite their sources.[62] Because of the effort he had put into creating a distinctive style, Buffon's words kept their integrity better than did those of many other authors. "The most noble conquest that man has ever made"—the description of the horse in the first sentence of the first animal article—must have been quoted hundreds of times in a multitude of contexts, but it never lost its association with Buffon, explicitly acknowledged or not. It is still repeated today by French writers with the assumption that readers will recognize its origin.[63] Compilers, abridgers, and reviewers were attracted to these lyrical and emotive passages, which came to have a much greater cultural presence than did the more technical sections.

Three thousand copies of the *Histoire naturelle* (vols. 1–3) reportedly sold out in six weeks, but a vast array of subsequent editions in all sizes and prices kept the work circulating, often in truncated form. In 1767, Buffon sold to the publisher Panckoucke a small-format edition of the *Histoire naturelle*, without Daubenton's anatomical descriptions.[64] Although some influential critics had declared Daubenton's work to be more valuable than Buffon's, few people read these dry details: "they seem to interest only scientists *[physiciens]*," remarked the *Affiches de Province*.[65] Three other small-format editions had already been printed, as had two editions published in the Netherlands, one with additions by the Dutch naturalist J. N. S. Allamand. Two editions of *Oeuvres complètes de Monsieur le comte de Buffon*—a fourteen-volume quarto edition and a twenty-seven-volume duodecimo edition, both without Daubenton's descriptions—were published between 1774 and 1789.[66]

According to one overworked bibliographer, an "avalanche" of extracts—for children, the public, or gens du monde—began after Buffon's death and continued through the nineteenth century.[67] One of the first was *Génie de M. de Buffon,* compiled by Giovanni Ferry de Saint-Constant. The preliminary discourse heaped the usual encomiums on Buffon, praising his vast genius (superior to that of Aristotle or Pliny), which enabled him to comprehend all of nature at a glance, and his poetic style, which appealed to both heart and mind by means of attractive descriptions, pleasant images, noble and touching sentiments, profound reflections, and sublime ideas. Replying to critics of Buffon's style, Saint-Constant retorted that Buffon had made the most abstract material accessible to the most ordinary minds: "Isn't it necessary for him to use brilliant and varied colors to keep the attention of readers unfamiliar with sublime objects and who become discouraged if it takes effort to understand them?" His book compiled picturesque descriptions and passages that had already become favorites in salons, such as "Love and liberty! What blessings!"[68]

Most books that recycled Buffon's words set them into a melange of excerpts from other authors. Small-format or abridged versions of Buffon were not adequate for those who wanted an overview of natural history;

Panckoucke's duodecimo edition ran to thirteen volumes and covered only man and quadrupeds. Enterprising compilers could pillage from Réaumur for insects, Brisson for birds, Buffon for quadrupeds, Valmont de Bomare for everything else, and put together complete natural histories in one or two volumes. A review of one of these texts, the *Cours d'histoire naturelle*, explained that "the public has been waiting for a long time for an elementary work that would include all of the most interesting details of natural history: all of the groups that make up the animal realm are either isolated in dictionaries or spread out in various specific treatises."[69]

The choice of texts depended on the targeted audience. Authors offered their works for teaching, for learning facts of practical use, for getting up to speed so as to be able to participate in learned conversations, or simply for pleasure. Thus, Wandelaincourt recommended natural history for students because it is "the foundation of commerce, the economy, and medicine, and the source of all of the pleasures that delight the sensitive and honest man." In choosing extracts from "the best writers of our day," he focused on the useful and familiar aspects of the "brilliant spectacle of nature": students would not broach more abstract notions of cause and effect until later. More oriented toward social uses, Chevignard de la Pallue stated that he had assembled the principal phenomena of nature that are frequently topics of conversation, because to participate in society "it is necessary to be a so-called educated man *[homme instruit]*."[70]

Cutting and pasting passages from other works also allowed authors to put their own philosophical or scientific spin on the material, and the excerpted words took on a different cast in their new contexts. The section on animals in Savérien's *Histoire des progrès de l'esprit humain dans les sciences naturelles*, for example, changed the order in which the quadrupeds were treated, abandoning Buffon's human-centered scheme for an arrangement based on anatomy. Hennebert and Beaurieu's *Cours d'histoire naturelle*, by juxtaposing Buffon's words with those of Virgil, Pliny, and La Fontaine, brought out his literary and pastoral side.[71] The abbé Sauri put a natural theological twist on Buffon in his *Précis d'histoire naturelle*, a five-volume work that began with insects and ended with birds and quadrupeds.[72] Duchesne and Le Blond created a work for children, the *Portefeuille des enfans*, published in installments, by juxtaposing illustrations from the *Histoire naturelle* with brief paragraphs describing each animal and bird. Children were to cut them out and paste them into their own scrapbooks. The physician Pierre-Joseph Buc'hoz reworked Buffon in a utilitarian mold, shamelessly copying extensive passages although he constantly claimed originality, maintaining that Buffon's works "have no relation to mine, and one can acquire them, along with mine, without running a risk of repetition."[73]

One of the more colorful authors to appropriate Buffon's animal descriptions, and so disseminate them to an audience that might not otherwise have encountered them, was Restif de la Bretonne, fan of Buffon and author of the *Lettre d'un singe,* as well as pornographic novels, memoirs, and philosophical jottings. In Restif's utopian novel *La découverte australe* (1781), a Frenchman, Victorin, flies (with wings of his invention) to an island in the South Seas that he names île Christine in honor of his wife, where he establishes an ideal community. He then travels with his two sons (both of whom have married Patagonian giants) to a series of islands, each one inhabited by a different race of animal-human hybrids: elephant-men, frog-men, horse-men, lion-men, and so on. In describing each type, Restif closely followed Buffon's descriptions (Buffon makes an appearance as the character Noffub on an upside-down island called Mégapatagonie). Victorin establishes colonies on the islands and aerially transports a pair of young hybrids back from each one to the île Christine, where they are tamed and civilized (fig. 6.2).[74] The *Affiches de Province,* although it dubbed the work "the strangest production imaginable," full of bizarre ideas and grotesque engravings, praised the author for his knowledge of natural history.[75]

Despite their varied goals, almost all of these compilers picked out the most arresting passages from the *Histoire naturelle,* helping to raise a handful of phrases to the level of cultural mantras. Popular periodicals compounded this process when they chose excerpts from the excerpts to quote in their reviews. "Buffon" for many readers, then, was the Buffon of evocative tableaus, tableaus that often contained moral lessons much like those of fables.

Natural History and Fables

Historians of Western civilization often portray the seventeenth century as a transitional period away from an older view of nature as full of meaning and allegory and toward a modern view of nature as a brute realm to be

FIGURE 6.2. Elephant-man and elephant-woman fight back against winged
humans who are trying to capture them. Illustration from Nicolas-Edmé
Restif de la Bretonne, *La découverte australe par un homme-volant*
(Leipzig, 1781), book 12. Courtesy of Special Collections Library,
University of Michigan, Ann Arbor.

investigated and manipulated with dispassion. In a rather extreme version of this view, Charles Raven claimed that "little by little nonsense was recognised, fables were exploded, superstitions were unmasked, and the world-outlook built up out of these elements fell to pieces."[76]

Allegories and symbols did not exit with the Renaissance, however, for most people. Buffon did not eliminate meaning from the animal world, nor did many others. The "slippage between the natural and the social," to use E. C. Spary's words, is now beginning to be explored, especially in works on popular science. This "slippage" is especially evident in the connections between natural history and fables.[77] Examining these connections offers a way to understand how people found meaning in animals during the Enlightenment.

One has to be careful with the word *fable*, though. In eighteenth-century French, *fable* could mean either a false report or a moralistic story. When naturalists such as Buffon said that they were eliminating fables from their work, they meant that they were discarding falsities such as unicorn sightings, not that they were refusing to draw moral lessons from nature. It is true that explicit moralizing diminished in some scientific texts and that this trend continued into the nineteenth and twentieth centuries. Moralizing did not disappear from popular works, however (nor has it yet): many familiar species continued to be associated with traditional lessons, and meanings concerning everything from slavery to motherly love were easily grafted onto newly discovered species. Many of these same connections were made (and still are being made) more subtly in scientific works not intended for a popular audience. Nature continued to be a source for lessons, but in a different way than it had been in earlier periods. This shift is seen most clearly in the relationship between fables and natural history. The mingling of fable and natural history occurred in both directions. Just as fables incorporated facts of natural history, natural history, both on display and in print, continued to blend scientific displays or descriptions with mythical, fabulous, or moralistic elements.[78]

The prince de Condé's menagerie and the duc de Montmorency's natural history cabinet were two locations where spectators could marvel at erudite collections of native and exotic species presented in a context of familiar animal stories. The prince de Condé resided at Chantilly, an expansive estate north of Paris that rivaled Versailles in opulence. The prince was well known for his art collection, stables, menagerie, and natural history cabinet. The cabinet became one of the most extensive in France when the prince bought Valmont de Bomare's collection in 1786 (along with his services as keeper and guide), and the menagerie consistently garnered rave reviews from visitors and guidebooks, some of which declared it superior to the one at Versailles. The menagerie, constructed in the last two decades of the seventeenth century and later expanded, consisted of enclosures for animals and birds arranged around a series of courtyards. In the middle of each courtyard was a rococo fountain with a sculpted animal figure in the center and a painting around the base of a fable from La Fontaine featuring the same animal. A visitor from the provinces noted that "everywhere in the chateau, as in the garden, we saw subjects taken from La Fontaine's fables, with the verses underneath; he seems to be the favorite author at Chantilly."[79]

Themes from fables also graced the extensive natural history cabinet belonging to the duc de Montmorency. The collection, spread out through several rooms of a Parisian mansion and containing a huge variety of species displayed in glass cabinets surmounted by ancient Japanese porcelain vases, had been prepared by a taxidermist named Desmoulins. In 1783, Desmoulins described the cabinet in an advertisement in the *Affiches de Paris:* "Natural history cabinet composed of 16 armoires containing, conserved under glass, the most beautiful birds of the Indies and of our continent, several quadrupeds, including a tiger with the traits of Bacchus, and several other groups also having analogical attributes; all of these animals are prepared in a manner to protect them forever from insects, and each armoire offers a feature *[trait]* that describes their moral character."[80]

A description of the cabinet in a contemporary guidebook also mentioned the cat (identified as a panther rather than a tiger), which was displayed in a large armoire along with a grapevine and other traditional trappings of Bacchus. According to the guidebook, all the specimens were grouped in a picturesque way, in attitudes that corresponded to their characters. The fable of the crow and the fox appeared above the entryway. A footnote explained, "The birds and quadrupeds that one sees in the cabinet were arranged and injected by M. Desmoulins, painter. This artist combines the talent of the painter with the ability to stuff all kinds of animals, birds, reptiles, etc. so as to portray their characters, to make them subjects of fables; which arouses interest, and seems to animate the groups that he creates. Because of his talents he has been chosen by M. de Buffon for the maintenance and arrangement of this part of the Cabinet du roi."[81]

Natural history texts often featured the same mixture of scientific facts about animals combined with fable-like moral messages or traditional tales as did these sorts of displays. People read fables for the pleasure of the poetry and the wisdom of the moral, and many read nonfiction works about animals for the same reasons. Moral messages emerged in these works in several ways. One was the textual juxtaposition of fables with naturalists' descriptions, as in Hennebert and Beaurieu's *Cours d'histoire naturelle*. The authors explained in the preface that they decided to temper the "somewhat cold but necessary" facts with the charms of poetry by citing Lucretius, Virgil, Vaniere, and La Fontaine in appropriate places. The article on the elephant, for example, began with a fable from La Fontaine. The effect of flowing from La Fontaine to Valmont de Bomare to Buffon in one page is to merge all of these sources into a single commentary on animals and their meanings. Reviewers seemed to have no problem with this hybrid genre: the *Affiches de Province* praised it as a work of instruction and amusement, accessible to many readers, and drawing on the most accredited and recent authors.[82]

Moral lessons were also conveyed by attaching value to particular species or behaviors: dogs are good, for instance, and so is loyalty, but wolves are bad, and so is unprovoked violence. The abbé Sauri, after describing the lion's benevolence to feeble creatures, commented that not all *grands* make such good use of their power. Buffon accorded top ranking to four species—elephants, dogs, beavers, and monkeys—based on their ability to learn, their sociability, and their possession of admirable character traits. Of these, monkeys scored the lowest because of their nastiness and sexual immodesty.[83] Good animals were often paired with their opposite, evil doubles: lions versus tigers; dogs versus wolves (or cats); eagles versus vultures. Later in the eighteenth century, some authors, most notably Rousseau, contended that anything that was natural was good. Social critics used this argument to fault overly refined humans for no longer engaging in natural (good) activities such as breast-feeding.[84] Not everyone agreed with this argument, however. In a letter to the *Journal de Paris*, for example, a subscriber reported seeing a swallow steal mud from another swallow's nest and asked why we should look to animals as moral models.[85]

Buffon was a master of the new fable. Although he engaged in some old-fashioned moralizing, holding certain species up for emulation or inserting a pointed comment, he also developed the naturalistic fable to its peak. The article on the elephant displays both of these techniques. Following Pliny, Buffon praises the elephant as the best of the animals. His attributes, according to Buffon, include the following: he approaches man in intelligence; is dexterous (with the trunk); docile; recognizes and becomes attached to humans; submits to good treatment but not force; serves with zeal, fidelity, and intelligence; likes the society of his own species; is powerful, courageous, prudent, moderate even in the strongest passions, and constant in love; recognizes his friends even when angry; attacks only those who have wronged him; remembers favors as well as injuries; is vegetarian; and is loved and respected by the other animals because they have no reason to fear him. Finally, his modesty exceeds that of humans: he never mates in the presence of witnesses, especially not when in captivity. The elephant is at the opposite end of the moral scale from the bloodthirsty tiger:

> The conformation of the body usually matches the disposition. . . . The tiger, too elongated, too low to the ground, his head naked, his eyes haggard, his tongue the color of blood and always hanging out of his mouth, has nothing but features of gross wickedness and insatiable cruelty; his only instincts are a constant rage, a blind fury, which knows no distinctions and often makes him devour his own children and tear their mother to pieces when she tries to defend them. May he thirst excessively for his own blood! May the thirst not be allayed until he has destroyed at birth the entire race of monsters he has produced![86]

Such passages do not seem far away from Renaissance animal books, presenting moral models for the reader, abounding in stereotypes, and evoking folk traditions that set up opposing pairs of species, one created by God and the other by the Devil.[87] This seems odd considering Buffon's disparaging comments in his "Discourse on the Nature of Animals" about those who have made a moral example of the so-called perfect republic of beehives. One should observe and describe bees, he writes, "but it is the morality, the theology of insects that I cannot stand to hear preached; they are marvels that the observers put there themselves and about which they then make a fuss, as if they really existed." He concludes the section, "Isn't nature astonishing enough herself, without our trying to surprise ourselves by becoming dazed with marvels which don't exist and which we put there ourselves?" In introducing the elephant, he criticizes moral virtues that the ancients accorded the animal, such as worshiping the sun. However, he continues, "in dismissing the fables of credulous antiquity, in also rejecting the puerile fictions of superstition that still exist, there still remains enough about the elephant, even in the eyes of the philosopher, that one must regard him as a being of the first rank; he is worthy of being known, of being observed; we will endeavor, therefore, to write the history impartially, that is to say, without admiration or scorn."[88]

Buffon believed that his description of the elephant was based on accurately assessing its capabilities rather than on attributing spurious marvels to it. Although an infinite distance separates men, who have souls, from animals, who do not, he contends, animals have memory and ideas, they share their material nature with humans, and they are endowed with varying degrees of *sentiment* (sense perception). Thus they can experience feelings that are the product of the material body: these include the passions of fear, horror, anger, courage, love, jealousy, and the desire for enjoyment. Through education, example, imitation, and habit, they can also develop attachment, pride, and ambition. The more *sentiment* a species has, the closer it is to humans and the more qualities it shares with them.[89]

Now it is clear why Buffon scorned those who lauded bees while not hesitating to praise the morals of other animals himself. Bees have a very low level of *sentiment* and so could not possibly have much in common with humans. Elephants, on the other hand, are as close to humans as it is possible for soulless matter to be, and so they can share all of the highest material characteristics of human beings.[90] Elephants do not simply represent a symbol or allegory of modesty—they *are* modest.

Buffon's search for characteristics common to more than one species—say humans and elephants—was an important element of his overall philosophical project. He stated at the outset of the *Histoire naturelle* that the natural historian should proceed by first acquainting himself fully with the details of the diverse objects of nature and then raising himself up to attain more general views that take in a variety of objects.[91] Buffon looked not only for shared traits like modesty, but also for general patterns or laws of behavior. Thus, for instance, he noted that in almost all species, males are furious during mating season, and females become ferocious when defending their young.[92]

One of Buffon's most astute readers, Mme Necker, appreciated this method of finding truths in nature. In her writings she praised his descriptions in contrast to those of his collaborator, Guéneau de Montbeillard, whom she faulted for making links between humans and animals that did not relate to some essential comparison between the two. Here she contrasts Guéneau de Montbeillard's article on the nightingale with Buffon's on the warbler:

> The great fault of the article on the nightingale . . . is that what it says has more to do with a musician than a bird. Compare the article on the warbler . . . the latter is much shorter, because there is less to say about a bird than a man, but all of its features are charming. M. de Buffon keeps our uninterrupted interest in the little bird; he reminds us constantly of man, in making us regret not having the morals of birds; but, in M. Guéneau, we only see the man in the bird: it is no longer a connection *[rapport]*, it is the thing itself.[93]

Drawing lessons from nature was not limited to authors: readers could engage in *rapport*-making as well, using the raw material of the natural history text. Mme Necker did just that when she remarked that after she read in Buffon that beavers develop their sense of instinct only in the company of other beavers, it made her think about what you see in society every day: witty people *(gens d'esprit)* are witty only when they are with others of their kind.[94]

Buffon's treatment of the myth of the swan song shows his remarkable ability to retain the lyricism of ancient fables while discrediting them. The article on the swan, one of the favorites for salon readings, exploited even as it demolished the fable that swans sing a beautiful song just before they die. Buffon concluded the article, which he wrote after an almost fatal bout with kidney stones, with this expressive passage:

> Moreover, the ancients were not content to make the swan a marvelous singer; alone among all of the beings that shudder in the face of destruction, it still sang at the moment of its last agony, and heralded its last sigh with

harmonious sounds; it was, they said, when the swan was about to expire and was saying a sad and tender goodbye to life that it rendered its strains so sweet and so touching. . . . One heard this song at dawn, when the wind and waves were calm, and could even see swans musically expiring while singing their funereal hymns. No fiction in natural history, no fable of the Ancients, has been more celebrated, more repeated, given more credit. . . . One must pardon them their fables: they were appealing and touching; they were worth as much as sad and arid truths; they were sweet symbols for sensitive souls. Swans, of course, do not sing when they die; but when speaking of the final flight and last burst of a great genius about to be extinguished, one will remember with emotion that touching expression: *it is the song of the swan!*[95]

Paying Buffon the compliment he had set up for himself, the chevalier de Cubières wrote these lines after Buffon's death:

> One day, I remember, he read me the history
> Of the harmonious swan. One saw his eyes
> Shining with the light of creative talent.
> But soon his career was to end,
> And already the tomb had opened beneath his feet.
> Alas! Who could have told me that in that fatal moment
> Death was not far, and that, swan himself,
> The unfortunate approached his final hour![96]

Buffon's musical strains touched the hearts of a public fond of sentiment but thirsting for knowledge and utility. His works and other books and articles on natural history provided readers with scripts for understanding the strange animals in their midst, as well as their own kind.

CHAPTER 7

Elephant Slaves

BUFFON'S article on the swan ended with the dusty myth of the swan song, but it began with a statement that would have resonated with contemporary issues: "In all societies, whether animal or human, violence makes tyrants, gentle authority makes kings."[1] Although this statement supposedly referred to the swan's gracious behavior toward other birds, any reader would have immediately connected it with debates on the tyrannical nature of Louis XVI's reign.

Many naturalists incorporated morals or observations applicable to human society in their writings about animals—often, in the late eighteenth century, on just such contentious matters as the nature of tyranny. Questions about forms of government and grounds for political authority emerged in accounts of the structure of animal societies; concerns about the social role of women were reflected in descriptions of animal families; anxiety about loss of religious grounding underlay interest in natural laws of morality; and debates about colonialism and slavery played out in stories about the subjugation of wild and exotic animals. The commercial success of these works probably had much to do with their attention to themes of current interest.

I focus here on the last of these issues, debates about colonialism and slavery; it is the one most closely associated with the presence of exotic species in France, and it lays the groundwork for an understanding of attitudes and actions concerning animals during the French Revolution. The language of "animal slavery" appeared in many natural history texts and other writings about animals, but in a way that initially seems puzzling; even within the same text, human domination of animals is sometimes praised and at other times criticized. Why is there this ambiguity? And what do the animals have to do with their metaphorical human counterparts? Analysis of this language illustrates how concerns about human affairs suffused writing that was ostensibly about animals, it shows how people looked to nature for laws that they could apply to social relations, and it allows exploration of some conflicting views about linking animal and human slaves.

Man as Rightful Master

Most writings on animals in eighteenth-century France (and England) accepted the traditional position that man has both the ability and the right to dominate nature and put it to his use. Those who looked to the Bible as an authority routinely cited as justification God's orders to Adam and Eve to "be fruitful and multiply, and fill the earth and subdue it; and have dominion over the fish of the sea and over the birds of the air and over every living thing that moves upon the earth" (Gen. 1:28). After the flood, God gave Noah essentially the same command, with the difference that all of the animals would fear and dread him (Gen. 9:1–3). Natural theologians explained that just as the first couple had rebelled against God, so the animals had rebelled against them. "No animal would be wild, if Adam and Eve hadn't sinned," explained Eraste to his student in Fillassier's *Ami de la jeunesse.* "Their revolt against God brought about that of the animals against man." But, explained another author, no matter how bothersome the insects and rebellious the animals, "we can, through work and thought, derive use from all of those, and convert the most pernicious into beneficial remedies."[2]

In addition to, or instead of, God's imprimatur, some naturalists identi-

fied man's superior abilities as justification for dominating nature. The abbé
Pluche contended in *Spectacle de la nature* that because man is the only animal endowed with reason, he claims both mastery and awareness of its
legitimacy: "[Reason] teaches him that everything animals have is meant for
him, that they are his veritable slaves; and that he can avail himself of their
lives or their service. . . . Animals provide his food, guard his door, fight for
him, cultivate his fields, carry his burdens." For Buffon, conscious thought
itself was sufficient to set man apart and above: "It is because of his natural
superiority that man reigns and commands; he thinks, and consequently he
is master of those beings that do not think." As mankind's influence spread,
claimed Hennebert and Beaurieu, more and more species would be under
his control until the world had become one vast menagerie and garden.[3]
The ability to govern others required first the ability to govern oneself,
however, and so the less advanced the culture the less the ability to subjugate other species; Buffon, for example, correlated American Indians' lack
of domestic animals with their low level of civilization. In his hypothetical
reconstruction of human history, man first formed societies, then tamed
animals, and only after that could begin the work of civilization.[4] Citing
travelers' reports, naturalists identified dozens of species like the trumpeter
and the zebra as untapped resources awaiting domestication.[5]

Human mastery of animals would have been obvious to most readers of
this literature, who daily used animals to pull their carriages, sing for them
in their salons, entertain them at fairs, and nourish them at dinner. But enough
reminders of rebellious species remained to keep them from complacency.
Travel books featured colorful accounts of attacks by fierce beasts. At fairs,
exhibitors emphasized the wildness of their animals, hinting at the brutal
rampages that would occur if not for the restraining iron bars (see chap. 3).

Rumors of just such a scary breakout gripped the province of Gévaudan
in the 1760s. Reports circulated that an escaped hyena from a traveling fair
had been attacking and killing people, especially women and young girls,
whom it mutilated and sometimes decapitated. Later reports blamed a wolf,
and the supposed culprit was shot and taken to Versailles to be displayed to
the king.[6] The incident spurred a flurry of gruesome engravings as well as
sales opportunities: in 1782, dame Fournier, a tobacco merchant, offered for
sale to "des curieux naturalistes" some stuffed hyenas, described as *animaux
du Gévaudan,* suitable as garden ornaments.[7] Periodicals breathlessly fol-

FIGURE 7.1. "La bête du Gévaudan dévorant une femme" (The Gévaudan beast devouring a woman). The beast (a wolf, originally thought to be an escaped hyena) supposedly devoured many people before it was hunted and killed. BnF Estampes Jb mat 1a. Photo: Bibliothèque nationale de France, Paris.

lowed the campaign to track and kill the animal and used the event as an occasion to philosophize about the "war of the animals against us" that heats up every now and then (fig. 7.1).[8] People feared the "bête du Gévaudan" and immediately called for its destruction, but they would not have wanted replicas of the beast in their gardens had they not also experienced a thrill from this eruption of wildness. This fascination for untamed nature pervaded natural history texts as well.

Animal Victims

Alongside language lauding domination and domestication, a contradictory strain became widespread in the eighteenth century—often in the very same texts—that portrayed animals as victims of the human race and exalted the freedom and independence of wild creatures. Sympathetic language about animals was not new (appearing, for instance, in Montaigne's essays),[9] but it took on a distinctive form in the later eighteenth century, particularly in

France. It is striking to see how the phrase *animal slave* shifted from a positive descriptive term as employed in the 1730s by Pluche to a critical one. Natural history books and literary works alike portrayed heartrending images of mistreated animals. The most common vignettes were the domestic animal as overworked slave, the wild animal as mercilessly harassed fugitive, and the captive animal as unhappy prisoner. Species that continued to resist human domination often appeared as noble adversaries.

Readers of Buffon's *Histoire naturelle* encountered the theme of exploited animal slaves in the very first lines of the discourse that introduced the domestic animals: "Man changes the natural state of animals in making them obey him and making them serve his needs: a domestic animal is a slave that we enjoy, use, abuse, alter, uproot, and denature, whereas a wild animal, obeying only Nature, knows no other laws than those of need and liberty."[10] Buffon proposed that man's first conquest was taming the dog; then with this loyal servant's help, he proceeded to enslave other species. Among animals that Buffon referred to as man's slaves were the horse, the ox, the camel, the canary, and the falcon, all of which he described as having become altered and degraded from their original nature.[11] In looking at these changes that man has produced, he wrote, "we will be astonished at the degree to which tyranny can degrade and disfigure Nature; we will perceive the marks *[stigmates]* of slavery, and the prints of her chains; and we will find that these wounds are deeper and more incurable the older they are."[12]

Even stronger language pervades Hennebert and Beaurieu's *Cours d'histoire naturelle*, in which the pitiful lives of horses, oxen, sheep, and camels are portrayed in purple prose that must have wrung a few tears from sensitive readers. The poor harnessed horse with drooping head, plodding along exhausted, barely able to put one foot in front of another, they wrote, "is the symbol of slavery and misery." Even in the pasture the horse bears the marks of abuse in its scarred flanks and feet pierced with nails. The only moment of relief comes at the end of its life, when, no longer of use to its cruel master, it is taken out to die in the field it plowed a thousand times. There it consoles itself with the reflection that "my troubles are over, I will soon be returned in a different form to the [mass of beings]. I will metamorphose into earthworms and flying insects, two opposite states, but free, and consequently happy."[13]

In most instances, writers identified mankind in general as responsible for the slave animal's torment. But particular groups rated special blame. In the article on oxen, Buffon pointed a finger at overfed, wealthy landowners. Meat eating in itself is not bad—it is important for good health, and, more generally, destruction is necessary for life—but some men go too far. The rich man, instead of renewing what he destroys, "derives all of his glory from consuming," loading his table down with enough provisions to sustain several families: "he abuses animals and men alike."[14]

Arabs also came in for censure. Buffon's description of cruel camel-raising techniques repeated the well-worn stereotypes of Middle Eastern societies as tyrannical and authoritarian. Camels, he explained, are the oldest and most hard-worked of man's slaves. Unlike horses, dogs, cattle, pigs, and sheep, which have a few moments of pleasure, if only when being fattened for slaughter, camels' lives consist of unrelenting work. They are forced at a young age to carry heavy loads and endure extreme thirst; their humps and callosses constitute the "marks of their servitude and stigmata of their suffering."[15] (No doubt this suggestion of Christ-like martyrdom inflicted by Muslim camel-tormenters was intentional.)

Cousin to the domestic animal as suffering slave was the wild animal as harassed native. Numerous authors recounted episodes of man's war against animals, presenting a view bluntly summed up by Antoine de Rivarol: "Man destroys everything."[16] Species that held out against this onslaught were represented as having a tragic nobility. Buffon, in his discourse on wild animals, exalted the lives of these still free species, "which we call savage *[sauvages]* because we have not yet subdued them." Neither slaves nor tyrants, living in peace, they enjoy equality, liberty, and love. No wonder they flee at the sight of man:

> Some of them, the most gentle, the most innocent, and the most peaceful, are content to distance themselves, and pass their lives in our countryside; those that are more defiant, more fierce, disappear into the woods; others, as though they knew there were no safe place on the surface of the earth, dig underground caves, or climb to the heights of the most inaccessible mountains; finally, the most ferocious, or rather the proudest, live only in the deserts and reign as sovereigns in those torrid climates where even humans as wild as they cannot challenge their empire.[17]

Ostriches, for instance, had fled to the desert, but even there man continued to search them out to take their eggs, blood, flesh, and feathers, and to attempt to add them to his stock of animal slaves.[18]

Critics identified the destruction of animal societies as one of the worst effects of man's depredations. Hennebert and Beaurieu described a scenario almost as heart-wrenching as that of the dying horse, of a world where solitary, sad, fearful animals, like rebels fleeing their master, eke out a precarious existence, venturing out of their hiding places only at night, when the fierce attack the timid in a "theater of carnage and destruction."[19] The North American beaver, avidly hunted for pelts used to make hats, provided the quintessential illustration of social destruction. Only in countries where they do not fear encountering humans, claimed Buffon, do beavers get together to build their lodges, "which represent . . . the first effort at a nascent republic."

> In contrast, in countries where humans are widespread, terror seems to live among them. There is no more society among the animals, all industry stops, all the arts are snuffed out, they no longer think about building, they neglect all comforts; always troubled /pressés/ by fear and necessity, they can only try to stay alive, they think only of fleeing and hiding; and if, as seems likely, the human species continues in the future to populate the entire surface of the earth, it may be that, in several centuries, the history of our beavers will be regarded as nothing but a fable.[20]

Naturalists foretold extinction of other species as well. They were already aware that the dodo had disappeared from the earth, and Sonnerat predicted that the penguin *(manchot)* would ultimately be eradicated from any place that became settled by *"homme destructeur,* who lets nothing subsist if he can annihilate it."[21]

Captivity was portrayed as an even worse fate than the precarious life of a fugitive for animals that prized liberty and independence. Reporting that a cat will ignore mice if they are put into a cage with it, Savérien proposed that when deprived of its liberty, the cat is indifferent to everything except being free. The monkey-human hybrid author of Restif de la Bretonne's *Lettre d'un singe* described his brother monkeys as "either free in the forests of their native country or reduced to a state of sad slavery among humans." Canaries, wrote Buffon, become sick in captivity, where their inability to satisfy the urge for love constitutes one of the worst evils of their enslavement. He and Daubenton painted a grim picture of life for wild animals in the menagerie at Versailles. Neither canaries nor parrots should be kept in small cages, scolded Hennebert and Beaurieu.[22]

Although this was an age of great optimism concerning domestication, many authors identified one or more species that would never become subservient. For Savérien, the tiger was the only completely untamable species. Buffon identified several, like the pheasant, which, even when raised from a chick, always tried to escape and could never be anything but a prisoner. He claimed that the majority of bird species would always fly from us, remaining with Nature "as witness of her independence."[23] Some animals earned respect because they preferred death to captivity. Travelers, as we saw (chap. 1), liked to ascribe noble motives to such sensitive species, like the white monkeys of Galam that could not bear to leave their homeland. Orangutans were also said to die of boredom and melancholy when they were chained up.[24] Buffon witnessed such behavior firsthand when he kept a coot in a cage. The bird did not struggle, even when chased and prodded, but simply wasted quietly away. "The tragedy of slavery is thus even greater than we think," he concluded, "since there are beings from whom it removes even the ability to complain."[25] Resistance to domestication sometimes indicated a forcefulness associated with virility; easily tamed animals, by contrast, were seen as pliant and effeminate.[26]

Buffon claimed that some animals could be tamed as individuals, but not as a species. Most monkeys fell into this category, as well as falcons, but the elephant provided the prime example.[27] The elephant held a special place in the *Histoire naturelle* as the pinnacle of the animal hierarchy, brushing its broad back up against the lowest rung of the human species on the ladder of beings (see fig. 7.2). In a frequently quoted phrase, Buffon claimed that the elephant unites "the intelligence of the beaver, the dexterity of the monkey, and the sensitivity of the dog" with its own particular qualities. Because the elephant is gentle, obedient, and loyal, once tamed he becomes firmly attached to his master: he cares for him and caresses him, learns his signs and voice, anticipates his wishes, and obeys all orders promptly, prudently, and attentively. But content as he is to serve, "the disgust for his situation lodges in the bottom of his heart"; he refuses to mate when in captivity despite the intensity of his passion. "He thus differs from all of the

FIGURE 7.2. The elephant, which, wrote Buffon, "surpasses all other terrestrial animals in size and approaches man in intelligence, at least as much as matter can approach spirit." From Buffon et al., *Histoire naturelle*, vol. 11 (1764), plate 1. Reproduced by courtesy of the Department of Special Collections, General Library System, University of Wisconsin–Madison.

other domestic animals that man treats or manipulates like beings with no will; he is not among those born slaves that we propagate, mutilate, or multiply for our use. Here only the individual is the slave; the species remains independent and consistently refuses to reproduce for the profit of the tyrant." The elephant's refusal to submit completely demonstrated its noble character. Two groups came off badly in comparison: servile animals like dogs, which gave up everything for their masters, and humans who could not muster enough authority to make elephants obey them. Elephants scorned Africans, according to Buffon, for their inability to reduce them to servitude.[28]

Buffon's treatment of elephants is not easy to characterize. He praises their submission, but also their resistance. He seems critical of humans for exploiting them, but honors those who tame them. He implies a criticism of the institution of human slavery, but at the same time denigrates Africans. These seeming contradictions were not peculiar to Buffon; they pervaded the literature on animals.[29] One way to understand them is to recognize that Buffon and others presented another view of animals that resolved some of these contradictions—what could be called the "happy servant" view, which I discuss below. The fact remains, however, that shame and pride at exploiting animals existed uneasily side by side. This schizophrenic situation also characterized contemporary texts about native peoples, thus reinforcing the parallels between animals and humans; I will return to this point later.

Whatever the large-scale reasons for this change in writings about nature, one thing is certain—the ways in which people referred to animals were inextricably bound up with social and political matters. Thus, criticism of "animal slavery" seems to have been weaker in England than it was in France, where political criticism was much more widespread.[30] Expressions of concern about mistreated animals went hand in hand with growing expressions of sympathy for subordinate human groups.

Happy Animal Servants

A partial answer to the puzzle of how animals could be praised simultaneously for their wildness and for their submissiveness lies in a solution proposed by some naturalists: the good master and happy servant model. According to this model, animals would be content in a subservient position as long as they were properly handled. Some liberty might have to be sacrificed, but the result would be beneficial overall. Cruelty and exploitation could be minimized by eating moderately, slaughtering humanely, pulling gently on the bit, taming wisely, and hunting only when necessary.[31] Not only would this diminish suffering, but many unhappy animals could actually be quite content if they were merely treated better, and they might even be happier under man's protection than fending for themselves in the tough world of raw nature.

The ornithologist Mauduyt, in a discourse urging the naturalization of exotic birds, contended that the man who pursues and destroys animals for his own needs gives proof only of his force, while he who subjugates and imports useful species demonstrates his industry.[32] There was no need, in the 1780s, to argue that industry had a higher value than force. Buffon described an example of how such subjugation could be attained. Waterfowl, he wrote, have been tamed by bringing their eggs into the farmyard. Although at first wary and wild, after the birds mate they become attached to the place of their first love, and their slavery becomes as sweet as liberty. Their young, "citizens by birth" of their parents' adopted country, know no other life, and "their natal country becomes dear even to those who live there as slaves." The trick is to convince the animal that it *wants* to be a slave, so that it enjoys a feeling of freedom even while doing its master's bidding. Buffon illustrated this in more detail in an explanation of how one could tame red partridges. "There are means for taming and subjugating the wildest animal, that is the one most enamored of his liberty, and that method is to treat them according to their nature, in leaving them as much liberty as possible." The animal's relation to its master must be based on sympathy and choice, not need, interest, or passivity. If constrained, even in a large aviary, the partridge refuses to mate—like the elephant, not wanting to "perpetuate a race of slaves"—and even loses its instinct for self-preservation. But if it is attached to its master by choice and knows that it can leave at any time, then the relation between the partridge and the man who "knows how to make himself obeyed" is attractive *(interessant)* and even noble.[33]

Hennebert and Beaurieu painted similar scenarios in their *Cours d'histoire naturelle*, which promoted improved treatment of animals as one of its major themes. Man destroys nature by modifying and changing it too fast, they

stated in the preface, but careful, slow changes "contribute to the beautifi-
cation of Nature and the happiness of sensitive souls [êtres sensibles]."[34]
Wild animals might seem to live in enviable conditions, free from man's
tyranny, and indeed their freedom could compensate for many hardships.
But overall, domestic animals are happier, especially when treated well (2:113).
It is actually a good bargain for them if they provide us with service in
exchange for protection, shelter, and food (2:60): "any animal that is well
tamed and well treated prefers (at least after one or two generations) our
paddocks, our parks, to the immense expanse of the forests" (2:105).

The authors provided specific guidelines for how to ease the lives of animal servants. Horses should be left as free as possible, not closed up in dirty
stables but lodged "under the sky, under the sky!" (1:417). Of course horses
must be accustomed to the harness; but they will give up a portion of their
liberty for "the pleasure of living with us and serving us" (1:410). Young
horses can be gradually introduced to the harness and eventually will enjoy
working. Similarly, oxen should be tamed when young by gradually accus-
toming them to the touch of a man's hand and the weight of a light chain.
They should be rewarded with a little salted wine, because "for oxen as for
men, it is wine that accustoms them to slavery" (2:7). When their days of
service are over, they should be turned out to pasture to enjoy a brief time
of liberty before they are slaughtered. This may seem cruel, but killing an
animal is legitimate as long as it is done with commiseration and regret,
rather than to satisfy a desire for luxury (2:25–26). These sorts of techniques,
they contended, not only would improve the lot of already domesticated
animals, but would make possible the taming of wild species like the guinea
pig, the hedgehog, the zebra, and the monkey (2:256, 260, 376–77, 460–61).
They admitted that zebras were hard to tame, and, in a comment with a critical edge, suggested that their intractability might save them from a fate to
which their beauty predisposed them, since "we like the animals whom we
make our slaves to combine the ability to serve with an attractive form"
(2:376). Nevertheless, they advocated naturalizing zebras in Europe, for,
"despite their love of liberty, they are docile and can be conquered by good
treatment" (2:376–77). Mauduyt similarly mixed a tinge of criticism in with
his call to import the South American trumpeter to France to take the place
of dogs. "If man needs to become attached to an animal to which he con-
veys sentiments that should be reserved for his own kind; if inside his dwell-

ing place it is a consolation for him to have a faithful, intelligent, docile, loving slave," he begins dubiously, only to conclude that it would be better to replace the rabies-prone dog with a safer animal that has all of the same qualities.[35]

Linking Animals and Humans

All of the texts just surveyed were ostensibly about animals but used words referring to humans. If talk of animal fugitives or slaves or servants conjures up visions of Carib Indians or African captives or French peasants for a present-day reader, such visions must have been considerably more vivid for eighteenth-century readers. This feature of literature about animals raises two central questions. What sort of commentary were the authors making about human society? And were the animals in these texts simply vehicles for such commentaries?

In some texts, animals stood for humans in analogies that did not imply any fundamental equivalence between the two realms. Here, authors used animal actors primarily for rhetorical effect, to make a point about society or politics. Such was the case in many poems and fables based on the image of the birdcage, for example in an anonymous pamphlet entitled "The Escaped Birds, the Peacocks, and the Oiseleur: Fable." The fable tells of a rich man who owns an aviary full of little birds that he dotes on. He lets his peacocks guard the aviary, however, and the peacocks chain the birds, make them labor, and tell the owner that the birds are bad and rebellious. The birds escape with the help of their wild brethren. The master hires a oiseleur who assures him that because the birds love him they will come back, as long as they are not mistreated or confined. For anyone who did not get the message, a footnote explained, "There is such a one [aviary] called 'bastille,' feared, I believe, in hell itself. . . . O Louis, Louis! . . . do not be a half father of your little people; hear the cries of all of your children; tear down that infernal aviary: you do not need prisons; its ruins will serve us as altars where you will be worshiped."[36] The author's intention was obviously to make a political commentary, not to urge better treatment of birds in aviaries.

A less metaphorical comparison appeared in one of the most popular novels of the eighteenth century, Jean-Jacques Rousseau's *Julie, ou La nou-*

velle Héloïse. At the end of the book, Rousseau describes a utopian community, Clarens, overseen by the saintly Julie and her husband, M. de Wolmar. When Julie's former lover Saint-Preux visits, he is taken on a tour of the grounds and the aviary. Coming into an open area where lush vegetation surrounds a peaceful pool, he sees multitudes of birds flying to and fro, unafraid of his presence. The "aviary," he realizes, has no bars. M. de Wolmar explains that the birds gather there by choice because they are provided with food, water, and nesting materials, are never disturbed or hunted, and are protected from owls and other enemies. Only four people have keys to enter the grove, and for two months during the spring no one is allowed in. "Thus," observes Saint-Preux, "for fear that your birds would become your slaves, you have become theirs." "Here indeed," Julie replies, "are the words of a tyrant, who believes that he can enjoy liberty only if he diminishes that of others." Although he did not directly compare them, Rousseau supplied human counterparts for the happy birds in Julie's aviary in the servants and laborers at Clarens, who were thoroughly content working for their wonderful masters. At Clarens, there was neither perfect equality nor tyrannous authority, but rather a natural hierarchy of governing and governed—the happy servant model applied to both animals and humans.[37] The difference between this comparison and that in the fable of the escaped birds is that in Rousseau's novel the birds are not just convenient stand-ins for people; they are part of Rousseau's coherent vision of natural innocence.

In the *Histoire naturelle* and other natural history texts, talk of animal servants, slaves, and savages embedded the comparison to human society more subtly, because the animal and human worlds were linked through the use of anthropomorphic language rather than through explicit comparisons. This more ambiguous construction allows a greater latitude of interpretation. It is possible to read these passages as fables in the style of the "Escaped Birds"—that is, as having more to do with people than with animals. Clarence Glacken, for instance, called Buffon's praises of wild animals "in fact transparent criticisms of human society."[38] But, as I discussed in the preceding chapter, readers like Mme Necker took these kinds of comparisons as *rapports* that applied to both humans and animals, based on shared characteristics. Many people believed, along with Buffon, that animals experienced a wide range of emotions, from jealousy to gratitude. In this kind of comparison, captive animals, rather than standing metaphorically for captive humans, are portrayed as suffering from their loss of liberty in the same way as humans do.[39] The distinction between the literal and the metaphorical is nevertheless murky. When Buffon compared the swan to a king, he did not mean that swans literally ruled the other birds. But it is less clear whether he and his readers interpreted the concept of slavery so broadly as to apply literally to peasants and to elephants as well as to humans held in bondage. From the early seventeenth century, the term *slave* had been used in an extended sense, to apply to subjects of despotic governments as well as "slaves of love."[40] Late-eighteenth-century political criticism regularly referred to enslaved Frenchmen; for example, in Jean-Paul Marat's *The Chains of Slavery* (London, 1774) or the opening of Rousseau's *Social Contract*: "Man was born free and everywhere he is in chains."[41] For many readers, *slave* and *slavery* must have been concepts that hovered on the border between the literal and the metaphorical, evoking images of both exploited animals and humans. As we will see, however, some writers objected to the extension of slave status to animals.

If Mme Necker had written down her reaction to Buffon's description of how to tame red partridges, it might have read something like this: Partridges can be tamed and become attached to their masters if kept in large enclosures, in the same way that human slaves and servants become tractable and fond of their human masters when allowed small freedoms. What she might not have been able to explain, because it would have seemed so obvious to her, was why it made sense to compare servants to partridges. Like many people at the time, Mme Necker believed that the gradations found in nature had parallels in the human world. Concerning an animal whose characteristics were intermediate between those of two other species, for example, she wrote, "I am persuaded that the nuances that we perceive in the physical order also exist in the moral order, and that souls also have their gradations from vice to virtue, and from stupidity to genius."[42]

The basis for many of the slippages between human and animal realms in the eighteenth century was a postulated hierarchy of living beings running from worms to angels. Civilization and reason occupied the top rungs in the form of educated European men; below them came peasants, women, children, mentally unfit and mad people, and nonwhite people (not necessarily in that order); then animals, led by the most intelligent and capable (the dog, the elephant). Lines could be drawn through this hierarchy between any two groups. The break between all humans and all animals was only one possible place to locate a line. Move that line up a little, and certain people would be lumped together with animals. Or Frenchmen could set themselves apart from Spaniards and other nationalities they considered to be less civilized. Move the line down, and the higher animals (e.g., warm-blooded) could be grouped with humans. Take away the line altogether and humans became just another animal. Not all rungs on this ladder were equally spaced; everyone would no doubt have agreed that a larger gap existed between humans and animals than within either of these groups, but humans lower on the ladder were generally considered to have more in common with animals than did those higher up, and it was not uncommon to group "lower" humans together with animals.[43]

When pointing out the inferior nature of various human groups, authors invariably compared them to animals. Buffon, for example, described primitive men as "a species of animal incapable of commanding others," and in one passage placed American Indians even lower, as "a species of impotent automatons."[44] Buffon contended that although these races at the bottom of the ladder of the human species acted like animals in the sense of not using their superior capabilities, they still differed essentially from animals: "whatever resemblance there is between the Hottentot and the monkey, the interval that separates them is immense, because internally, he [the Hottentot] is furnished *[rempli]* with thought, and externally with speech."[45]

Sometimes people turned the ladder upside down; then stupid Indians metamorphosed into noble savages, deficient women into men's moral superiors, and slaves and peasants into virtuous primitives. The same people who were reading about noble elephants and fugitive beavers would have been reading elsewhere about Africans who refused to bear children into a state of slavery and Indians who had been driven out of their homes by cruel conquistadors.[46] Around the middle of the eighteenth century, the concept of the noble savage, which had been proposed by Montaigne and others, solidified into a cliché. Rousseau, Diderot, and scores of others lauded the Caribs and especially the Tahitians for their natural morality and happy-go-lucky lives. This image never disappeared, though it was tempered by the end of the century by reports of the killing of the French explorer Marion du Fresne and sixteen of his men by Maoris in 1772, the murder of Captain Cook in Hawai'i in 1779, and other tales of violence and even cannibalism in the South Seas.[47]

Simultaneously, colonialism began to be critically scrutinized in France (especially after the losses of the Seven Years War). The abbé Raynal's widely read *Histoire philosophique des deux Indes* (which included contributions by Diderot and others) deplored the Spanish invasion of America and English despotism in India and warned the Hottentots against the Dutch in an emotionally charged passage that sounds much like the descriptions of persecuted wild animals in natural history texts: "Fly, unlucky Hottentots, fly! Hide in the forests. The ferocious beasts that abide there are less fearsome than the monsters under whose empire you are about to fall. The tiger might tear you to pieces, but it will have taken only your life. The other will ravish your innocence and your liberty. Or if you have the courage, take up your axes, draw your bows, and make your poisoned arrows rain down on those foreigners [*étrangers*]."[48] Overall, however, the *Histoire philosophique* portrayed colonialism as beneficial, if undertaken gently and with the goal of advancing commerce, not national power. (First published in 1770, it reached most readers in an expanded third edition published in 1780.) The book described the Jesuit settlement in Paraguay as a model community— much like Rousseau's Clarens—where the Jesuits exerted a benevolent authority over a people who were neither lazy nor overworked, who were all equal, who were well lodged, fed, and dressed, who married whomever they liked, and who did not pay taxes.[49]

Like colonialism, slavery aroused ambivalent sentiments. Slavery was officially not allowed on French soil, but, conveniently, this law did not apply to the colonies.[50] Saint Domingue, France's richest possession, was widely regarded as essential for continued economic prosperity. By the end of the eighteenth century, reports of abuses and slave revolts spread, and criticism grew—but the solution was not clear. Critics disagreed about whether the slaves should be liberated gradually (few argued for immediate liberation), or whether already established slave-labor communities should be maintained while eliminating the trade. Some proposed that instead of transporting Africans elsewhere to do manual labor, the French should have Africans work for them in Africa (as the English were doing in India). A group of prominent people formed a *Société des amis des Noirs* in 1788, which initially called for the abolition of both the slave trade and of slavery but later, as a concession to colonial planters, rescinded the latter demand. Abolition did not come until 1794, after the slave rebellion in Saint Domingue made it inevitable.

Animals and certain humans, then, were linked through explicit comparisons and through a parallel language that pervaded both natural history texts and works on colonialism and slavery. Popular natural history books may have appealed to readers partly because they presented naturalized versions of comfortably reformist views. Reading Buffon or Hennebert and Beaurieu on animal slaves would most likely have been a satisfying experience for both pet keepers and absentee owners of slave plantations, who could have experienced sorrow over mistreated camels, celibate elephants, and wild animals forced into deserts and mountains, while at the same time feeling relieved that these creatures would be happy under the control of a benevolent authority.[51]

Historians who have assessed the late-eighteenth-century reform movement have disagreed about its significance as a humanitarian movement. Some see the Enlightenment as an important period of humanitarian advance on all fronts: treatment of animals, prisoners, and slaves, improvements in education, and so on.[52] Richard Grove, for instance, argues that colonial administrators like Pierre Poivre and Bernardin de Saint-Pierre combined humanitarian sentiments and recognition of environmental destruction in a view that provided a foundation for the modern environmental movement: "The two tasks of the French Utopians, one social, the other environmental, were . . . inseparable."[53] Some commentators have recognized the paternalistic tendencies of this movement (as evident in the "good master" attitude), but still rate it as positive overall.[54]

Others take a more negative view. Michèle Duchet has argued that despite their sometimes vehement denunciations of slavery and colonialism, Poivre, Bernardin de Saint-Pierre, Raynal, Buffon, and Diderot were still caught up in a colonial ideology centered on domination. She and other critics have pointed out that expressions of admiration and sympathy toward oppressed groups do not necessarily indicate sincere humanitarian sentiment, since it is common for people to regard groups they consider to be inferior with intense admiration and contemptuous scorn at the same time: noble or savage; madonna or whore.[55]

Many authors who presented sympathetic or positive portrayals of savages and slaves had little knowledge of or interest in these alien cultures (often represented in inaccurate caricatures) but used them as mirrors for self-criticism. Some were indulging in shallow sentiment: historian Yves Benot has dismissed weepy stories about miserable slaves as "pity à la mode."[56] This sort of sappy complacency characterized numerous poems such as "A Woman to Her Canaries," published in the *Affiches de Paris* in 1786, which plays off the widespread idea that (real) slaves could not be emancipated because they would not be able to sustain themselves:

> Charming little birds, with gold plumage,
> You are born among us in sad slavery.
> You sing even better in your native lands;
> But man, too jealous, loves only to ravage.
>
> .   .   .   .
>
> Ah! If liberty is offered to you one day,
> Do not accept it, little unfortunates.
> Flying under our skies would bring your destruction:
> Death and unhappiness have surrounded you.
>
> Liberty is only a word, a seductive fallacy *[flatteuse erreur]*;
> A gentle bond between hearts makes a narrow cage:
> In captivity you find the happiness
> That we often search for from shore to shore.[57]

Given the benevolent portrayal of slavery in poems like this, no wonder some abolitionists rejected the comparison between slaves and captive animals. Even in a context urging better treatment for animals, this metaphor naturalized the analogy between animals and slaves, linking them as groups that shared an inferior status. Perhaps for this reason, some proponents of abolition, rather than agitating for improved treatment of both animals and slaves, emphasized the distinction between them. They pulled human slaves up—rhetorically—into fellowship with other humans while cutting the connection to animals, thereby making it clear that people deserved more freedom than did animals. The author of one antislavery tract, for example, argued that "to treat humans tyrannically would be a reprehensible barbarity; to roughly treat animals that merely look like humans would be no worse than prodding cattle or whipping horses. Thus, to justify cruel injustices, . . . people have not hesitated to put men on a level with animals."[58] Another tract contended that all animals experience a "boiling ardor for liberty" but then explained why this urge could justly be curbed for all species but man. The lion strains against his chain, the bird struggles until he dies against the bars of his cage: "Timid or fierce, they all want to be free." However, the author continued, unlike fellow humans, animals are destined to serve us, and so "perhaps one shouldn't conclude [from their strong resistance to servitude] that we are doing them an injustice to subjugate them."[59] Condorcet also drew a firm line between human and animal laborers. Planters say they treat their slaves as well as they do their horses, he remarked. However, they don't treat their horses well, and even if they treated slaves as well as they should be treating their horses, the slaves would still be unhappy: "a man is not a horse."[60]

One important difference between human and animal slaves is in their ability to speak for themselves. It may be true, as some historians argue, that despite the racial prejudice and lukewarm antislavery sentiment of most Enlightenment thinkers, their focus on individual happiness, natural rights, and compassion helped to create an intellectual transition that became one of the factors (along with economic interests) that led to the abolition of slavery.[61] But human slaves were not just recipients of freedom; they also freed themselves. At least one call to arms went out to animals, from Restif de la Bretonne's monkey-human narrator César in *Lettre d'un singe*. César urged animals to revolt against their human masters, predicting that once they educated themselves as he had, they would break their chains. Bulls would chase, lions would attack, and elephants would trample humans.[62] But apparently the animals did not respond to César's exhortation. They did become involved in the French Revolution, but as symbols, not rebels.

CHAPTER 8

Vive la Liberté

DURING THE French Revolution, the catchword *liberté* appeared again and
again in orations, pamphlets, songs, and posters. Indeed, many of the re-
pressive features of the ancien régime were dismantled: the Bastille was
destroyed, the Third Estate won political representation, divorce was legal-
ized, and slavery outlawed. The symbols of the ancien régime were as thoroughly overhauled as were its institutions. In renaming streets and buildings (e.g., Palais Royal to Maison Égalité), effacing aristocratic symbols like
fleurs-de-lis, and establishing a new calendar, revolutionaries hoped to create pure ground on which to build a virtuous republic. They saw themselves
as throwing off accreted corruption and breaking the chains that had enslaved them.[1]

These reformers looked to nature as the source of freedom and to natural laws as the means of restoring it: "Nature has made only free beings,"
declared the *Journal de Paris* in 1790.[2] They renamed the months after the
seasons and advocated practices such as breast-feeding.[3] The unencum-

bered lives of wild animals often served as models. The oppressed and en-slaved people who had in earlier decades been depicted as caged or chained animals now appeared unfettered: a common image accompanying the slo-gan *Vive la liberté!* on posters and revolutionary pottery was that of birds bursting out of a cage or sitting cockily on top of their former prisons (fig. 8.1).

Just as in the period before the Revolution, this language of enslaved and liberated animals balanced on the boundary between the literal and the metaphorical. Sometimes people talked about freeing real animals (particu-larly those belonging to the king), and the most utopian-minded advanced specific proposals for improving the circumstances of both show and work-ing animals. But in the long run, animal liberation foundered both because of practical considerations and because of competing symbolism. On the practical side, liberating lions and parrots, or even giving them large cages, posed difficulties. On the symbolic side, carnivorous wild animals suffered from their association with violence. They were frequently portrayed not as unjustly imprisoned victims but as bloodthirsty creatures that had to be restrained. Caged animals also had a trophy value that revolutionaries were as susceptible to as were kings. What is surprising is not so much that the animals remained in their cages as that they aroused such impassioned dis-cussion.

Inciters of the Mob or Henchmen of the Nobles?

Wild animals first caught the attention of the authorities during the Revo-lution not as creatures to be liberated but as potentially dangerous inciters of popular violence. In March 1790, Jacques Peuchet, director of the de-partment of police, published an article in the daily paper the *Moniteur* rec-ommending that public animal fights be outlawed because of the threat that the violent performances would incite the ferocity of the Parisian masses. Anything that foments cruelty, atrocity, and destructive feelings should be prohibited, he wrote, and he went on to show why the animal fights fit into this category: "This horrible entertainment consists of slaying sometimes a dog, sometimes a bull, sometimes a bear by means of the murderous teeth of a multitude of animals; cries, howls, moans of pain and death accompany this fearful scene, where a blind multitude goes to take lessons in barbarism

VIVE LA LIBERTÉ

FIGURE 8.1. "Vive la liberté!" A revolutionary poster in which freedom is symbolized by a sans-culotte freeing birds. By Smeeton and Cosson. Musée Carnavalet. © Photothèque des Musées de la Ville de Paris / photo: Habouzit.

and to accustom itself to spill blood with the tranquility of a routine act and the calm of a satisfied taste." In this "school for murder," he contended, the physical impressions act directly on people's organs, and thus on their moral characters. In the past, we scorned the masses too much to care about educating them, but now we realize that it is our responsibility to civilize them and make them gentler. We should replace this bloody public entertainment with festivals that evoke sentiments of peace and kindness: dances, fireworks, and calm scenes.[4]

A M. Brisset seconded Peuchet's opinion in a letter to the same journal several months later. Brisset favored outlawing the animal fights and other animal-killing entertainments in all of France, contending that the best way to assure prosperity and loyalty among the people was to encourage the progress of reason and its essential characteristics: kindness, sensitivity, and humaneness.[5]

Why such concern about violence? Only nine months before Peuchet's letter was published, in July 1789, hundreds of angry Parisians had surrounded and captured the Bastille, a fortress-like prison that housed a few political prisoners and a stock of gunpowder. The governor of the prison and another official were killed and mutilated, and their severed heads were stuck on pikes and carried through the streets in a grisly parade. Nine days later, two other officials suffered the same fate; one of them, who was suspected of participating in a plot to hoard flour, had grass stuffed in his mouth. Political moderates and those charged with keeping order—the chief of police and the mayor—had good reason to try to identify and curb factors they believed might incite such bloody mob justice.

The mayor of Paris, Jean Sylvain Bailly, wrote a brief letter to the *Moniteur* in August 1790 in which he agreed with Peuchet that all enlightened people must wish for the suppression of the animal fights and announced that a decree to that effect would be issued the following month.[6] In September, the Paris journals announced that the police had banned the animal fights, "which dishonored the laws and morals of a free people."[7] The people, however, were not very cooperative. They preferred using their freedom to watch animals fight than to take lessons in civilization and gentility. In the spring of 1791, Peuchet wrote again to the *Moniteur,* complaining that the animal fights were still going on, "by permission of the mayor," according to the poster advertising it. Didn't the mayor agree with him that it was particularly important to encourage gentleness and humanity in the people at this time?[8] The mayor did agree, he stated in his reply, but his hands were tied: the proprietor had moved the spectacle to Belleville, just outside the boundaries of Paris, and it was the mayor of that community who had given permission for the fights.[9]

Peuchet was undeterred. He wrote to the *Moniteur* again, stating his case even more strongly and suggesting that the administrators of the entire department of Paris (a larger administrative unit that included Belleville) outlaw the animal fights. This time he addressed the vexing problem of how to justify inhibiting the activities of a supposedly free people, a dilemma that arose in many other contexts, as well. One can act completely freely only when one's actions stay in the personal realm, he wrote. But when they affect groups of people, then they must be subject to laws. Public entertainments in particular must be regulated so as to allow those that encourage sensitivity, greatness *(grandeur)*, and courage, but not those that arouse ferocity and barbarity. Activities like animal fights directly impress themselves on the soul via the senses and thus affect the spectators' subsequent actions: people might think they were acting of their own free will when they were actually responding passively like machines. The less enlightened the person, the stronger the effect. The government must act like a father to the citizens who, in time, will thank it for having done what was necessary for their happiness. Whether or not the intendant of the department followed Peuchet's recommendation is unknown; he may have been distracted by the king's attempted escape from Paris, which occurred a few days later.[10] In any case, the fights were back in business in Paris by 1797.[11]

In the meantime, wild animals had grabbed the attention of Paris's municipal authorities for another reason, this time in connection with rumors of an aristocratic plot. Once again the animals were linked with violence, but from the elite rather than the masses. It was July 1790, and the first anniversary of the storming of the Bastille was coming up. Elaborate preparations were under way for the Festival of Federation, to take place at the Champ de Mars, a vast parade ground where, for another festival ninety-nine years later, the Eiffel Tower would be erected. The Festival of Federation, which attracted close to half a million spectators, was organized to celebrate and cement the relationship between national guardsmen, legislators, and their king, now in his new role as a constitutional monarch. Throngs

of people joined in the preparations, which included excavating and level-
ing the parade ground and constructing an amphitheater, a triumphal arch,
an altar where the oath of allegiance would be sworn, and a royal pavilion
to house the king. Among the niggling details that had to be attended to was
what to do about the displays of wild animals that had been installed in the
parade grounds, presumably by entrepreneurs anxious to make some money
from spectators looking for diversion. On 8 July, the council of the City of
Paris, over the objections of two "proprietors of cages destined for these
establishments," passed a decree that all of the "ferocious beasts" that had
been brought to the area around the Champ de Mars were to be removed
from there, and the police were to assure that the establishments not be set
up again. Two days later, the administrator of the Department of Public
Works reported that the orders had been carried out.[12]

The movement to banish these animals seems to have come from the public as well as the municipal authorities. When the *Chronique de Paris* recommended removing the animals from a building at the École militaire where
they were being housed, it referred to them as "great living aristocrats, that
is to say a well-stocked menagerie of lions, tigers, leopards, and monkeys
of the largest size." The word *aristocrat* alludes to some chilling rumors that
had been circulating. The *Chronique* warned of the danger of an involun-
tary accident, but apparently fears of intentional mischief were rife. Several
people had communicated their fears to the committees of their districts:

> When going to the Champ de Mars to check on the preparations for the cer-
> emony, the public had noticed on the avenue that leads from the hôtel des
> Invalides to the École militaire several stalls full of lions, tigers, leopards,
> etc. Who brought these animals? For what purpose were they gathered
> there? Could it be that the aristocrats, who had good reason to disturb patri-
> otic festivals, had come up with the idea of releasing on the crowd these
> beasts, transformed into henchmen of the counterrevolution *[la réaction]*, to
> devour the spectators?[13]

Indeed, one of the animal shows at the Champ de Mars was called "Great
Aristocrats of Africa"[14]—but it is hard to imagine that the owner was an
aristocrat. It is more likely that these animals were being kept and shown by
the same sorts of entrepreneurs who displayed animals at the city fairs.

But strong associations existed between nobles and fierce carnivores, partly because of the tradition of princely menageries. Caricatures and satires often described the king and queen and their friends as lions, tigers, and panthers. Two pamphlets published in 1789 exploited this imagery by announcing a hunt for "stinking and ferocious beasts" and offering huge rewards for the extermination of well-known nobles and members of the royal family in the guise of exotic animals such as a panther, a tiger "raised in the Versailles menagerie," an "orang-outang," a porcupine, a Mexican monkey, and a parrot, along with native pests like wolves, crows, and rats. They may appear to be tame, the second pamphlet declared, but they are monstrous despots who must be destroyed.[15] Although the king himself was absent from these early pamphlets, he appeared in a later one (probably from 1792) as the "Royal-Veto," who inhabited a menagerie at the Louvre. He was described as a voracious pig-like animal, stupid as an ostrich, timid and fat, who drank all day and whose food cost 25 million to 30 million livres per year.[16]

The police could chase wild animals and their keepers away from certain areas of the city, but they don't seem to have been able to extirpate them. Paris during the revolutionary years, with its excitement and roiling crowds, provided a rich ground for hawkers and entrepreneurs of all types. Concerns about wild animals cropped up again in 1793, when the city again outlawed their public display. This time the decree applied to all of Paris. "All dangerous animals, such as leopards, lions, and others that are being shown in public places, will be killed or taken to the menagerie at Versailles, and their proprietors compensated."[17] The sweep of animal shows stimulated by this decree netted a couple of dozen animals that made up the core of the new menagerie at the Jardin des plantes (formerly the Jardin du roi).

Conditions continued to favor street entertainment, and wild animals continued to be favorites. At the height of the Terror, in prairial, Year II (May 1794), two citizens complained to the Paris Commune about the charlatans and dishonest proprietors of games who entrapped good citizens, as well as the "disgusting displays of wild animals at the place de la Révolution and along the boulevards."[18] At this time, the guillotine, slicing up to twenty heads per day, was in the center of the place de la Révolution, later renamed the place de la Concorde. Entrepreneurs setting up there would have been sure of large crowds looking for excitement between beheadings.

Questions about the relation between cruelty to animals and cruelty to humans continued to be discussed in subsequent years, especially in relation to revolutionary violence. In 1802, for example, one of the prize questions posed by the Institut national was, "What effect do barbaric treatments of animals have on public morality? And should laws be instituted in this area?" Most of the submitted essays discussed the bad influence of brutality to animals on the masses (with reference to events of the Revolution), although a few connected hunting with cruel behavior on the part of nobles.[19]

Proposals for a New Menagerie

In May 1794, about the same time that the city government was dealing with the proliferation of wild animals around the guillotine, the primary governing body of the National Convention, the Committee of Public Safety (whose most famous member, Maximilien Robespierre, would be deposed and executed two months later) approved the establishment of a more orderly display of animals, the menagerie at the Jardin des plantes. Animals had been accumulating there since the October 1793 decree to get wild animals off the streets, but the menagerie had not yet been formalized as an institution or accorded any funding. Money was in short supply during this time of war against domestic and foreign enemies, and those who wanted to establish a menagerie had to make arguments for its importance. The director of the Jardin, Jacques-Henri Bernardin de Saint-Pierre, had to justify why the state should support an institution that had been decried as a symbol of monarchist tyranny when it was at Versailles. Bernardin de Saint-Pierre and other proponents argued for the menagerie with high-flown rhetoric that contrasted with the sordid reality on the ground.[20]

In the years between the Festival of Federation and the approval of a new menagerie, the political situation changed; the people began to take the reins away from the hands of the aristocrats and turn the Revolution in a more radical direction. August and September 1792 saw the attack against the king's guard at the Tuileries, replacement of the Legislative Assembly by the National Convention, establishment of the French Republic, and preparations for the trial of Louis XVI. During the September massacres, mobs invaded Parisian prisons and killed fourteen hundred aristocrats and clergy, after condemning them in mock trials as enemies of the people. Some nobles' estates were ransacked, including, in a few cases, their animal collections.

One of the locations pillaged during the uprisings in September was Chantilly, north of Paris, former home of the prince de Condé. The prince, who had already emigrated and was forming a counterrevolutionary army, was one of the most reviled enemies of the republic. On 15 August 1792, a group of national guardsmen from Paris went to Chantilly, where they first killed a local miller and paraded his head on a pike. The prince's retainers fled. On the next day, the soldiers ransacked the chateau and bombarded the menagerie with cannons. Only one enclosure, containing a tiger, a sheep, a civet, two eagles, and five peacocks, remained intact. All of those surviving animals, except the eagles, which were adopted by a sieur Colmache, were later executed. An official report on the state of the menagerie after the assault described one after another the broken statues and fountains, including a figure of Narcissus that had fallen into the basin it had been gazing into.[21]

The menagerie at Versailles may have been ransacked sometime between 1789 and 1792, but almost certainly not in the manner described in 1868 by historian Paul Huot. Huot claimed that a band of Jacobins had marched up to the director of the menagerie and declared that they had come "in the name of the people and in the name of nature to order him to liberate the beings that had emerged free from the hands of the Creator and had been unduly detained by the pomp and arrogance of tyrants." The director supposedly replied that he was perfectly willing to accede to their request, but that some of the inhabitants were insensible enough to feelings of gratitude that they might make use of their freedom to devour their liberators: "as a consequence, he declined performing this role personally, and offered the society the keys to the cages enclosing the lions, tigers, panthers, and other large carnivores." After reflecting and taking a vote, the society decided that the dangerous animals should be taken to the Jardin des plantes in Paris and that the harmless ones be released. Although this story has been repeated a number of times, it is most likely an antirevolutionary tall tale, perhaps based on an earlier, more haphazard pillaging.[22] For one thing, no animals were taken to the Jardin until 1794; for another, a visitor in April 1791 found it already reduced to almost nothing: he remarked that the menagerie, which had deteriorated a great deal since 1785, now housed only a rhinoceros, a lion with its dog companion, a hartebeest, and a "hybrid derived from a zebra and an ass" (the quagga).[23]

The few remaining menagerie residents almost contributed to science as skeletons rather than living specimens. In September 1792, the general overseer for Versailles wrote to Bernardin de Saint-Pierre, director of the Jardin des plantes, informing him that "the menagerie is going to be destroyed." He asked the intendant to come to Versailles to indicate which of the animals he would like for the natural history cabinet, noting especially the presence of a "superb rhinoceros." Bernardin de Saint-Pierre's trip to the menagerie along with the botanists André Thouin and René Desfontaines spurred him to write the impassioned *Mémoire sur la nécessité de joindre une ménagerie au Jardin national des plantes de Paris*, which he printed at his own expense and sent to the minister of the interior in December.[24] Although the animals remained at Versailles for another year and a half (during which time the rhinoceros died), this memoir probably saved them from the taxidermist. More important, in the long run, it played a part in convincing members of the revolutionary government that an institution that had been excoriated as an example of tyrannical luxury and waste should be supported by the new republic. Converting this scornful attitude to one that would allow scarce resources to be accorded to a new menagerie required breaking both the connection between exotic beasts and aristocratic decadence and the connection between caged animals and slavery. Bernardin de Saint-Pierre's memoir, along with the work of several other proponents of the menagerie, helped cement the new imagery.[25]

Bernardin de Saint-Pierre's memoir is a masterpiece of persuasive prose. He presents and then demolishes one by one all of the arguments against maintaining a menagerie, subtly appeals to feelings of national pomp and pride, and weaves a thread of populism through the whole. Using the same sentimental language that earned devoted fans for his widely read novel *Paul et Virginie*, he exalts the tenderness of natural virtue and beauty in a way guaranteed to appeal to Rousseau-loving revolutionaries. He dispenses easily with two objections to keeping a menagerie in Paris—its expense and its possible danger. Funds should be derived partly from the state and partly from public dues for courses, he recommends, remarking that people always give greater esteem to things they have to pay for. Escaped wild animals do not need to be feared; this has never been a problem with animals shown on the streets, and besides, the animals will become more gentle during their sojourn in the menagerie. Observing this effect on the animals' temperament would be good for the people, Bernardin de Saint-Pierre explains, turning to the major objection to the menagerie, that of its utility.

The menagerie at Versailles was indeed a useless luxury, admits Bernardin de Saint-Pierre, but he argues that the new menagerie would be useful for the nation, for naturalists, and for the people. As a national institution, a menagerie is necessary for one very practical reason: foreign dignitaries like to give live animals as diplomatic gifts. Such animals can neither be killed nor refused; therefore, "foreign relations require the existence of a menagerie" (402). The menagerie would also draw visitors. Artists would come to use the exotic species as models, and crowds would be attracted. People are naturally curious about animals, he remarks; anyone who wants to make a fair successful knows that all they need to do is bring in some exotic *(étrangers)* animals (403). These arguments would have been attractive to the National Convention, eager to see itself as the government of an established and proud nation.

But live animals are also important for scientific study, Bernardin de Saint-Pierre argues. Here he pits himself against those who claimed that naturalists could do their work from preserved specimens. The study of animals' tastes, passions, and instincts is in fact the most important part of natural history, he claims. "It was this study that made Buffon interesting, not only to scholars, but to all people" (397). Just think what Buffon could have done if there had been a menagerie at the Jardin des plantes! The new director also has an answer for those who argue that naturalists should study animal behavior in the animals' native habitats, because they lose their character when displaced and confined. "If animals lose their character in captivity, they lose it even more in death," he retorts, but he then goes on to argue that their characters are not in fact ruined by captivity, if they are treated right (415).

Here, as elsewhere, Bernardin de Saint-Pierre evokes the concept of the good master that I described in the preceding chapter. "The animal under man's power still shows his instinct; if it is altered by bad treatment, it seems to become perfected by kindness *[bienfaits]*" (416–17). As evidence he points to the lion and the rhinoceros that he observed at the Versailles menagerie.

The lion showed tenderness toward the dog friend with whom he was confined, and the rhinoceros came up to the bars of his enclosure for a pat when he saw visitors coming (400–401). (Presumably the menagerie animals had been receiving good treatment since the departure of their former tyrannous master.) Furthermore, traveling naturalists are not a solution. Is it right for only a few people to observe real animals, while the nation that is paying their expenses has to be content with pictures and descriptions? Besides, these voyagers would not observe the animals' natural habits anyway: since they would be hunting the animals to acquire specimens, they would see them only "fleeing and trembling" (416). Better to bring the animals to France, along with their native vegetation. There, in an Eden-like setting, the trust between animals and man would be reestablished, and the animals would forget their captivity so thoroughly that they would happily begin to reproduce (408–9).

Critics of the Versailles menagerie in the 1760s had suggested that menageries should be used for acclimatizing useful exotic species and breeding new varieties, not for superficial display of rarities. Bernardin de Saint-Pierre echoes those recommendations, suggesting that, to start with, mates could be acquired for the African antelope and the Asian pigeon still at Versailles. The menagerie could also be used for research on nutrition and disease in domestic species. Slipping momentarily away from the practical, he suggests that "charming" species, like the hummingbird, might also be acclimatized and become lovely new denizens of the French countryside (407–10).

It was a little harder to justify keeping animals that could clearly never become naturalized in Europe, like carnivorous lions and tigers or spectacular but unwieldy rhinoceroses. But even these species would be useful, Bernardin de Saint-Pierre argues, for they would provide lessons in philosophy and moral training for citizens. In observing animals at the menagerie, the people would see how society softens even the fiercest character. They would see how wild animals are calmed by spending time with humans and even in the company of other animals (404 5). Indeed, he himself observed this phenomenon. He happened to have been a passenger on the ship from île de France that transported the rhinoceros to France, where he saw this intractable beast become friends with a goat, whom it allowed to eat hay from between its legs (406–7).

The lion's story was even more remarkable. It had arrived from Senegal in 1788 with a dog companion. "Their friendship is one of the most touching sights that nature can offer to the speculations of a philosopher," writes Bernardin de Saint-Pierre, "Never have I seen such generosity in a lion, and such amiability in a dog" (400, 401). This story of the friendship between the lion and the dog was often told during this period, in a way that clearly connected the lion with a repentant, benevolent king and the dog with the loyal people.[26] In late 1792, however, when this memoir was composed, the king's reputation had plummeted (he would be placed on trial in December), and the king of animals seems to have suffered along with him. While watching the lion play with his dog friend, Bernardin de Saint-Pierre noticed that the dog had an inflamed scar on its side, which it kept licking "as though to show us the effects of an unequal friendship" (401). Still, the lion was much more friendly than solitary ones are and thus suggested a new model for the relationship between man and animals: although man initially used the dog to subjugate other species by force, he might now use the dog and other domesticated species to attract them to him by kindness. Such a gentle relationship between animal and human would provide a good example for the people, whose violent tendencies are often aroused by watching the cruel treatment of animals (419). Finally, as a place where natural laws are both studied and displayed, the menagerie would promote universal brotherhood. For "nature alone brings together people . . . who are divided by religion and patriotism" (424).

The memoir ends on a flourish that masterfully appeals both to the delegates' authority and to their populist sentiments. Support the Jardin des plantes, he urges, and "the people will regard you as gods who hurl lightning bolts with one hand and sprinkle fertile dewdrops with the other" (425).

As Yves Laissus and Jean-Jacques Petter noted in their history of the menagerie, it is unlikely that many members of the Convention read the sixty-three-page memoir cover to cover.[27] Its major messages, however, could have come to them through other, more easily traversed avenues. A review of the memoir that appeared in November 1792 summarized its main points and quoted crucial passages. It went even further than Bernardin de Saint-Pierre had in unlinking aristocrats from their animals: "Common observers scarcely considered the menagerie at Versailles as anything other than one of those objects of luxury and curiosity with which kings love to surround the pomp of the throne. It was necessary for tigers and lions to live next to the sanctuary of despotism and for monkeys to associate with courtiers; but just because we have thrown down tyrants of the human species, do we also need to abolish those other kings of the animals, who are much less terrible than the first?"[28]

Another report that came out in December 1792 was written by a committee of three members of the recently formed Société d'histoire naturelle de Paris. The authors reported that they had been asked by the society to give their opinion on Bernardin de Saint-Pierre's proposal. They admitted that royal menageries were nothing but costly and useless imitations of Asiatic ostentatiousness, but argued that a "ménagerie sans luxe" would be important for advancing the study of natural history. Although less certain than was Bernardin de Saint-Pierre that all of the animals would become tame and happy in their new home, they believed that observations on captive animals could still be useful. One could note how they drank and ate, when their colors changed, and so forth. And if they were slaves, well, that allowed for study of the effect of slavery on their characters. It had already been observed, for instance, that lions and tigers can be tamed, but jaguars cannot.[29]

One person who responded to Bernardin de Saint-Pierre's memoir was Couturier, the official who had warned that the menagerie was to be destroyed. In January 1793, he reported that the minister of the interior had suggested he consult with the director of the Jardin about the fates of the remaining animals at Versailles, in particular the rhinoceros. Someone had offered to buy it, he wrote, but "I would rather it become an object of public instruction in the hands of a philosophe like you."[30] Transporting a rhinoceros was a major undertaking that may have been impossible during this chaotic time; in any case, the animal died nine months later and arrived at the museum as a corpse to be dissected by the anatomists. The professors had to ask the commissioner of public instruction for extra funds to pay for preparing and mounting the animal's bones and hide, estimated to amount to 3,000 livres.[31] The funds were found, the animal was preserved, and it can be seen today, looking smaller than one expects, staring impassively out of a glass case at the Muséum national d'histoire naturelle in Paris.

In the meantime, the trial and execution (on 21 January 1793) of the tyrant of Versailles were occupying the Convention, and the members paid little attention to the fate of his animal slaves. A few months later, however, in June 1793, the Convention established the Muséum d'histoire naturelle, which included the Jardin national des plantes, and established several professorships. A supplement to this decree declared that when the funds were available, a menagerie should be maintained where the professors would study, acclimatize, breed, and distribute useful species.[32]

The new museum and menagerie, as imagined by Bernardin de Saint-Pierre and others, fit well within the general tenor of revolutionary schemes for reforming science. No longer the frivolous preserve of the elite, science would be democratic and utilitarian: membership in academies would be determined by merit rather than birth, museums would welcome and educate the public, and scientists would carry out projects that would produce useful results. Buffon's bewigged, powdered head, laid to rest with perfect timing in 1788, was rhetorically buried as a symbol of the ancien régime by advocates of the new science. Condorcet, in a letter written while he was preparing an *éloge* (eulogy) of Buffon that he would read at the Académie des sciences, complained, "Here I am occupied with another charlatan, the great Buffon. The more I study him, the more I find him empty and puffed up." Apparently unimpressed with Buffon's paeans to liberty in the *Histoire naturelle,* Condorcet accused the naturalist of not having participated in the general movement during the century to release the human spirit from its chains.[33]

If the menagerie was conceived with starry eyes, it was born in gritty reality. Animals started to accumulate at the Jardin des plantes at the end of 1793 as a result of the police orders to clear the streets of animal displays, although the menagerie was not accorded official status until the spring of 1794, when the remaining Versailles orphans arrived. The menagerie developed higgledy-piggledy, in response to unplanned circumstances, and in a form quite different from that in the idealized memoirs and plans.

The Animals Arrive

Animals started arriving at the Jardin des plantes during the fall and winter of 1793–94, following the order by the Commune of Paris to round up all of the street animals, compensate their owners, and take them either to the

Jardin or to Versailles. The commissioners who undertook the operation naturally conveyed the animals to the more convenient location, where they probably expected them to end up as specimens in the natural history cabinet at the Jardin.[34] From the stalls of the entrepreneurs who made their livings from Parisians' curiosity about strange animals, they gathered together a ragtag collection for the professors.

Police Commissioner Charbonnier brought in not only some impressive animals, but a keeper, as well. In November 1793, he went to the shop of C. Dominique Marchini, near the place de la Révolution, where he reported finding a sea lion worth—according to Marchini—4,000 livres, a leopard worth 10,000 livres, a civet worth 2,000 livres, and a little monkey worth 200 livres, along with cages, carts, and tools worth a total of 700 livres. (The commissioner's knowledge of zoology must have been rather limited, for it turns out that the "sea-lion" he mentioned was actually a polar bear.) Marchini was not about to give up his business just like that for a one-time payment, however. According to the report, he pleaded to be allowed to stay on as guardian of the animals, with his assistant Remi Amet to groom them: "[Marchini] observed . . . that it was the only resource [they] had for making a living . . . ; both of them offering to care for the said animals, being very familiar with them and knowing perfectly everything necessary to maintain them, and asked us to request that the citizen administrators of the police department accept them to care for the animals."[35] Amet's fate is unknown, but Marchini was hired.

A less prepossessing flock arrived the following February, shepherded by a commissioner named Baudin. At six in the evening, Baudin went to the boutique of citoyen Cochon at the Maison Égalité. There he found and later carefully described a bull, twenty-and-one-half months old, with a multicolored growth on its left horn, a ram with four horns and a double eyelid, a sheep with two mouths under its eyes, another sheep with a mouth containing two teeth under its left ear, and a third sheep "having nothing remarkable about it." After a stop at Cochon's residence, they all trooped over to the Jardin des plantes, doubtless attracting a few curious glances from people out in the winter evening. The professors, apparently not interested in studying deformed sheep, tried to get rid of them.[36]

This flood of the exotic and the bizarre flowed into a Jardin not at all prepared or outfitted to receive it. Desfontaines, the secretary at the Jardin, wrote to the Committee of Public Instruction on 6 November, two days after Marchini arrived with his four animals, asking what to do with them. He noted that the animals could be kept provisionally in an area underneath the museum galleries, and that it would cost approximately 12 livres a day to feed them and pay someone to care for them. The professors, who wouldn't have known the first thing about what to do with a polar bear, must have been glad to accept Marchini's services. An addendum to Desfontaines's letter leaves to the imagination the chaotic conditions at the Jardin: "P.S. The moment I was about to seal this letter, I learned that a serval, a male polar bear, two mandrills, and three eagles just arrived at the museum."[37]

How much to pay for the animals was another quandary. Naturally, the owners were inclined to estimate the value of their property on the high side. A local expert was called in to do the assessments. Sieur Martin, the proprietor of the combat d'animaux, provided estimates for animals confiscated from (or offered by) five different proprietors, including Marchini, and two others, Bernard Louzardy and Félix Cassal, who, like Marchini, would become animal keepers.[38]

In response to a request from the Committee of Public Safety, Desfontaines estimated how much money was needed to pay for the animals, their food, and a building to house them.[39] While waiting for a reply, the professors prepared to welcome the stragglers from Versailles, which arrived on 26 April 1794. The antelope died a few days later, but the quagga and the lion lived for another few years. The pigeon, which is not mentioned, seems to have succumbed before the move.[40] A month later, a much larger group of thirty-six animals arrived from Raincy, the former estate of the duc d'Orléans. Some of the animals at Raincy had already been sold, and the Committee of Public Safety had to step in to rescue the rest of them for the state; other than some parrots and a camel, these were mostly native species.[41] Along with the animals came cages, bars, and other materials from Versailles, Raincy, and Chantilly.

In May, Étienne Geoffroy Saint-Hilaire, the zoology professor who had been put in charge of the menagerie, listed the animals that had accumulated at the Jardin during the preceding six months: thirty-two mammals and twenty-six birds.[42] On 16 May, the committee approved funds for a provisional menagerie.[43] The menagerie finally opened to the public, who came in droves.[44] A painting representing the scene shows well-dressed citizens

FIGURE 8.2. Provisional menagerie at Jardin des plantes, 1794. The menagerie provided temporary housing for animals that had been confiscated from street displays or transferred from the Versailles menagerie. © Bibliothèque centrale M.N.H.N. Paris 2000.

in tricorne hats, accompanied by many children, peering into murky barred cells (fig. 8.2). An architect designed plans for a spacious permanent menagerie, but they stayed on the drawing board. The animals continued to languish in their prison-like accommodations, far from the peaceable kingdom envisioned by Bernardin de Saint-Pierre.

Advocates for the menagerie reiterated appeals for support, but with little success. Antoine-Laurent de Jussieu, professor of botany at the museum, penned a justificatory document in the winter of 1793–94. He echoed many of Bernardin de Saint-Pierre's arguments and listed the benefits that would come from importing and domesticating exotic species (including zebras) that could be used for transportation, food, and as sources for raw materials.[45]

In January 1794, a list of articles concerning establishment of the menagerie was presented to the Committee of Public Instruction.[46] The following December, Antoine-Claire Thibaudeau, a member of the committee, spoke up for the menagerie as part of his recommendations for enlargement of the museum and increased salaries for the professors. He pointed out that it was the duty of the committee to make sure that the menagerie animals were not kept as prisoners, as they had been in the past. "Until now," he reminded the deputies, using language that paralleled the critical observations Buffon had made several decades before, "the most beautiful menageries were nothing but prisons where the cramped animals displayed the physiognomy of sadness, lost their fur, and remained permanently in positions that attested to their listlessness." To be useful for public instruction, he continued, the animals in the menagerie must enjoy as much freedom as possible. "In bringing close to us all of the productions of nature," he concluded, "let us not make them prisoners. An author has said that our [natural history] cabinets were tombs: well then! Let everything enjoy new life as a result of our attentions, and let the animals destined for our pleasure and for the instruction of the people no longer wear on their brows, as in the menageries built by the pomp of kings, the brand of slavery." It must indeed have been embarrassing to be displaying animals in bondage during a period when nature and freedom were being publicly honored. Little could be done to remedy the situation, however, since money for renovations was lacking.[47]

The following winter of 1795 was grim for everyone, including the animals, as famine gripped Paris. The professors at the menagerie warned the Commission on Provisioning that the animals would starve, but the commissioners remained adamant that what little stores of bread they had on hand must go to the people first; they recommended that the professors try to buy potatoes or whatever they could get their hands on in the open market. The professors replied that they did not have enough money in the budget to make such purchases. They tried to get horsemeat from the veterinary school at Alfort, but even that was scarce, because it was being reserved for people's dogs, or possibly for the people themselves, and for Martin, the proprietor of the animal fights. In the end more than half of the animals died; their bodies then became useful scientific specimens for dissection and display.[48]

After that terrible winter, the conditions in the menagerie generally improved (with the exception of 1799, when famine prevailed again and the professors kept the more valuable animals alive by feeding the less valuable ones to them).[49] As the civil war drew to a close and France expanded outside her borders, new sources of animals became available. In the winter of

1794–95 the French invaded Holland. In 1797, troops led by General Bonaparte occupied northern Italy. The following year, a French expedition embarked for Egypt, complete with 167 savants.[50] In 1800, Captain Baudin led two ships on an exploring mission to Australia. On each excursion, the grasping hands of empire gathered up live animals to be returned to the capital. Always justified in terms of public instruction and utility, the living booty also boosted national pride and brought crowds to the Jardin to marvel at the new acquisitions.

The Dutch raid was exciting for the naturalists, for it yielded the fruits of a competing colonial enterprise. Geoffroy Saint-Hilaire, the young director of the new menagerie, dashed off an excited letter urging quick retrieval of the animals at the stadtholder's menagerie: "Holland is taken: several measures must be taken without losing any time: there are lots [une foule] of living animals in Amsterdam, at the Hague; two elephants, a zebra, an onager at Loo, two leagues from Amsterdam."[51] Elephants and a zebra! The very animals that the court had so ardently desired in the 1780s were now within reach. The zebra turned out to be a chimera, but the elephants later made a triumphal entrance to Paris. The stadtholder's natural history cabinet promised equal riches. Here were specimens from Africa and the East Indies that the French knew only from the works of Dutch naturalists. The combination of the Dutch and French collections "will form a museum the likes of which all of the efforts of all of the naturalists of Europe during twenty years could hardly have created," wrote one of the French naturalists in charge of confiscating the goods.[52]

The naturalists understood both the scientific and patriotic worth of their harvest. Thouin, the botanist at the Jardin who led the scientific expedition to Holland, recommended that when the natural history objects were displayed in Paris, their provenance should be indicated "so as to render justice to the valor of the French armies and to perpetuate the memory of their victory."[53] Thouin and his crew expeditiously packed up thousands of preserved specimens to send by boat along the North Sea coast and up the Seine, but conveying the live animals required more preparation. Bernard Louzardy, one of the former exhibitors who now worked at the Jardin, traveled to the Hague to help escort them. In August 1796, a procession of carriages containing several Indian deer and goats and a collection of exotic birds rolled into Paris.[54] The elephants, Hans and Parkie (later renamed Marguerite), were not among them: cages had to be built to order for them—and rebuilt after the enraged male twice broke his—and extrasturdy carriages constructed and repaired en route. The elephants entered Paris two years later, in March 1798, to great excitement. Hopes that they would propagate ran so high that a book about them included an imaginary scene of their coupling (fig. 8.3).[55] Given the reputed refusal of elephants to mate in captivity, a willingness to reproduce in their new home would have had great symbolic importance. They left no offspring behind, however, when they died a few years later.

Just a month after the Dutch animals arrived in Paris, Félix Cassal, who, like Louzardy, had found employment at the menagerie with his animals, returned from a trip to Tunis with an assortment of North African animals that had been lacking from the menagerie: lions, camels, gazelles, and ostriches. A popular book about the Jardin des plantes published in 1797/98—a source of dubious authenticity—quoted Cassal as telling a visitor before he set off on his mission to Africa:

> I would give the little I own to see this menagerie filled with all kinds of ferocious animals. When the public comes to visit, I am ashamed to be able to show them only ordinary animals. I would like to be able to say to you: "There is the tiger, there the panther, there the leopard, there the jackal, there the hyena, there the lion." Do you know what I dream of now? What my ambition is? To be sent to the deserts of Africa to snatch away all of the most redoubtable animals that I just named and bring them to France. My trip would be sure to succeed because I have spent all my life with wild animals. I know how to subdue them. And besides; it is less difficult to tame tigers than to tame certain men.[56]

Soon after the African contingent settled in, more spoils of war filed into the menagerie from Italy. Diplomatic gifts arrived from Algeria and the United States, fulfilling Bernardin de Saint-Pierre's prophecy that the new republic would need a place for such gifts. Part of the former menagerie of the king of Mysore, Tippoo Saib, which had been purchased by a Mr. Penbrock in England and comprised tigers, lynxes, a panther, a mandrill, and several birds, was purchased and conveyed to the menagerie personally by Delaunay, Geoffroy Saint-Hilaire's replacement as director of the menagerie.[57] The first naval expedition to be undertaken after the Revolution,

FIGURE 8.3. Imaginary mating scene depicting Hans and Marguerite, the pair of elephants brought to Paris from Holland in 1798, in a tender (if biologically inaccurate) embrace. From J. P. Houel, *Histoire naturelle des deux éléphans mâle et femelle* (Paris, 1803), plate 16. Bancroft Library #fQL 737 U8 H64. Courtesy of the Bancroft Library, University of California, Berkeley.

that of Baudin to Australia, was in most respects a disaster—the captain died, Australia did not become a French colony, and most of the scientists died or deserted—but it did furnish the menagerie with another large group of exotic animals, a few from Australia and many, including a zebra, from South Africa (1804).[58]

Still Behind Bars

Just before these new animals arrived, Bernard de Lacépède, a professor of zoology at the museum who was continuing Buffon's *Histoire naturelle* with volumes on reptiles and fishes, spelled out a new vision for the menagerie even more idyllic than Bernardin de Saint-Pierre's. In a letter published in the journal *Décade philosophique*, Lacépède wrote that, up until now, most

European menageries had been like those of the Roman despots, characterized by "cramped cages where the successors to these despots locked up,
mutilated, degraded, [and] denatured the individuals; and far from wanting
to pluck useful secrets from nature, they could conceive only of enchaining
her, and man as well." An enlightened society in which liberty and reason
reign, he argued, should have a menagerie that focuses on utility and shuns
spectacle and barbarity. Rare animals should be collected judiciously for
study, not simply to excite "the useless curiosity of the vulgar." Like Bernardin de Saint-Pierre, he stressed the importance of public education: it
would be most beneficial for the people to see the useless and dangerous lion
give way to the gentle, humble, peaceful, and useful cashmere ram *[belier de
Cachemire]*. These useful species could be kept in large, open enclosures
where "images of constraint or the evidence of slavery would be removed
as far as possible from the eyes of a free people."[59]

According to this vision, the animals in the menagerie were to instruct
rather than to arouse the people's passions or ornament aristocrats' estates;
well-treated and happy, they would provide models of natural morality for
the people. In theory. The Revolution had indeed made a significant change
in the place of exotic animals in Paris by moving them from the streets and
the king's palace to a state-sponsored institution. The grounds of the menagerie were enlarged in the first decades of the nineteenth century, in an
English-garden type of design, with paddocks for benign species and scat-
tered rustic huts. Many species remained in rows of small cages behind bars,
however, and throughout the nineteenth century critics periodically decried
the prison-like conditions.[60]

Lacépède's vision never materialized in part because of competing visions of the menagerie by different groups. One of the groups, the keepers,
were responsible for much of the day-to-day running of the menagerie;
they knew how to drum up a crowd but their ideas about the scientific and
symbolic value of the animals were different from the professors'. The professors needed the keepers' expertise with both the animals and the public,
and they gave their entertainment-oriented helpers wide latitude.

This attitude comes through in a letter that the librarian and menagerie
director Delaunay sent to the museum when he was on his way back from
England to Paris in 1800 with a caravan of animals that had originally be-

longed to Tippoo Saib. In order to make money to cover the expenses of the trip, the animals' conductors, including the former exhibitor Bernard Louzardy, displayed them in Brussels, Lille, and other towns they passed through.

Delaunay explained the procedure somewhat sheepishly: the menagerie's arrival was announced by a drummer and by "somewhat inflated" notices *(annonces un peu charlatanes)* in journals. The setup—"pretty crude," but cheap, he wrote—consisted of a stage of pine boards set on barrels, with wooden barriers to separate the first- and second-class seating. A crier called the crowds in, and a guard made sure that those with second-class tickets didn't try to sneak into first-class seats. They attracted large audiences in good weather, though more unusual offerings would have brought in even greater numbers, Delaunay surmised: "If there were kangaroos our fortune would be made." He noted that they could also have done better if they had been willing to put on more of a fairground-type sideshow *(prendre l'air de forains)*. But then he himself would have had to stand at the door and enforce order; not only did that role not appeal to him, he wrote, but he was afraid that if people recognized him later it would harm the dignity of the museum. Delaunay's dignity intact, the show went on: "Bernard [Louzardy], a real showman, does the demonstration, in a manner as amusing as it is bizarre." Perhaps a little too much street talk crept in, however, for Delaunay then inquired, "Is he going to do it [the demonstrating] in Paris, too?"[61] Keepers like Louzardy were more interested in providing the public with entertainment than in creating a peaceable kingdom.

Some of those responsible for accumulating and displaying animals at the menagerie were intent mostly on portraying wild animals as symbols of national strength and as magnets for the public—they did not seem to share Lacépède's concern about displaying them as symbols of imprisonment and slavery. The president of the Commission temporaire des arts (the committee responsible for gathering up treasures from émigrés, executed aristocrats, and conquered lands) argued for the establishment of a menagerie on the grounds that it would exhibit a zeal for education that would be impressive in the eyes of France's enemies.[62] The entry into Paris of wild animals, paintings, and other confiscated "monuments" from Italy took place in an atmosphere of national triumph. A report noted that the animals in the procession had been in cages, but that "it would have been better to show them bare [à nu], chained to chariots, as the Romans did."[63]

The fierce wild animal also remained a popular metaphor. Where blood-thirsty carnivores had previously been used as metaphors either for members of out-of-control mobs or for despotic aristocrats, in a pamphlet published after the Terror they stood for guillotine-happy Jacobins. The pamphlet listed a mock sale of ferocious animals from the menagerie of the Jacobins: the tiger Carrier (Felix Sanguinolenta), which eats twenty-five hundred young girls and boys a day, on sale for 500,000,000,000 livres; the "orang-outang" Collot-d'Herbois (Simia Jacobina) for 190,908,700,437 livres, and so on.[64]

The menagerie itself, although it provided some heartwarming stories of interspecies friendships like that between the lion from Senegal and its dog companion, also generated stories that did not harmonize so well with the vision of its founders. In a children's guide published in Year VI (1797/98), Voyage au Jardin des plantes, the narrator takes a child to visit the menagerie. When the child, Gustave, asks whether the hyenas and leopards will eat out of his hand, the narrator explains that they are still somewhat wild, and "to keep them from doing harm, they are not allowed complete liberty: they are not tied up, but they are in cages [loges] with bars, through which one can look at them without fear." Later the keeper, Félix Cassal, shows Gustave some of the animals, including the friendly cockatoo. "Buffon claimed that they cannot stand to be in a cage," says Cassal; "this one, however, seems to put up with captivity quite well." When Gustave returns on the next day, the animals confiscated from Holland have just arrived. Cassal explains that the peaceful ones have been placed in the garden ("this is what you earn when you have gentle ways and do not do harm to anyone"), but the porcupine, the cassowary, and other dangerous ones are lodged in the menagerie, under his care. "The menagerie is a prison," he explains, "where the only animals that are locked up are those that would abuse their liberty. These cages are a type of dungeon." Gustave, impressed, tells Cassal "Your position is quite dangerous. You are the guard of a formidable prison; because indeed, these vicious, ferocious animals could revolt and play a nasty trick on you."[65]

Epilogue

IN VOLTAIRE'S satirical tale *Candide ou l'optimisme* (1759), as Candide and his companion approach the Dutch colony of Surinam they encounter a black man lying on the ground with his left leg and right hand missing. The slave explains that his master has treated him according to standard custom. If your finger gets caught by the grindstone in the sugar mill, your hand is cut off, and if you try to escape, your leg is cut off; both happened to him. "It's at this price that you eat sugar in Europe," he tells the horrified travelers. "Dogs, monkeys, and parrots are a thousand times less unhappy than we are."[1]

In the introduction to the edition of *Candide* that I read in high school, the editor notes that because Voltaire was writing for his contemporaries, many in-jokes are lost on the modern reader.[2] Though not exactly a joke, the reference to monkeys and parrots certainly escaped me then, and it continued to do so even after I learned something about eighteenth-century history and philosophy. Now this passage makes sense to me and, I hope, to readers of this book. I can visualize an eighteenth-century Parisian sitting in a comfortable chair reading *Candide*, taking in these words, glancing up at the Amazon parrot in an ornate cage by the window, looking at the sugar bowl on the tea tray, thinking about slavery and pets, cages and liberty, commerce and luxury. The words Voltaire attributed to the slave would not have worked if readers had not been familiar with monkeys and parrots (and dogs, too) as household pets.

We have seen in the preceding chapters how parrots, monkeys, elephants, and other creatures traveled to France via ships plying trade routes, especially those of the slave trade. We have observed them and their observers at the king's menagerie, street fairs and fights, oiseleurs' shops, and private homes. And we have seen the different meanings they took on for the people who encountered them, meanings that can conveniently if somewhat artificially be divided into four categories: commodities, scientific objects, sources of emotional connection, and symbols.

As commodities, foreign species in eighteenth-century Paris, because of their rarity and exotic origin, had a high value that enabled some people to make a living from charging to see them and classed them among luxury items for personal ownership and monarchical display. They became part of the consumer revolution. Caught up in the network of consumption, they were bought and sold, advertised, lost and reclaimed, flaunted as signs of wealth, and criticized as symptoms of decadence.

As scientific raw material, exotic animals became objects of scrutiny for eighteenth-century naturalists, who sought them out in all of their Parisian habitats and benefited from the knowledge of their owners and keepers. They helped to fuel and at the same time provided fuel for the growing interest in science among a larger public; both authors and exhibitors increasingly touted their value for the amateur of natural history. To reformers late in the century and during the revolutionary period, they became potential sources of utilitarian domestication schemes. Even in their new scientific garb, though, they continued to sport elements of their older dress from fables and emblem books, speaking to humans about human, not animal, affairs. In Buffon's descriptions, in particular, naturalistic fables conveyed lessons about human society through descriptions of animal behavior.

Because they are alive, exotic animals elicit different sorts of emotional responses than do inanimate consumer or scientific objects. Pet owners, fair exhibitors, sailors—all developed bonds with the living creatures they encountered. Sympathy toward animals became a common theme in writings of many kinds, but in practice, tenderness existed side by side with brutality and indifference. The Buffon who wrote movingly about his parrot and monkey pets, for example, was the same Buffon who performed experiments on living animals and watched captive birds starve to death, and many spectators at animal fights no doubt kept pets in their homes.

Many questions concerning the moral dimensions of animals—both exotic and native—were popular topics for discussion in the eighteenth century. Do animal lovers ignore human misery? Is sentiment for animals a low-level emotion? Does sympathy for animals translate into sympathy for humans? Can exposure to nature through observations of animals (for example, in a menagerie) inculcate beneficial values in the populace? Some of these questions had been debated for centuries, but the increasing size of the reading public along with the growing presence of different species gave them new reach and significance.

As symbols, caged and chained exotic animals became especially significant through their metaphorical representations as slaves, prisoners, native people, or the oppressed masses. In an era when the nature of domestic political authority, relations with indigenous cultures, and slavery were all being questioned, these constrained creatures became apt vehicles for discussions of tyranny and slavery. The complex language that elided metaphorical and literal meanings in images such as that of the animal slave conveyed conflicting messages that reflected tensions and uncertainties about the legitimacy of cultural hierarchies. In France, where criticism of political tyranny was widespread even in elite circles, pity for animal slaves and praise of independent wild creatures became standard language in both natural history books and fictional works.

As we saw, though, many authors, and presumably readers, too, favored replacing cruel tyranny with benevolent paternalism, a model that envisioned both animals and subordinate humans in a state of voluntary but contented servitude. Even radical reformers in the revolutionary period, although they used animal liberation for rhetorical purposes, did not do much to alter the conditions of the animals themselves, partly for practical reasons, and partly because of conflicting symbolism that connected wild animals with uncontrolled violence or used trophy animals to display national pride. Looking at these complicated symbolic uses of animals, it is obvious how much talk about animals was either really, or additionally, talk about something else; this should make anyone reading historical documents, even scientific ones, think twice before using them to draw straightforward conclusions about the authors' attitudes toward animals or nature.[3]

The place and meanings of exotic animals in eighteenth-century Paris seem familiar in some ways and very foreign in other ways. Today, exotic animals in most Western countries are both more and less accessible: we can no longer go to the park to buy a pet monkey from a vendor, and we would not come across an elephant on a city street; but circuses, zoos, television shows, and magazines and books all capitalize on the public's continued interest in curious stories about unusual species. The underground trade in exotic animals thrives, despite strict regulations, and the rich continue to flaunt unusual pets.

In the realm of the sciences, most zoologists today prefer to observe animals in their natural habitats, not in zoos or circuses, and wildness has largely overshadowed tractability as a laudable trait. Serious scientific literature about animal behavior often puts on an objective front, using unpoetic, technical language and claiming to present only scientifically gathered data. When science writers and even many scientists write for the general public, though, they often use the very same style as popular eighteenth-century writers of natural history, appealing to interest about human nature through descriptions of animal behavior. The justification for linking animal and human behavior has changed somewhat during the past two centuries, however, since the development of the theory of evolution. Although there is considerable disagreement about how much of human behavior is affected by genetic predispositions, many biologists, especially those in the fields of sociobiology and evolutionary psychology, claim that human behavior is, at bottom, animal behavior.[4]

Whether people feel the same range of emotions about animals today as they did in the past is impossible to know. Although the animal-rights movement has certainly grown in strength, many of the discussions about attachment to animals and the effect of animals on human morality seem very similar to those that were current in the eighteenth century. Analysts of contemporary culture, for example, have pointed out how often attitudes of sympathy toward animals coexist with cruel or indifferent behavior.[5] Although scientific studies demonstrating the health benefits of pet ownership are widely reported, many people still criticize excessive attention lavished on animals. And the relation between animal and human cruelty is still a matter of debate: violent criminals are often described as having enjoyed torturing animals when they were young, and people worry about the effect on children of watching nature documentaries showing gruesome scenes of animals attacking and killing each other.

We continue to use animals as symbols in all kinds of ways. Interestingly, the metaphorical link between animals and slaves, which was such a distinctive feature of ancien-régime France, has recently reemerged in the animal-rights movement. A number of twentieth-century critics of Western imperialist ideology have linked exploitation of nature with exploitation of women and native peoples and have argued that liberation movements for racial minorities, women, and animals should be carried on together.[6] It may be easier to talk metaphorically about slavery now that human slavery is practiced on a vastly smaller scale than it was in the eighteenth century. Still, just as in the past, there are those who have doubts about making connections between animal and human rights. Primatologist Frans B. M. de Waal, for example, argues that "rights are part of a social contract that makes no sense without responsibilities. This is the reason that the animal rights movement's outrageous parallel with the abolition of slavery—apart from being insulting—is morally flawed: slaves can and should become full members of society: animals cannot and will not."[7] The concept of animal slavery, however, has different associations now than it did two hundred years ago.

The places occupied by exotic animals in eighteenth-century France, both material and conceptual, reflect the specific historical circumstances of that culture, in all of its diversity, as well as more universal aspects of the relation between humans and animals. The elephant, for instance, embodied the most prevalent and even contradictory strains of thinking about exotic animals. It intrigued naturalists, as a scientific novelty, but at the same time it resonated with other cultural preoccupations. In its obedience, loyalty, and intelligence, the elephant epitomized the good servant, representing the proper subservient role both for animals and for their metaphorical human counterparts. In earning praise for its resistance to complete domination, though, it reflected late-eighteenth-century guilt about exploitation of indigenous people and of nature. A big burden to carry for the little elephant on the rue Dauphine that spent its days swigging beer and picking rice grains from ladies' hands.

NOTES

Abbreviations

A.-c.	*Avant-coureur*
Aff. de Paris	*Affiches de Paris*
Aff. de Prov.	*Affiches de Province*
Alm. for.	*Almanach forain*
AN	Archives nationales
AN Col	Archives nationales, Colonies
BHVP	Bibliothèque historique de la ville de Paris
BnF	Bibliothèque nationale de France
doc.	document
f.	folder
fol.	folio
HN	Georges-Louis Leclerc, comte de Buffon, *Histoire naturelle, générale et particulière,* 15 vols. Paris: Imprimerie royale, 1749–67. With descriptions by Louis Jean-Marie Daubenton.
HNO	Georges-Louis Leclerc, comte de Buffon, *Histoire naturelle des oiseaux,* 9 vols. Paris: Imprimerie royale, 1770–83. With contributions by Philibert Guéneau de Montbeillard and abbé Gabriel-Léopold Bexon.
HNS	Georges-Louis Leclerc, comte de Buffon, *Supplément à l'Histoire naturelle,* 7 vols. Paris: Imprimerie royale, 1774–89. Vol. 5 contains *Époques de la nature;* vol. 7 edited by Bernard de Lacépède.
J. de Paris	*Journal de Paris*

Mauduyt, *Encyc. méth. ois.*, 1	Mauduit [Pierre-Jean-Claude Mauduyt de la Varenne], *Encyclopédie méthodique: Histoire naturelle des animaux,* vol. 1. Paris: Panckoucke; Liège: Plomteux, 1782; second half of volume, "Ornithologie," beginning on 321.
Mauduyt, *Encyc. méth. ois.*, 2	Mauduit [Pierre-Jean-Claude Mauduyt de la Varenne], *Encyclopédie méthodique: Histoire naturelle: Oiseaux,* vol. 2. Paris: Panckoucke; Liège: Plomteux, 1784.
MNHN	Bibliothèque centrale du Muséum national d'histoire naturelle.
PUF	Presses Universitaires de France
subf.	subfolder

Preface

1. [Luc-Vincent] Thiéry, *Guide des amateurs et des étrangers voyageurs à Paris* . . . (Paris: Hardouin & Gattey, 1787), 1:35. See also Félix Nogaret, *Apologie de mon goût* (Paris: Couturier, 1771), 21, 42, who defined *exotique* (describing a plant) as "étranger." The 1778 edition of the *Dictionnaire de l'Académie française* (Nîmes: Beaume) defined *exotique* as "Qui ne croît point dans le pays," and gave as examples *plante exotique* and *terme exotique.* The word does not appear in the 1694 edition.

2. Jean le Rond d'Alembert, *Preliminary Discourse to the Encyclopedia of Diderot,* trans. Richard N. Schwab (Indianapolis: Bobbs-Merrill, 1963), 146–48. See also Thomas Hankins, *Science and the Enlightenment* (Cambridge: Cambridge UP, 1985), 10–13.

3. Sources for species names and translations: Buffon et al., *HN, HNO, HNS;* Georges Cuvier et al., *Dictionnaire des sciences naturelles* . . . (Strasbourg: Levrault; Paris: Le Normant, 1816–30); M. G. Wells, *World Species Checklist* (Herts., U.K.: Worldlist, 1998) (for names of birds); *Grzimek's Encyclopedia of Mammals* (New York: McGraw-Hill, 1990) (for names of mammals). For list of species translations, see Louise E. Robbins, "Elephant Slaves and Pampered Parrots: Exotic Animals in Eighteenth-Century France," Ph.D. diss., Univ. of Wisconsin–Madison, 1998, app. A.

Introduction

1. *A.-c.*, 14 Jan. 1771 (no. 2), 23–24. The only reference I have seen to the elephant's place of origin is in a source of dubious accuracy: a poem accompanying a print of the elephant, which calls the animal Asian (see chap. 3).

2. Buffon, "L'Éléphant," *HN*, 11 (1764), 16, 17.

3. Tim Ingold identifies the transition from hunting-gathering to pastoralism as the beginning of a dangerous detachment from nature in "From Trust to Domination," in *Animals and Human Society*, ed. Aubrey Manning and James Serpell (London: Routledge, 1994), 1–22. Lynn White Jr. blames Judeo-Christian attitudes in "The Historical Roots of Our Ecologic Crisis," in *Machina Ex Deo* (Cambridge: MIT P, 1968), 75–94. Among many identifying Bacon or Descartes or both as culprits are Carolyn Merchant, *The Death of Nature* (1980; paper., San Francisco: Harper & Row, 1983) and John Berger, "Why Look at Animals?" in *About Looking* (1980; New York: Vintage Int'l., 1991).

4. Claude Lévi-Strauss, *Le totémisme aujourd'hui*, 2nd ed. (Paris: PUF, 1965), 128; in this passage, Lévi-Strauss is describing Radcliffe-Brown's conception of totemism (with which he agrees). See also Lévi-Strauss, *La pensée sauvage* (Paris: Plon, 1962).

5. Mary Douglas, "The Pangolin Revisited: A New Approach to Animal Symbolism," in *Signifying Animals*, ed. R. G. Willis (London: Unwin Hyman, 1990), 25–36 (quote at 25). Douglas claims that the practice of relating two objects metaphorically precedes the recognition of their similarity. See also Lévi-Strauss, *Pensée sauvage*, 73–89, on the relation between animal and symbol.

6. Lévi-Strauss, *Pensée sauvage*, 82.

7. Chevalier de Jaucourt, "Sciences," in *L'encyclopédie*, ed. Denis Diderot and Jean le Rond d'Alembert, 14 (Neuchâtel: Faulche, 1755), 788. On connections between science and literature, see also Thomas L. Hankins, *Science and the Enlightenment* (Cambridge: Cambridge UP, 1985), 7–10, and my discussion in chap. 6. E. C. Spary recognizes the importance of Buffon's literary side in *Utopia's Garden* (Chicago: U of Chicago P, 2000).

8. Jeffrey Moussaieff Masson, *The Emperor's Embrace* (New York: Pocket Books, 1999); Rupert Sheldrake, *Dogs That Know When Their Owners Are Coming Home* (New York: Crown, 1999); Katy Payne, *Silent Thunder* (New York: Simon & Schuster, 1998).

9. See, e.g., Peter Singer, *Animal Liberation* (New York: New York Review, 1975); Marjorie Spiegel, *The Dreaded Comparison* (New York: Mirror, 1988); Steven M. Wise, *Rattling the Cage* (Cambridge, Mass.: Perseus, 2000).

10. George Page, review of *Wild Minds*, by Marc D. Hauser, *New York Times*

Book Review, 12 March 2000, 25–26. Page, creator of the PBS series *Nature*, was executive editor of the *Nature* series *Inside the Animal Mind*, aired Jan. 2000.

Chapter 1. Live Cargo

1. Dom [Antoine-Joseph] Pernetty, *Histoire d'un voyage aux isles Malouines . . .*, new ed. (Paris: Saillant & Nyon; 1770), 1:308–9. For more on this expedition and on Pernetty, see Jean-Étienne Martin-Allanic, *Bougainville navigateur et les découvertes de son temps* (Paris: PUF, 1964), 1:70–201. On Pernetty's account, see Pierre Berthiaume, *L'aventure américaine au XVIIIe siècle* (Ottawa: P de l'Univ. d'Ottawa, 1990), 130–66. The word *tigre* was often used for any large patterned cat, although many naturalists tried to restrict it to the Old World tiger. See Buffon, "Le Tigre," *HN*, 9 (1761), 52–56.

2. Pernetty, *Voyage aux isles Malouines*, 1:178–80 (at 180). Pernetty noted that only five to six parrots per hundred were said to survive the color-producing injections (the liquid being a sort of "poison"). On this process, see Buffon, "Le Crik à Tête Violette," *HNO*, 6 (1779), 235–36; "Perroquet tapiré," in Mauduyt, *Encyc. méth. ois.*, 2 (1784), 327. Mauduyt wrote that these multicolored parrots were sometimes seen in Parisian bird shops, where they had a high value.

3. See, e.g., Numa Broc, *La géographie des philosophes* (Paris: Ophrys, 1975), 294. A French expedition tried to take Aotourou back to Tahiti, but he died on the way, in 1771, of smallpox he probably contracted during a stopover in île de France. Martin-Allanic, *Bougainville navigateur*, 2:1307–28.

4. "Journal de Louis Caro, lieutenant sur *l'Étoile*," in Étienne Taillemite, *Bougainville et ses compagnons autour du monde, 1766–1769* (Paris: Imprimerie nationale, 1977), 2:362. Caro identified the birds as "des lory et cracatois [cacatois]," or lories (a type of small parrot) and cockatoos. For other references to acquiring birds, see Taillemite, ibid., 1:335, 411, 414 (Bougainville's journal), 2:279 (journal of François Vivez, surgeon on *l'Étoile*), 2:358, 359 (journal of Louis Caro).

5. For a similar argument from silence, see Robert Louis Stein, *The French Slave Trade in the Eighteenth Century* (Madison: U of Wisconsin P, 1979), 103–6; Stein argues that slave revolts on French slave ships were so common that captains' reports might mention them only briefly or not at all.

6. Catherine Gaziello, *L'expédition de La Pérouse, 1785–1788* (Paris: CTHS, 1984), 185.

7. [Robert Challe,] *Journal d'un voyage fait aux Indes orientales . . .* (Rouen: Machuel le jeune, 1721), 2:301. For a dissenting view, see Jean Baptiste Thibault de Chanvalon, *Voyage à la Martinique . . .* (Paris: Bauche, 1763), 98: "people eat them, their flesh has always seemed to me to be tough and rather bad tasting. I've seen very few Europeans set much store by them."

8. Jean de Léry, *Histoire d'un voyage faict en la terre du Brésil (1578)*, 2nd ed. (1580), ed. Frank Lestringant (Paris: Lib. générale française, livre de poche, 1994), 514, 526–27, 530, 536–37 (chaps. 21 and 22).

9. "Journal de Bougainville," in Taillemite, *Bougainville et ses compagnons*, 1:389; *The Journal of Jean-François de Galaup de la Pérouse, 1785–1788*, trans. and ed. John Dunmore (London: Hakluyt Soc., 1995), 2:384. La Pérouse's name appears in different forms (see Dunmore's introduction); I have used the one that is most common in the secondary literature.

10. Florian [Jean-Pierre de Claris], "Le perroquet confiant," in *Les fables de Florian*, ed. Jean-Noël Pascal (Perpignan: PU de Perpignan, 1995), 181.

11. Léry, *Histoire d'un voyage*, 526–27.

12. N. A. M. Rodger, in *The Wooden World* (London: Collins, 1986), 70, notes that exotic animals might be on British navy ships as pets, on commission, for future sale in London, or sometimes for scientific purposes. Also noting multiple reasons for bringing home exotic pets is Barbara T. Gates, *Kindred Nature: Victorian and Edwardian Women Embrace the Living World* (Chicago: U of Chicago P, 1998), 217.

13. Frézier, *Relation du voyage de la Mer du Sud* (1716), cited in Anne-Marie Brenot, "Les voyageurs français dans la vice-royauté du Pérou au XVIIIe siècle," *Revue d'histoire moderne et contemporaine* 35 (1988): 240–61, on 246. Legal trade in Pacific Spanish America was prohibited in 1724. See E. W. Dahlgren, *Les relations commerciales et maritimes entre la France et les côtes de l'océan pacifique (commencement du XVIIIe siècle)* (Paris: Champion, 1909).

14. Théophile Malvezin, *Histoire du commerce de Bordeaux depuis les origines jusqu'à nos jours* (Bordeaux: Bellier, 1892), 2:369, 3:204; Paul Butel, "France, the Antilles, and Europe in the Seventeenth and Eighteenth Centuries," in *The Rise of Merchant Empires*, ed. James D. Tracy (Cambridge: Cambridge UP, 1990), 153–73 (at 162).

15. Jean Meyer et al., *Histoire de la France coloniale des origines à 1914* (Paris: Colin, 1991), 101, gives population numbers for Saint Domingue, Martinique, and Guadcloupe from the 1680s to 1789. In 1789 the population of Saint Domingue (rounded to the nearest thousand) consisted of 31,000 whites, 28,000 free blacks, and 465,000 slaves.

16. Philip D. Curtin, *The Atlantic Slave Trade* (Madison: U of Wisconsin P, 1969), 75–85; Serge Daget, *La traite des Noirs* ([Rennes]: Ouest-France Univ., 1990), 151–73.

17. Patrick Villiers, "The Slave and Colonial Trade in France Just before the Revolution," in *Slavery and the Rise of the Atlantic System*, ed. Barbara L. Solow (Cambridge: Cambridge UP; Cambridge, Mass.: Du Bois Institute for Afro-American Research, Harvard Univ., 1991), 210–36.

18. Jean Meyer, "La France et l'Asie: Essai de statistiques (1730–1785), état de la question," *Histoire, économie et société* 2 (1982): 297–312; Philippe Haudrère, *La Compagnie française des Indes au XVIIIe siècle* (Paris: Lib. de l'Inde, 1989), 4:1215; Philippe Haudrère, *L'empire des rois, 1500–1789* (Paris: Denoël, 1997), 179–82.

19. The French, of course, were not alone in outfitting exploratory voyages; James Cook's voyages (1768–80), in particular, had a major impact on international relations and on geographical and botanical knowledge, especially through the work of the botanist Joseph Banks. See David Philip Miller and Peter Hanns Reill, eds., *Visions of Empire* (Cambridge: Cambridge UP, 1996).

20. Gaziello, *L'expédition de La Pérouse*, 217–18.

21. For Bougainville's voyage, see Martin-Allanic, *Bougainville navigateur*, 1:456–504 passim; for La Pérouse's voyage, see Gaziello, *L'expédition de La Pérouse*, 59–65, 288; and John Dunmore, introduction to La Pérouse, *Journal*, 1:cxli–cxlvi; for d'Entrecasteaux, see Hélène Richard, *Une grande expédition scientifique au temps de la Révolution française* (Paris: CTHS, 1986), 61–82, 122–25, esp. 124 ("There was never a question of bringing live animals on board. First, it was necessary to describe the living animals and their behavior. Then, dead, they had to be prepared to last until the return to France"). See Silvia Collini and Antonelle Vannoni, "La società d'histoire naturelle et il viaggio di d'Entrecasteaux alla ricerca de La Pérouse: Le istruzioni scientifiche per i viaggiatori," *Nuncius* 10, no. 1 (1995): 257–91, for text of instructions from Société d'histoire naturelle de Paris to naturalists on the expedition; and Lorelai Kury, "Les instructions de voyage dans les expéditions scientifiques françaises (1750–1830)," *Revue d'histoire des sciences* 51, no. 1 (1998): 65–91. On the expectation that specimens would not remain alive, see also Jean-Bernard Lacroix, "L'approvisionnement des ménageries et les transports d'animaux sauvages par la Compagnie des Indes au XVIIIe siècle," *Revue française d'histoire d'outre-mer* 65 (1978): 153–79, on 160–61.

22. [Pierre] Sonnerat, *Voyage aux Indes orientales et à la Chine . . .* (Paris: Author; Froulé, 1782), 1:viii–ix.

23. Richard W. Burkhardt Jr., "Unpacking Baudin," in *Jean-Baptiste Lamarck, 1744–1829*, ed. Goulven Laurent (Paris: CTHS, 1997), 497–514.

24. Marquis de Chastellux, *Travels in North-America in the Years 1780, 1781, and 1782*, "trans. . . . by an English gentleman," 2nd ed. (London: 1787; repr. n.p.: Arno, 1968), 2:426. I have not seen any reference in the *Histoire naturelle* to the animal.

25. *Correspondance inédite entre Réaumur et Abraham Trembley*, ed. Maurice Trembley (Geneva: Georg, 1943), 401–13. See also Daubenton, "Description de la partie du cabinet qui a rapport à l'Histoire Naturelle de l'Éléphant," *HN*, 11 (1764), 143; Rodger, *Wooden World*, 70. The *Aff. de Prov.* reported that the elephant's food

and drink, including wine and other liquor, cost 9 livres per day (12 Nov. 1755 [no. 46], 183).

26. Pernetty, *Histoire d'un voyage aux Malouines*, 1:190–91. For the fate of Pernetty's birds, see P. Fournier, *Voyages et découvertes scientifiques des missionnaires naturalistes français à travers le monde pendant cinq siècles* (Paris: Lechevalier & fils, 1932), part 1:68.

27. Richard, *Le voyage de d'Entrecasteaux*, 306.

28. [François] Le Vaillant, *Voyage de Monsieur Le Vaillant dans l'intérieur de l'Afrique* . . . (Paris: Leroy, 1790), 1:199.

29. See Léry, *Histoire d'un voyage*, 281–82, for a case in which an indigenous woman refused to sell her well-trained pet parrot.

30. On animals as diplomatic gifts, see Michael A. Osborne, "The Role of Exotic Animals in the Scientific and Political Culture of Nineteenth-Century France," in *Les animaux exotiques dans les relations internationales*, ed. Liliane Bodson (Liège: U de Liège, 1998), 15–32. The tradition of animal gifts is an old one and is common in many cultures.

31. La Pérouse, *Journal*, 2:357.

32. For references to monkeys and parrots given as gifts, see Tibierge, "Journal du sieur Tibierge, . . . sur le vaisseau 'Le pont d'or' au voyage de l'année 1692," in *L'établissement d'Issiny, 1687–1702*, ed. Paul Roussier (Paris: Larose, 1935), 54; Godefroy Loyer, *Relation du voyage du royaume d'Issyny, Cote d'Or, païs de Guinée en Afrique* . . . [1714] in ibid., 155.

33. [Michel Jajolet] de La Courbe, *Premier voyage du sieur de La Courbe fait à la coste d'Afrique en 1685*, ed. P. Cultru (Paris: Champion, 1913), 157–58. This incident, in slightly different form, was attributed to General Brue by Labat, who, according to Cultru, copied much of his *Nouvelle relation* from La Courbe. See intro., i–lviii (at v). For Labat's account, see Jean-Baptiste Labat, *Nouvelle relation de l'Afrique occidentale* . . . (Paris, 1728), 3:131–32. On Labat's reliability as a source, see Broc, *La géographie des philosophes*, 72 n. 55.

34. [François] Froger, *Relation d'un voyage fait en 1695, 1696, & 1697, aux côtes d'Afrique, détroit de Magellan, Brezil, Cayenne, & isles Antilles* . . . (Paris: Nicolas le Gras, 1699), 144. De Gennes was on a military reconnaissance mission in the South Atlantic.

35. Gustave Loisel, *Histoire des ménageries de l'antiquité à nos jours* (Paris: Doin & fils, 1912), 2:112–15. Documents listing Mosnier's expenses for 1679 and the animals he acquired in 1685 are reprinted in Loisel, 2:335–39. Loisel states (2:113) that Colbert requested savants going abroad to bring back rare animals (alive, presumably). I have not seen any such instructions in the eighteenth century.

36. Chandra Mukerji discusses formal gardens as symbols of power in *Territorial Ambitions and the Gardens of Versailles* (Cambridge: Cambridge UP, 1997),

and Harriet Ritvo discusses animals in this vein in *The Animal Estate* (Cambridge: Harvard UP, 1987), chaps. 5 and 6.

37. Extract of letter from B. Semaire, chancellor of the French consulate at Cairo, 1 Nov. 1714 (ostriches), and extract of letter from Le Maire, consul at Cairo, 26 Aug. 1715 (loss of swamphens, delayed trip), AN AJ/15/512, f. A 14, subf. "Ménagerie de Versailles," doc. 506.

38. M. Percheron, "Instructions pour servir à soigner les plantes et animaux destinés pour le Jardin du roi, et pour la Ménagerie du roi," July 1787, AN AJ/15/511, f. "Envois de graines, plantes, minéraux etc. des pays étrangers au Jardin du roi," subf. "Afrique," doc. 457.

39. J.-B. Lacroix, "L'approvisionnement des ménageries," 162.

40. Comte de Pontchartrain to consuls in the Levant, 21 Jan. 1711, reprinted in Loisel, *Histoire des ménageries*, 2:339–40; see discussion at 2:114.

41. Percheron to maréchal de Castries, 22 Mar. 1785, AN Col C/5B/7, no. 62. See also Percheron to Castries, 28 July 1786, ibid., no. 137.

42. [Quesnel?] to Castries, 22 July 1788, AN Col C/2/285, f. "Animaux," 192 (the letter states that it should take ten to twelve days for a cassowary and its two attendants to travel from Havre to Saint Germain en Laye). It took twenty-four days for the tigers to travel from Lorient to Versailles in 1770, according to J.-B. Lacroix, "L'approvisionnement des ménageries," 162.

43. Rouillé (ministre de la Marine) to ?, 8 Feb. 1751 and 22 Mar. 1751; copies of letters from Archives de la Chambre de commerce de Marseille, AA 63, in AN AJ/15/512, f. A 14, subf. "Ménagerie de Versailles," docs. 507 II, III.

44. La Courbe, *Premier voyage . . . à la coste d'Afrique*, 47, 135. La Courbe identified the birds as "peignées." The ministers were Seignelay and Pontchartrain (probably Louis de).

45. Le Juge to "Monseigneur," île de France, 26 Mar. 1751 and [Imbert?] to "Monseigneur," Lorient, 5 Aug. 1752, MNHN MS 293, f. C. This exchange shows the importance placed on giving a live animal as a gift.

46. J.-H. Bernardin de Saint-Pierre, "Voyage à l'île de France," in *Oeuvres complètes de J. H. Bernardin de Saint Pierre* (Paris: Armand-Aubrée, 1834), 2:249.

47. Letters from the Archives de la Marine at Lorient, as extracted by J.-B. Lacroix, "L'approvisionnement des ménageries," 157–59.

48. Quoted in J.-B. Lacroix, "L'approvisionnement des ménageries," 159. Lacroix suggests that the naturalist Michel Adanson, who was in Senegal at the time, might have been involved in collecting the birds to fill this commission, which took three years: Jean-Bernard Lacroix, *Les Français au Sénégal au temps de la Compagnie des Indes de 1719 à 1758* (Vincennes: Service historique de la Marine, 1986), 20.

49. [Jean Bernard] Bossu, *Nouveaux voyages dans l'Amérique septentrionale . . .* (Amsterdam: Changuion, 1777), 390.

50. Joseph-François-Charpentier de Cossigny to Louis-Guillaume Le Monnier, Paris, 24 April 1775, and Cossigny to Le Monnier, Lorient, 5 May 1775, MNHN MS 1995, letters 78 and 79 (quote from letter 79). The "Princesse" is probably the princesse de Marfan (see ibid., letters 72 and 90).

51. On Boufflers, see Paul Bouteiller, *Le chevalier de Boufflers et le Sénégal de son temps, 1785–1788* (Paris: Lettres du Monde, 1995).

52. Journal of the comtesse de Sabran, 29 June 1787, *Correspondance inédite de la comtesse de Sabran et du chevalier de Boufflers, 1778–1788*, 2nd ed. (Paris: Plon, 1875), 252–53; ibid., 5 July 1787, 259–61. The prince had come to France from Cochinchina to gain support from Louis XVI.

53. Boufflers to Sabran, 9 [19?] July 1787 [1786?], *Correspondance inédite*, 513–14. The letter is headed "Ce 9" but comes after one dated "Ce 18" and before one dated "Ce 20." In the 1875 edition of the correspondence, the letter is identified as having been written during Boufflers's second return from Senegal in 1787, but later scholars have suggested that it dates from the first trip, in 1786, in which case it would have been written before Sabran's letters about the escaped birds (see Little, below). Nesta Webster assumes that the yellow parrot was for Sabran (*The Chevalier de Boufflers* [New York: Dutton, 1924], 217), but since Boufflers referred to the recipient as "quelqu'un qui est seul" (not "quelqu'une qui est seule"), I presume the recipient was a male (the king?). The "little captive" was a Senegalese child who lived with the prince and princesse de Beauvau until her death at the age of sixteen; her story became the basis for *Ourika* (1824), a best-selling novel by Claire de Duras (Roger Little, "Madame de Duras et *Ourika*," in Claire de Durfort, duchesse de Duras, *Ourika* [Exeter: U of Exeter P, 1993], 27–67, esp. 37–40).

54. Boufflers to Sabran, 30 July 1787 [1786?], *Correspondance inédite*, 519–20.

55. Wilma George, "Sources and Background to Discoveries of New Animals in the Sixteenth and Seventeenth Centuries," *History of Science* 18 (1980): 80, citing Samuel Eliot Morison, *The Great Explorers* (Oxford: Oxford UP, 1978), 454.

56. Claude Lévi-Strauss, *Tristes tropiques*, trans. John Russell (New York: Atheneum, 1968), 87.

57. Alfred Franklin, *La vie privée d'autrefois: Les animaux, du XVe au XIXe siècle* (Paris: Plon, 1897–99), 2:194. The only primates on Madagascar are lemurs; the animals in question might, however, have been monkeys from Africa or India transported via Madagascar.

58. Ibid., 2:57–59. Also mentioned in Loisel, *Histoire des ménageries*, 1:274–75.

59. From the travels of Thomas Platter, as cited in E.-T. Hamy, "Le commerce des animaux exotiques à Marseille à la fin du XVIe siècle," *Bulletin du Muséum d'histoire naturelle* 7 (1903): 316–18. Merchants from La Rochelle who furnished goods for René de Laudonnière's voyage to Florida in 1572 (originally destined to Peru) demanded a half share of the birds and animals to be acquired, although they

supplied only one-third of the merchandise to be traded. Historian G. Musset concludes that the merchants must have expected a good return from the animals, but he does not mention whether any animals were acquired: "Les collectionneurs de bêtes sauvages, 1047–1572," *Bulletin du Muséum d'histoire naturelle* (1902): 242–43.

60. [François Alexandre Stanislaus] baron de Wimpffen, *Voyage à Saint-Domingue* . . . (Paris: Chocheris, an 5 [1799]), 1:6–7 (emphasis and dots in original).

61. [Pierre] Sonnerat, *Voyage à la Nouvelle Guinée* . . . (Paris: Ruault, 1776), 152.

62. For Bougainville's expedition, see above; for La Pérouse, see *Journal*, 2:393, 394; for d'Entrecasteaux, see Richard, *Voyage de d'Entrecasteaux*, 157.

63. Crew members on d'Entrecasteaux's expedition were prohibited (apparently with little effect) from trading with local people for fear that the price of food would be driven up (Richard, *Voyage de d'Entrecasteaux*, 153). Johann Reinhold Forster complained about sailors on Cook's expedition buying up natural curiosities for sale in England (the gunner and carpenter had accumulated several thousand shells): Nicholas Thomas, *Entangled Objects* (Cambridge: Harvard UP, 1991), 140–41.

64. Wages for ordinary sailors on Bougainville's ships ranged from 12 to 20 livres per month (Taillemite, *Bougainville et ses compagnons*, 1:160–75); average wages on slave ships were 22 to 30 livres per month in 1770 (Jean-Michel Deveau, *La traite rochelaise* [Paris: Karthala, 1990], 148). For prices of parrots (which averaged around 100 livres) see chap. 5.

65. "Mémoire sur les Lamas," by abbé Beliardy, 21 Aug. 1779, MNHN MS 864, f. III D, quotations on 5v. The memoir is quoted in Buffon, "De la Vigogne," *HNS*, 6 (1789), 211.

66. Le Page du Pratz, *Histoire de la Louisiane* . . . (Paris: De Bure l'aîné, 1758), 2:128 (parrots), 2:139–40 (cardinal). Also critical of the cardinal's "overly loud voice" were [Henri-Gabriel Duchesne and Pierre-Joseph Macquer,] *Manuel du naturaliste* (Paris: Desprez, 1771), 96.

67. François-Xavier de Charlevoix, *Journal d'un voyage fait par ordre du roi dans l'Amérique septentrionale*, ed. Pierre Berthiaume (Montreal: P de l'Univ. de Montreal, 1994), 1:373 (from letter dated 1 April 1721). Charlevoix noted that there were already cardinals in Paris that had come from Louisiana. Mauduyt also remarked that cardinals were often brought to Paris and "during peacetime" could be found in bird shops. He found their song "very loud and very pleasant." Mauduyt, *Encyc. méth. ois.*, 1 (1782), 570. The name *cardinal* was applied to several species, including a red South American tanager that Buffon dubbed the *scarlatte*. He surmised (wrongly) that when travelers described a cardinal with a pleasant song they were referring to the South American bird, because the North American species was a type of grosbeak and so would not sing. "Le Scarlatte," *HNO*, 4 (1778), 243–47.

68. [Nougaret Bourgeois,] *Voyages intéressans dans différentes colonies françaises, espagnoles, anglaises, &c. . . .* (Paris: Bastien, 1788), 286 (birds), 289 (squirrels). *Pape* often referred to the painted bunting, but here it seems to refer to the indigo bunting. Mauduyt noted, in 1784, that live *papes* were often brought back to France: *Encyc. méth. ois.*, 2 (1784), 299–300.

69. Mauduyt remarked that slave traders brought back with them large numbers of *grenadins, veuves, bengalis* and *senegalis: Encyc. méth. ois.*, 2 (1784), 77. On parrots and monkeys, see Barbot's description, below, and also *Journal d'un voyage sur les costes d'Afrique et aux Indes d'Espagne . . .* (Amsterdam: Marret, 1723), 58, 124–25. The anonymous author of the latter described the parrots and monkeys as pets *(animaux domestiques)* of the native people (124).

70. Père Raymond Breton, *Relation de l'île de la Guadeloupe* (1647; repr. Basse-Terre: Société d'histoire de la Guadeloupe, 1978), 75. Pierre Barrère, a physician who served in Cayenne from 1722–24, noted that the Indians of Guyana sold live birds to the colonists: *Nouvelle relation de la France equinoxiale . . .* (Paris: Piget, 1743), 30–31.

71. [Jean Baptiste Thibault de Chanvalon,] *Voyage à la Martinique . . .* (Paris: Bauche, 1763), 98.

72. Buffon, "La Perruche à Collier Couleur de Rose," *HNO*, 6 (1779), 152. The practice of transporting species from one place to another could cause difficulties for naturalists trying to determine the geographic origin of the species. Buffon claimed that Brisson had wrongly identified the rose-collared parakeet as a New World species. Buffon also mentioned another parrot often imported on slave ships that he suggested might be native to Guinea rather than to Brazil, as other naturalists had claimed. "Le Paragua," ibid., 248–49.

73. Mauduyt, *Encyc. méth. ois.*, 2 (1784), 335.

74. Médéric Louis Élie Moreau de Saint-Méry, *Description topographique, physique, civile, politique et historique de la partie française de l'isle Saint Domingue*, new ed. [Philadelphia, 1797], ed. Blanche Maurel and Étienne Taillemite (Paris: Société de l'histoire des colonies françaises, 1958), 301, 1069–70 (aviary of the surgeon Le Sage).

75. *Barbot on Guinea*, ed. P. E. H. Hair, Adam Jones, and Robin Law (London: Hakluyt Soc., 1992), 1:319 (Bissago Is.), 2:362 n. 40, 416 (Gold Coast), 476–77 (Príncipe Is.), 694 (Calabar) (quote at 694). The description of Calabar is by James Barbot and was included in the 1732 edition of Barbot's voyages.

76. William Bosman, *A New and Accurate Description of the Coast of Guinea . . .*, new ed. (1705; repr. London: Cass, 1967). On bird observations by Bosman and others, see Alan Tye and Adam Jones, "Birds and Birdwatchers in West Africa, 1590–1712," *Archives of Natural History* 20, no. 2 (1993): 213–27.

77. *Barbot on Guinea*, 2:474 n. 10. The editors note that the animal was probably a Diana monkey.

78. Ibid., 2:421 n. 8 (parakeets), 2:477–78 (Príncipe Is.), 2:473–74 (monkeys), 2:471 (civet), 2:478 (guns), 2:623 (rats). Bosman mentioned similar practices and mortality rates: of the small parakeets from the Gold Coast, he wrote: "They are very beautiful little Creatures; and daily, or when-ever opportunity offers, we send great numbers of them to *Holland;* where they bear a good value. We generally buy them here at the rate of a Rycksdollar *per* Dozen: But most of them commonly die in their Passage to *Holland* . . . at present of one hundred that we send over scarce ten survive" (Bosman, *New and Accurate Description*, 270, emphasis in original).

79. Mauduyt, *Encyc. méth. ois.*, 1 (1782), 430.

80. The *Encyclopédie méthodique,* which was divided by subject, was a successor to d'Alembert and Diderot's *Encyclopédie.* See Robert Darnton, *The Business of Enlightenment: A Publishing History of the "Encyclopédie,"* 1775–1800 (Cambridge: Harvard UP, 1979); Christabel P. Braunrot and Kathleen Hardesty Doig, "The *Encyclopédie méthodique,*" *Studies on Voltaire and the Eighteenth Century* 327 (1995): 1–152. Braunrot and Doig (at 47 and 137) mistakenly identify the author of the two half-volumes on ornithology as Antoine-René Mauduyt, who was a mathematician; on the title page, the author is identified simply as M. Mauduit *[sic].* For Mauduyt's controversy with other naturalists about how best to preserve birds, see Paul Lawrence Farber, "The Development of Taxidermy and the History of Ornithology," *Isis* 68 (1977): 550–66.

81. MNHN MS 352: "Notice des oiseaux qu'on desire recevoir de Cayene vivans." The memoir was probably written between 1764 and 1776, because it identifies M. [Bertrand] Bajon as a surgeon in Cayenne (fol. 7r) and these are the dates of his stay there (Alfred Lacroix, *Figures de savants,* vols. 3 and 4: *L'Académie des sciences et l'étude de la France d'outre-mer* [Paris: Gauthier-Villars, 1938], 3:71–74). E. C. Spary identified the handwriting as Daubenton's: *Utopia's Garden* (Chicago: U of Chicago P, 2000), 83; and pers. comm.

82. MNHN MS 352, fols. 1r–1v.

83. Ibid., fol. 8r ("Suplement a l'instruction pour recevoir des oiseaux").

84. See Mauduyt, *Encyc. méth. ois.*, 1 (1782), 425–68 (fourth discourse) (general comments on importing birds), 2 (1784), 477 (toucans), 1 (1782), 428, 471–72 (trumpeters).

85. For an overview, see "Le rôle des ménageries dans l'acclimatation et la zoologie économique," Loisel, *Histoire des ménageries,* 2:322–34. On Daubenton's sheep-raising project, see Camille Limoges, "Louis-Jean-Marie Daubenton," *Dictionary of Scientific Biography,* ed. Charles Coulston Gillispie, 15, supp. 1 (New York: Scribner's, 1978), 111–14. On llamas, see Beliardy, "Mémoire sur les Lamas," MNHN MS 864, f. III D; and Buffon, "De la Vigogne," *HNS,* 6 (1782), 211. Exam-

ples of recommendations to import and breed species include the paca (*HNS*, 3 [1776], 210) and the zebra (*HN*, 12 [1764], 10). Plant introductions, esp. of crops such as coffee and sugar cane, were more often undertaken and more successful.

86. J.-B. Lacroix, "L'approvisionnement des ménageries," 158, states (without documentation): "It is necessary . . . to note the presence of little animals like the monkey . . . as pets of long-distance sailors. . . . Parrots and little birds were also common on ships." See Richard, *Le voyage de d'Entrecasteaux*, 151–52, and Haudrère, *Compagnie française*, 2:678–87 for descriptions of shipboard diversions. For a study of modern sailors that discusses pets as a source of affection, see Knut Weibust, *Deep Sea Sailors* (Stockholm: Norstedt & Söner, 1969), 428. On sailors' mortality rates, see Curtin, *Atlantic Slave Trade*, 282–86; Stein, *French Slave Trade*, 99; Deveau, *Traite rochelaise*, 154; Haudrère, *Compagnie française*, 2:687–89.

87. AN O/1/597, f. "Mémoires concernant la Marine, les Colonies, la Compagnie des Indes, XVIIIe s.," doc. 11: "Mes reflexions sur la marine" (anonymous and undated but written after 1757, a date mentioned in the MS); Yves Laissus, "Les voyageurs naturalistes du Jardin du roi et du Muséum d'histoire naturelle," *Revue d'histoire des sciences* 34 (1981): 259–317 (at 273).

88. La Courbe, *Premier voyage . . . à la coste d'Afrique*, 157–58 (eagle), 164 (guineafowl). He was still on shore when he wrote about the guineafowl, but he noted elsewhere (at 47) that people often took them back to France.

89. Ibid., 273.

90. Richard, *Voyage de d'Entrecasteaux*, 317. De d'Auribeau reported that the natives *(naturels)* solemnly asked the monkey and a young goat that was also taken ashore to sit down next to them.

91. Buffon, "Les Orang-outans, ou le Pongo et le Jocko," *HN*, 14 (1766), 55–56, and [Alexandre] Savérien, *Histoire des progrès de l'esprit humain . . . Histoire naturelle* (Paris: Humblot, 1778), 279–80, both of whom cite de la Brosse, *Voyage à la côte d'Angola* (on chimpanzees); Nicolas Restif de la Bretonne, *Lettre d'un singe aux animaux de son espèce*, ed. Monique Lebailly (1781; Levallois-Perret: Manya, 1990), 62–63, citing père Le Comte, *Mémoire de la Chine*, 2 (on monkeys from Malacca).

92. Wimpffen, *Voyage à Saint-Domingue*, 2:184–85, emphasis in original.

93. Pierre Berthiaume, in *L'aventure américaine*, compared manuscript and printed versions of French accounts of voyages to America, to see how authors elaborated or altered their texts. He contended that personal and especially sentimental passages were more common later in the century and were often additions made for publication (see esp. 130–66, 189–205).

94. Challe, *Journal d'un voyage fait aux Indes orientales*, 2:302–3. After recounting this episode, Challe went on to claim that animal instinct is no different from reason in man and that animals are not simply machines.

95. Froger, *Relation d'un voyage*, 201–3. Berthiaume *(L'aventure américaine)*

cites this story as an example of the way voyagers used their adventures to express their emotions: this minor incident, he writes, "finds a place at the heart of the account because it derives its meaning from the narrator's emotion" (193).

96. [Charles-Marie de] La Condamine, *Relation abrégée d'un voyage fait dans l'intérieur de l'Amérique méridionale* . . . (Paris: Veuve Pissot, 1745), 165–67 (at 166). The skin of the monkey (probably a tamarin) ended up in the Jardin du roi. Buffon quoted La Condamine's description in "Le Mico," *HN*, 15 (1767), 121–22.

97. MNHN MS 1765, "Fragmens copiés d'après les rélations de M. de Maudave et ses lettres à M. de Voltaire," sec. 4: "Oiseaux et autres productions de L'isle grande," 58–59 (or 31v–32r in second numbering system on MS). The comte de Maudave (or Modave) led an ill-fated attempt to colonize Madagascar in the 1770s.

98. [Jean-François de Galaup de] La Pérouse, *Voyage de La Pérouse autour du monde,* . . . *rédigé par M. L. A. Milet-Mureau* . . . (Paris: Plassan, an VI [1798]), 3:227. The recent English edition translates this passage: "it was unlikely that we would succeed in keeping it until Europe where all we could bring back was its plumage which had lost all its brilliance" (*Journal,* trans. John Dunmore, 2:390); the difference between my translation and Dunmore's could possibly reflect a difference in the source, since Dunmore used La Pérouse's original journal, to which Milet-Mureau made a few stylistic changes (see ibid., 2:571–76). The text in Milet-Mureau's version (the one that contemporary readers would have seen) is "il n'était guère vraisemblable qu'il pût arriver vivant en Europe: en effet, sa mort ne nous permit que de conserver sa robe, qui perdit bientôt tout son éclat."

99. [Bertrand] Bajon, *Mémoires pour servir à l'histoire de Cayenne, et de la Guiane françoise* . . . (Paris: Grangé, 1777), 1:401–2 (at 402).

100. Léry, *Histoire d'un voyage,* 536–37. Léry was taking the parrot back as a present for "the Admiral" (Admiral Gaspard de Coligny).

101. [Bourgeois,] *Voyages intéressans dans différentes colonies,* 288. I use *hummingbird* for both *oiseau-mouche* and *colibri,* terms that were sometimes used interchangeably, although Buffon used *colibri* for slightly larger birds with curved bills.

102. Breton, *Relations de l'île de la Guadeloupe,* 36. Breton explained that the natives caught the birds *(colibris)* alive with a gummed pole. On giving up, see Pernetty, *Histoire d'un voyage aux Malouines,* 1:190–91; Wimpffen, *Voyage à Saint-Domingue,* 1:249–50.

103. Chastellux, *Travels in North America,* 2:173; Le Page du Pratz, *Histoire de la Louisiane,* 2:142–43. Both stories were told to the authors by friends.

104. Bernardin de Saint-Pierre, "Voyage à l'île de France," in *Oeuvres complètes,* 2:231. This bird, too, was a colibri.

105. Le Page du Pratz, *Histoire de la Louisiane,* 2:143. See also Charlevoix, *Journal d'un voyage,* 1:375, who says he gave a hummingbird to a friend as a present but it died the next day.

106. MNHN MS 369, f. "Ornithologie de Saint Domingue, 1780," doc. "Oiseau-mouche. Duplicata," 13–14. On the front of the document someone has penciled *Sonnini (?)*. The reference is to Charles-Nicolas-Sigisbert Sonnini de Manoncourt, a frequent contributor of information on Guyana to Buffon; see Elizabeth Anderson, "La collaboration de Sonnini de Manoncourt à l'*Histoire naturelle de Buffon*," *Studies on Voltaire and the Eighteenth Century* 120 (1974): 329–58.

107. [Jean-Baptiste-François Hennebert and Gaspard Guillard de Beaurieu,] *Cours d'histoire naturelle* . . . (Paris: Lacombe, 1770), 4:91–92.

108. [Duchesne and Macquer,] *Manuel du naturaliste*, 146. They also claimed that the birds could be preserved with no loss of color.

109. Buffon, "L'Oiseau-Mouche," *HNO*, 6 (1779), 8; "Le Colibri," ibid., 42–43.

110. Le Page du Pratz, *Histoire de la Louisiane*, 2:143.

111. Chastellux, *Travels in North America*, 2:173.

112. Buffon, "Les Orang-Outangs," 56, citing Henri Grosse, *Voyage aux Indes orientales;* Buffon, "Le Solitaire," *HNO*, 1 (1770), 490. Two solitaires being sent from île de Bourbon to the king died on the ship because they would not eat or drink; Buffon cites *Voyage de Carré aux Indes.* Two species of solitaires, large birds related to the dodo, lived on two of the Mascarene islands; they are now extinct. See H. E. Strickland and A. G. Melville, *The Dodo and Its Kindred* (London: Reeve, Benham, & Reeve, 1848).

113. E.g., see [Pons Augustin Alletz,] *Histoire des singes, et autres animaux curieux* . . . (Paris: Duchesne, 1752), 23–24; Restif de la Bretonne, *Lettre d'un singe*, 80.

114. Labat, *Nouvelle relation de l'Afrique occidentale*, 4:192.

Chapter 2. The Royal Menagerie

1. "Mémoires pour servir à l'histoire naturelle des animaux," in *Mémoires de l'Académie royale des sciences*, 1 (The Hague: Gosse & Neaulme, 1731), 15.

2. R. J. Hoage, Anne Roskell, and Jane Mansour, "Menageries and Zoos to 1900," in *New Worlds, New Animals*, ed. R. J. Hoage and William A. Deiss (Baltimore: Johns Hopkins UP, 1996): 8–18 (at 15).

3. Some work on botanical and zoological displays has begun to focus on aspects of audience and conflicting interpretations that I emphasize here; e.g., Éric Baratay and Élisabeth Hardouin-Fugier, *Zoos* (Paris: Découverte, 1998). Robert W. Jones discusses the variety and contradictory responses of visitors in "'The Sight of Creatures Strange to our Clime': London Zoo and the Consumption of the Exotic," *Journal of Victorian Culture* 2.1 (1997): 1–26.

4. Jeffrey W. Merrick, *The Desacralization of the French Monarchy in the Eighteenth Century* (Baton Rouge: Louisiana State UP, 1990).

5. Chandra Mukerji, *Territorial Ambitions and the Gardens of Versailles* (Cambridge: Cambridge UP, 1997), 250.

6. On fables and Versailles, see Aurélia Gaillard, *Fables, mythes, contes* (Paris: Champion, 1996), esp. 280–81.

7. Regular visitors may have entered along a pathway from the road, rather than through the chateau and observatory. For literature on the menagerie, see note on secondary sources.

8. Delamarre, *Traité de la police,* quoted in Alfred Franklin, *La vie privée d'autrefois: Les animaux* (Paris: Plon, 1897–99), 2:191. On the symbolic importance of Mediterranean plants in the gardens, see Mukerji, *Territorial Ambitions,* 179–81.

9. Gustave Loisel, *Histoire des ménageries de l'antiquité à nos jours* (Paris: Doin & fils, 1912), 2:118–19. On the menagerie as the king's theater of power, see Baratay and Hardouin-Fugier, *Zoos.* See also Peter Burke, *The Fabrication of Louis XIV* (New Haven: Yale UP, 1992).

10. [Jean de] La Fontaine, "Les Amours de Psyche," in *Oeuvres complètes,* 2: *Oeuvres diverses,* ed. Pierre Clarac ([Paris]: Gallimard, 1958), 128; Claude Denis, *Explication de toutes les grottes, rochers, et fontaines du chasteau royal de Versailles, maison du soleil, et de la menagerie, en vers heroique* (Paris: n.d.), quoted in Masumi Iriye, "Le Vau's Menagerie and the Rise of the *Animalier,*" Ph.D. diss., Univ. of Michigan, 1994, 42. For other descriptions, see Iriye, "Le Vau's Menagerie," 39–41; Franklin, *Les Animaux,* 2:191–92.

11. "Mémoires pour servir à l'histoire naturelle des animaux," 2:324.

12. In the later seventeenth century, the menagerie was open to the public when the king was not present, although at some periods the king allowed access only to courtiers. During the eighteenth century access was unlimited. Baratay and Hardouin-Fugier, *Zoos,* 73–74.

13. J[oachim] C[hristoph] Nemeitz, *Séjour de Paris* . . . (Leiden: Van Abcoude, 1727), 486–508 (at 493). The fare was 25 sous each for the coach.

14. Loisel, *Histoire des ménageries,* 2:115, 140–41, 147–48.

15. Alice Stroup, *A Company of Scientists* (Berkeley: U of California P, 1990), 39, 41, 293 n. 29 (bear dissection). Based on this use of menagerie animals, Chandra Mukerji claims that the menagerie "replace[d] one order of nature based on bestiality with another order of nature derived from science" (*Territorial Ambitions,* 282). Many visitors, however, may neither have known nor cared about the dissections. On the Academy, see Roger Hahn, *The Anatomy of a Scientific Institution* (Berkeley: U of California P, 1971).

16. The dissection occurred in 1681, but the fight between the elephant and the tiger put on for the Persian ambassador supposedly took place in 1682. Could there have been two different elephants? Or is one of the dates wrong? See Loisel, *Histoire des ménageries,* 2:298.

17. "Mémoires pour servir à l'histoire naturelle des animaux," vols. 1 and 2. The memoirs were published first in Latin in two volumes in 1671 and 1676, then translated into French; they were later published in 1737 in French at Amsterdam and also by the Académie at about the same time in three volumes. The list of species observed includes, in volumes 1 and 2: lion, chameleon, camel, bear, gazelle, serval, shark, lynx, beaver, otter, civet, elk, coati, seal, *vache de Barbarie,* cormorant, chamois, African porcupine, hedgehog, monkeys, red deer, axis deer, guineafowl, eagle, *cocq indien,* bustard, demoiselle crane, ostrich, cassowary, and land tortoise; in volume 3 (which I have not seen): tiger, panther, spoonbill, marmot, dormouse, flamingo, purple swamphen, white ibis, stork, salamander, *lézard écaillé,* elephant, crocodile, pelican, crowned crane, and griffon vulture. See Loisel, *Histoire des ménageries,* 2:296–300. On Perrault, see Antoine Picon, *Claude Perrault ou la curiosité d'un classique* (Paris: Picard, 1988).

18. Iriye, "Le Vau's Menagerie," 51.

19. Nemeitz, *Séjour de Paris,* 506. A book from 1736 praised the "large and magnificent aviary" at Versailles as "the most beautiful to be seen anywhere," but this description might have been left from an earlier edition: [Louis] Liger, *La nouvelle maison rustique,* 4th ed. (Paris: Veuve Prudhomme, 1736), 2:796.

20. Loisel, *Histoire des ménageries,* 2:141, 298–99.

21. *Journal inédit du duc de Croÿ, 1718–1784,* ed. le vicomte de Grouchy and Paul Cottin (Paris: Flammarion, 1906–7), 1:135, 148, 264.

22. A ninth painting, *Chinese Hunt,* which was displayed for only three years, showed a variety of large cats and an African hyena being attacked by a group of Chinese men. See Xavier Salmon, *Versailles: Les chasses exotiques de Louis XV* (Amiens: Musée de Picardie; Versailles: Musée national des châteaux de Versailles et de Trianon, 1996). As Salmon points out, a few of the artists may have gone to the menagerie to do sketches, but Boucher copied a painting by Oudry of the tiger at the menagerie, and Charles Parrocel's elephant was probably modeled on sixteenth-century Italian works (44–45, 70, 105–6).

23. *Mémoires du duc de Luynes sur la cour de Louis XV, 1735–1758* (Paris: Firmin-Didot frères, 1862), 10:317. The duke simply described the bird—the size of a turkey, with a hooked beak and fire-red head; Loisel identified it as a condor (*Histoire des ménageries,* 2:143). Either Nemeitz was wrong about the animals having been sold or the menagerie was restocked after the 1720s. A Swedish visitor who toured the menagerie in 1754 also left an inventory of its contents; see Loisel, *Histoire des ménageries,* 3:562–63.

24. [Antoine-Joseph Dezallier d'Argenville,] *Voyage pittoresque des environs de Paris . . .* (Paris: De Bure l'aîné, 1755), 123–24.

25. *Aff. de Prov.,* 18 June 1760 (no. 25), 99–100.

26. Jean-Bernard Lacroix, "L'approvisionnement des ménageries et les trans-

ports d'animaux sauvages par la Compagnie des Indes au XVIIIe siècle," *Revue française d'histoire d'outre-mer* 65 (1978): 154.

27. Ibid., 162 (tigers), 164–70 (rhinoceros). On the rhinoceros, see also L. C. Rookmaaker, "Histoire du rhinocéros de Versailles (1770–1793)," *Revue d'histoire des sciences* 36, nos. 3–4 (1983): 307–18; and T. H. Clarke, *The Rhinoceros from Dürer to Stubbs, 1515–1799* (London: Sotheby's, 1986), 69–70.

28. J.-B. Lacroix, "L'approvisionnement des ménageries," 174.

29. Loisel, *Histoire des ménageries,* 2:342–45.

30. Prince de Poix to Castries, 30 Oct. 1782, AN AJ/15/512, f. A 14, subf. "Ménagerie de Versailles," doc. 508.

31. *Mémoires de la baronne d'Oberkirch sur la cour de Louis XVI et la société française avant 1789,* ed. Suzanne Burkard (Paris: Mercure de France, 1970), 160–61; [Anna Francesca] Cradock, *Journal de Madame Cradock: Voyage en France (1783–1786),* trans. and ed. Mme O. Delphin Balleyguier (Paris: Perrin, 1896), 80.

32. See, e.g., "Du Tapir ou Maipouri, Addition de l'éditeur hollandais (M. le Professeur Allamand)," *HNS,* 6 (1782), 17–19; "Du Gnou ou Niou," ibid., 89; "Addition à l'histoire du Condoma ou Coësdoës par M. Le Professeur Allamand," ibid., 127–33.

33. Noailles [prince de Poix] to Castries, 17 Jan. 1783, and unaddressed note, AN Col C/2/285, f. "Animaux," 167–68.

34. Animal acquisitions: AN O/1/807, f. "Fournitures: Versailles. Pourvoierie, Menagerie. 1780–1785," subf. "Ménagerie 1785," doc. 17 (hyena and panther); f. "Fourniture: Ménagerie 1786," doc. 78 (badgers), 79 (*tigre, coaïta* [spider monkey?]), 80 (mandrill and barbary ape), 85 (ostriches and *pithèques* monkeys), 87 (hyena and jackal). The prince de Poix also asked Castries's help in acquiring animals from northern Africa. In October 1782 he asked for ostriches, a camel, a dromedary, a tiger, and a pair of lions from Algeria (prince de Poix to Castries, 30 Oct., 1782, AN AJ/15/512, f. A 14, subf. "Ménagerie de Versailles," doc. 508), and in 1783 he sent a wish list that included two pairs of ostriches, a jackal, a hyena, and two *babales* (probably bubal, or hartebeest, an African antelope) (prince de Poix to Castries, 9 Dec. 1783, ibid., doc. 509).

35. Buffon, "Le Zèbre," *HN,* 12 (1764), 1. There was also an attempt to get zebras in 1777: see Lacroix, "L'approvisionnement des ménageries," 176–77.

36. Castries to Souillac and Chevreau, 14 Feb. 1783, AN Col C/2/285, f. "Animaux," 165 (no. 122). When writing to the prince de Poix explaining that his request had been attended to, Castries made no mention of economizing (Castries to prince de Poix, [?] Feb. 1783, AN Col C/2/285, 164).

37. Figures for other countries during the same period include 334 Dutch, 84 English, and 62 American ships. George M'Call Theal, *History of South Africa under*

the *Administration of the Dutch East India Company (1652–1795)* (London: Swan Sonnenschein, 1897), 2:236.

38. Percheron to Souillac and Chevreau, [4?] Feb. 1784, AN Col C/2/285, f. "Animaux," 170. Another slightly different version of the letter is in AN Col C/5B/8, 382.

39. Percheron to Castries, 17 Feb. 1784, AN Col C/5B/7, no. 17 (another copy in AN Col C/5B/8, 388).

40. Castries to Percheron, 12 June 1784, AN Col C/5B/7, no. 34.

41. Percheron to Castries, 21 Dec. 1784, AN Col C/5B/8, 427. Percheron may have meant that the zebra had been kept in a stable by the Dutch for a year; according to his previous letters, he had it for only a couple of weeks before loading it on the ship.

42. Percheron to Souillac and Chevreau, 26 Jan. 1785, AN Col C/5B/8, 440; Souillac and Chevreau to Castries, 10 May 1785, AN Col C/2/285, f. "Animaux," 173. See also ibid., 171 (copy of part of letter from Percheron to Souillac and Chevreau, 16 Mar. 1784), for Percheron's explanation about the zebra's death.

43. Percheron to Souillac and Chevreau, 16 Mar. 1784, C/5B/8, 395. Copy of part of this letter in AN Col C/2/285, f. "Animaux," 171.

44. Souillac and Chevreau to Castries, 10 May 1785, AN Col C/2/285, f. "Animaux," 173, 2nd page of letter.

45. Percheron to Castries, 22 Mar. 1785, AN Col C/5B/7, no. 62.

46. Ibid.

47. Percheron to Castries, 26 Nov. 1785, AN Col C/5B/8, 484. On the governor's extravagant lifestyle, see Theal, *History of South Africa*, 2:226–29. The French may also have benefited from the prince of Orange's troubles: in a letter to Arnout Vosmaer, 25 Jan. 1786, the prince declared that for financial reasons he no longer planned to keep mammals at his menagerie: Florence Pieters, with Kees Rookmaker and Corinne Sieger, "La ménagerie du stathouder Guillaume V dans le domaine Het Kleine Loo à Voorburg," in *Een vorstelijke dierentuin / Le zoo du prince*, ed. B. C. Sliggers and A. A. Wertheim (n.p.: Walburg Instituut, 1994), 49.

48. Percheron to Castries, 25 Feb. 1786, AN Col C/2/285, f. "Animaux," 175; Percheron to Clouet, 25 Feb. 1786, AN Col C/5B/8, 510 (the meaning of the letter is not entirely clear).

49. Percheron to Castries, 25 Feb. 1786, AN Col C/2/285, f. "Animaux," 175; Percheron to Souillac, 6 Mar. 1786, AN Col C/5B/8, 514.

50. Extract of letter from Castries to prince de Poix, 14 Mar. 1786, AN Col C/2/285, f. "Animaux," 177; Castries to prince de Poix, 31 Mar. 1786, ibid., 183; prince de Poix to Castries, 18 June 1986, ibid., 178.

51. Lardy received 400 livres upon his return to Lorient. Baras received room and board, 3 livres per day, and 25 livres for the cost of his trip from Versailles to

his home in Marne. The coach fare was 90 livres, plus a tip of 6 livres for each coachman: order signed by Jean-Charles Clouet, king's councillor, 24 June 1786, AN Col C/2/285, f. "Animaux," 180–81. Percheron described the starling(s) (two were embarked, one arrived) as "extraordinary" in his letter to Castries, 25 Feb. 1786, ibid., 175; receipt signed by L'Aimant, 12 July 1786, ibid., 182. I do not know if the captain presented the animals to Castries. On use of the coach service, see J.-B. Lacroix, "L'approvisionnement des ménageries," 163.

52. Trublet to Monseigneur [Castries?], 11 May 1789, AN Col C/5B/6, no. 4. The cost of the zebra and its food was 689 livres, 15 sous, 6 deniers.

53. On historical accounts of the quagga, see P. Tuijn, "Historical Notes on the Quagga Comprising Some Remarks on Buffon-Editions Published in Holland," *Bijdragen tot de dierkunde* 36 (1966), 75–79. On the quagga's role in nineteenth-century hereditary theories, see Richard W. Burkhardt Jr., "Closing the Door on Lord Morton's Mare," *Studies in the History of Biology* 3 (1979), 1–21. The British expressed (belated) regret about their role in the disappearance of the quagga: Ritvo, *Animal Estate*, 283–84. See also Reay H. N. Smithers, *The Mammals of the Southern African Subregion* (Pretoria: U of Pretoria, 1983), 566–77, 696–98. The quagga is now classified as a subspecies of the plain's zebra; the Quagga Project at the South African Museum is trying to re-create the quagga (i.e., a genetically similar variety) through selective breeding.

54. Castries to Percheron, 30 June, 1787, AN Col C/5B/5 bis, no. 198 (1). Comparison with a draft of this letter (ibid., no. 198 [2]), which shows deletions and insertions, suggests that Castries may have vacillated about how strongly to stress economizing (e.g., "not always" in the quoted passage is an insertion).

55. Percheron to Castries, 28 July 1786, AN Col C/5B/7, no. 137.

56. Cassowary: Mistral to Monseigneur [Castries?], 11 July 1788, AN Col C/2/285, f. "Animaux," 190, and subsequent letters, 191–200. Serval: Poulletier to Monseigneur [Castries?], 25 May 1789, ibid., 202, and subsequent letters, 203–6.

57. Noailles [prince de Poix] to Castries, 5 Aug. 1787, AN Col C/2/285, f. "Animaux," 199.

58. Noailles [prince de Poix] to M. le comte de la Luzerne, 8 June 1789, ibid., 206.

59. *L'encyclopédie*, ed. Denis Diderot and Jean le Rond d'Alembert, 10 (Neuchâtel: Faulche, 1765), 330.

60. Loisel, *Histoire des ménageries*, 2:159; Henri F. Ellenberger, "The Mental Hospital and the Zoological Garden," in *Animals and Man in Historical Perspective*, ed. Joseph and Barrie Klaits (New York: Harper & Row, 1974), 64–65; Bob Mullan and Garry Marvin, *Zoo Culture* (London: Weidenfeld & Nicolson, 1987), 107.

61. Chevalier de Jaucourt, "Versailles," *L'encyclopédie*, 17 (1765), 162.

62. Steven L. Kaplan, *The Famine Plot Persuasion in Eighteenth-Century France*

(Philadelphia: American Philosophical Soc., 1982; orig. publ. in *Trans. Am. Phil. Soc.*, 72), esp. 47–51.

63. [Louis-Antoine de Caraccioli,] *Dictionnaire critique, pittoresque et sentencieux* (Lyon: Duplain, 1768), 2:48.

64. Francis Bacon, *Essays and New Atlantis* (Roslyn, N.Y.: Black, 1942), 291–92.

65. *A.-c.*, 7 Dec. 1767 (no. 49), 777–78 (emphasis in original).

66. [Louis-Sébastien Mercier,] *L'an deux mille quatre cent quarante* (London: n.p., 1772), 250 (chap. "Le Cabinet du Roi"). On this book, see Robert Darnton, *The Forbidden Best-Sellers of Pre-Revolutionary France* (New York: Norton, 1996), 115–36, and excerpts, 300–336.

67. *La cour de Louis XV: Journal de voyage du comte Joseph Teleki*, ed. Gabriel Tolnai (Paris: PUF, 1943), 124–25.

68. *Aff. de Prov.*, 18 June 1760 (no. 25), 99–100.

69. Louis-Sébastien Mercier, *Tableau de Paris*, quoted in Daniel Roche, *La culture des apparences* (Paris: Fayard, 1989), 137–38 (men's fashions); [Henri-Gabriel Duchesne and Pierre-Joseph Macquer,] *Manuel du naturaliste . . .*, 2nd ed. (Paris: Rémont, an V [1797]), 4:310–11 (zebra in Cabinet du roi).

70. In his first article on the zebra, Buffon contended that the animal could be domesticated and could eventually replace the horse: Buffon, "Le Zèbre," *HN*, 12 (1764), 9–10. In a subsequent article he reported the rumor about the Dutch: "Addition aux articles de l'Âne, tome IV; & du Zèbre, tome XII," *HNS*, 3 (1776), 53–54. He rescinded the rumor in a final article, "Du Czigitai, de l'Onagre & du Zèbre," *HNS*, 6 (1782), 40.

71. Duc de Croÿ, *Journal inédit*, 2:485–86. The duke called the bird a *Tatoua;* the editor guessed that it must be a toucan. Bertin was *secrétaire d'État*, 1763–80. The rhinoceros was from India but sailed via the Cape of Good Hope. Others who remarked on the rhinoceros are Cradock, *Voyage en France*, 80; J. A. Dulaure, *Nouvelle description des environs de Paris* (Paris: Lejay, 1786), 2:309; François Cognel, *La vie parisienne sous Louis XVI* (Paris: Calmann-Lévy, 1882), 61.

72. Duc de Croÿ, *Journal inédit*, 4:16.

73. Loisel, *Histoire des ménageries*, 2:149–50.

74. *Correspondance secrète, politique et littéraire . . .* (London: Adamson, 1787), 13: 384–86 (14 Nov. 1782).

75. For references in guidebooks see [George-Louis Le Rouge,] *Curiosités de Paris . . . et des environs*, new ed. (Paris, 1778), 2:222; Dulaure, *Nouvelle description des environs de Paris*, 2:307–9; [Pierre-Thomas-Nicolas] Hurtaut and [Pierre] Magny, *Dictionnaire historique de la ville de Paris et de ses environs* (Paris: Moutard, 1779), 3:525. Reference to a tour group from Brittany is in duc de Croÿ, *Journal inédit*, 4:16.

76. On the concierge, see Bachaumont, *Mémoires secrets*, 23 July 1782, cited in

Yves Laissus and Jean-Jacques Petter, *Les animaux du Muséum, 1793–1993* (Paris: Muséum national d'histoire naturelle, 1993), 78–79; *Mémoires de la baronne d'Oberkirch*, 161; Loisel, *Histoire des ménageries*, 2:140–41, 147–48. The Hungarian count Teleki wrote in 1761: "A *Suisse* served as my guide. These men are forbidden to accept tips, but when you offer them something they turn their heads and hold out their hands" (Teleki, *La cour de Louis XV*, 125).

77. Duc de Croÿ, *Journal inédit*, 4:216. Subsequent page references are indicated in text.

78. During his visit to the prince de Condé's menagerie at Chantilly, he also marveled at its cost (25,000 livres per year, including 15,000 livres just for food); ibid., 4:132.

79. Four years later, however, after touring Mauduyt's cabinet of expertly stuffed birds, he concluded that a cabinet was more desirable than a menagerie (ibid., 3:311).

80. *Page à la cour de Louis XVI: Souvenirs du comte d'Hézecques*, ed. Emmanuel Bourassin (n.p.: Tallandier, 1987), 103–4; [Edward Bennett,] *The Tower Menagerie* (London: Jennings, 1829), intro.

81. Mauduyt, *Encyc. méth. ois.*, 1 (1782), 428, 432, 433.

82. Animals in the *Histoire naturelle* observed at the Versailles menagerie: *HN: castor* (beaver), 8 (1760), 310; *panthère*, 9 (1761), 160, 163, 176; *caracal*, 9 (1761), 266; 12 (1764), 442, 450–51; *buffle* (African buffalo), 11 (1764), 332; *axis*, 11 (1764), 401; *zébu*, 11 (1764), 439, 441; *zèbre*, 12 (1764), 3, 9–10; *serval*, 13 (1765), 233–34, 236–37; *HNO: sacre* (saker falcon), 1 (1770), 248; *pauxi, ou pierre* (curassow), 2 (1771), 383; *oie à cravatte* (Canada goose), 9 (1783), 83; *HNS: daim* (fallow deer), 3 (1776), 124; *rhinocéros*, 3, 297; *guénon à crinière* (crowned guenon?), 7 (1789), 81. I have not listed the menagerie specimens whose skeletons provided material for Buffon and Daubenton (some of which had been first dissected by Perrault). For some species, Buffon or Daubenton simply noted having observed a living one but did not specify where; some of them may also have been at the menagerie. Jacques Roger discusses Buffon's observation of live animals in *Buffon* (Paris: Fayard, 1989), 279–80, 357–59.

83. Daubenton, "Description du Caracal," *HN*, 9 (1761), 266; Buffon, "Le Serval," *HN*, 13 (1765), 234; Daubenton, "Description du Serval," ibid., 237.

84. Buffon, "Le Tigre," *HN*, 9 (1761), 133; Daubenton, "Suite de la Description du Caracal," *HN*, 12 (1764), 442, 450–51 (at 451). Buffon compares this degeneration to that produced by climate ("Le Tigre," 133–34), which he claims elsewhere explains the low fertility of animals and humans in the New World ("Animaux communs aux deux continents," *HN*, 9 [1761], 103–14).

85. Buffon, "L'Outarde," *HNO*, 2 (1771), 3–4. Richard W. Burkhardt Jr. discusses this passage and the general issue of studying animal behavior in "L'idéolo-

gie de domestication chez Buffon et chez les éthologues modernes," in *Buffon 88*, ed. Jean Gayon (Paris: Vrin, 1992), 569–82, esp. 573.

86. For other passages where Buffon discusses the altered behavior of captive animals, see "Les Orang-outangs, ou le Pongo et le Jocko," *HN*, 14 (1766), 55; "Oiseaux étrangers qui ont rapport aux Vautours, IV: Le Marchand," *HNO*, 1 (1770), 183; and references in Roger, *Buffon*, 373–74.

87. E.g., see "La Poulette d'Eau," *HNO*, 8 (1781), 177–78.

88. Plans for a renovation of the Jardin du roi in 1776 (which was only partially completed) included buildings for housing wild quadrupeds and birds (Loisel, *Histoire des ménageries*, 2:316–17), and, in 1782, Bachaumont, in his *Mémoires secrets* (23 July 1782), reported on talk of moving the menagerie animals from Versailles to the Jardin du roi (as quoted in Laissus and Petter, *Les animaux du Muséum*, 78–79). Richard Burkhardt Jr. contrasts Buffon's criticism of menageries with his attempt to establish one at the Jardin du roi in "La ménagerie et la vie du Muséum," in *Le Muséum au premier siècle de son histoire*, ed. Claude Blanckaert et al. (Paris: Muséum national d'histoire naturelle, *Archives*, 1997), 483. A transcription of Verniquet's mémoire is in Loisel, *Histoire des ménageries*, 2:320–21.

Chapter 3. Fairs and Fights

1. Gustave Loisel, *Histoire des ménageries de l'antiquité à nos jours* (Paris: Doin & fils, 1912), 2:275–76.

2. *Alm. for.* (1775), 104–6.

3. Exotic and bizarre animals were also popular in other countries, especially those with active overseas trade. In Britain, as in France, showmen often claimed that their animals did not fit into known categories, and there was a similar mutual exchange between naturalists and the public: Harriet Ritvo, *The Platypus and the Mermaid and Other Figments of the Classifying Imagination* (Cambridge: Harvard UP, 1997).

4. Quoted in John Lough, *France on the Eve of Revolution: British Travellers' Observations, 1763–1788* (Chicago: Dorsey, 1987), 214. For more on popular entertainment and animal shows in Paris, see note on secondary sources.

5. J[oachim] C[hristoph] Nemeitz, *Séjour de Paris . . .* (Leiden: Van Abcoude, 1727), 170–71, 180 (quote at 170).

6. Quoted in Lough, *France on the Eve of Revolution*, 216.

7. Information on the animal fights is from *Aff. de Paris, Alm. for., J. de Paris*, and additional references listed in note on secondary sources. The fights were not, as Robert Isherwood states, always fights to the death (*Farce and Fantasy: Popular Entertainment in Eighteenth-Century Paris* [Oxford: Oxford UP, 1986], 210). An

editor's note in the *Alm. for.* suggests that the court may have had ready access to animal fights in the 1770s:

> The Italian Nicolet, who showed a large menagerie at last year's St. Germain fair, put on combats d'animaux at Versailles at the same time. But no doubt he could no longer split his energies between the spectacle at Versailles and that of the St. Germain fair, and he decided to choose for the former an assistant, or perhaps a successor, since a new person came to open the Arena. Here is what appears in the *Affiches de Versailles* for Monday 25 March 1776: "Sieur Lensemberg, a bourgeois from Vienna in Austria, will present today a Combat of all kinds of unrestrained animals, &c. A Neopolitan mule will fight *in the new style*. . . . Then a large bear will be made to fight *in the new style*, &c." (*Alm. for.* [1778], 175–76, ellipsis in original)

In addition, the 1783 competition sponsored by the Académie royale d'architecture was for a "menagerie enclosed within the park of a sovereign's chateau . . . [with] an open amphitheater and arena suitable for combats des animaux" (*J. de Paris*, 23 Aug. 1783 [no. 235], 970). For full text of proposal and choice of winner, see *Procès-verbaux de l'Académie royale d'architecture, 1671–1793*, ed. Henry Lemonnier (Paris: Colin, 1926), 9:100–102, 111–12.

8. Loisel, *Histoire des ménageries*, 2:282–83.

9. The *Alm. for.* ([1776], 150) gives the fees as 3 livres for loges and gallery, 48 sous for amphitheater and balcony; 24 sous for parterre. De Jeze (*État ou Tableau de la ville de Paris*, new ed. [Paris: Prault père, 1763], part 3, 14) gives slightly different prices: gallery and first loge: 3 livres; loges: 2 livres, 8 sous; amphitheater: 1 livre, 10 sous; parterre under the gallery: 15 sous. For comparison, prices at the Opéra ranged from 2 to 7 livres, at the Comédie française from 1 to 4 livres (de Jeze, *État de Paris*), and at the fair theaters from 6 to 24 sous (Martine de Rougement, *La vie théâtrale en France au XVIIIe siècle* [1988; repr. Geneva: Slatkine, 1996], 277).

10. On animals for sale in Marseille in the sixteenth century (from the travel narrative of Thomas Platter), see E.-T. Hamy, "Le commerce des animaux exotiques à Marseille à la fin du XVIe siècle," *Bulletin du Muséum d'histoire naturelle* 7 (1903): 316–18. See also chap. 1.

11. Émile Campardon, *Les spectacles de la foire* (Paris: Berger-Levrault, 1877), 1:306; Buffon, "Le Choras," *HNS*, 7 (1789), 44; "Addition aux articles du Boeuf, tome IV, du Bison, du Zébu, & du Buffle, tome XI," *HNS*, 3 (1776), 57; "Addition à l'article de la Brebis, vol. V; & à celui du Moufflon & des Brebis étrangères, vol. XI," ibid., 66; "La Genette," ibid., 238.

12. The rhinoceros that toured Europe from 1741 to 1758 was chaperoned by Douwe Mout, a sea captain from the Dutch East India Company (T. H. Clarke,

The Rhinoceros from Dürer to Stubbs, 1515–1799 [London: Sotheby's, 1986], 48–49), and some ocelots were shown by a M. l'Escot who brought them to the Saint Germain fair from Cartagena (Buffon, "L'Ocelot," *HN*, 13 [1765], 239, 243).

13. *Alm. for.* (1775), 102.

14. Campardon, *Spectacles de la foire*, 2:312–13.

15. *Journal inédit du duc de Croÿ, 1718–1784*, ed. le vicomte de Grouchy and Paul Cottin (Paris: Flammarion, 1906–7), 4:150–51.

16. Campardon, *Spectacles de la foire*, 1:303–7.

17. Ibid., 2:106–7.

18. Ibid., 1:192.

19. Duc de Croÿ, *Journal inédit*, 4:151.

20. Campardon, *Spectacles de la foire*, 1:306.

21. *Singulier: Aff. de Paris*, 23 Sept. 1751 (no. 38), 306; 18 Feb. 1754 (no. 14), 111; *très singuliers: J. de Paris*, 3 Feb. 1777 (no. 34), 4; *fort rare: Aff. de Paris*, 18 Feb. 1754 (no. 14), 111; *n'a point encore paru en Europe:* ibid., 19 Feb. 1770 (no. 15), 162. Éric Baratay and Élisabeth Hardouin-Fugier also stress the exhibitors' emphasis on the bizarre and sensational: *Zoos* (Paris: Découverte, 1998), 77–82.

22. Campardon, *Spectacles de la foire*, 2:312–13.

23. *Alm. for.* (1775), 102.

24. Campardon, *Spectacles de la foire*, 2:312–13 (rhinoceros); *Aff. de Paris*, quoted ibid., 2:219–20 (pelican).

25. *Aff. de Paris*, 28 Feb. 1752 (no. 17), 135.

26. E.g., the *subsilvania:* "the singularity of its hide cannot be described." Ibid., 19 Feb. 1770 (no. 15), 162.

27. Beard: the Vcrus bubalus, or bison, ibid., 3 Feb. 1752 (no. 10), 78; tree-climber: *Alm. for.* (1775), 105.

28. Rhinoceros: Campardon, *Spectacles de la foire*, 2:312–13; bear: *Alm. for.* (1778), 31. A livre is about 1.1 pounds.

29. *Alm. for.* (1777), 48.

30. *Aff. de Paris*, 18 Feb. 1754 (no. 14), 111 *(belzebut, satyre,* and *couxcousou);* 19 Feb. 1770 (no. 15), 162 *(subsilvania);* 23 Sept. 1751 (no. 38), 396 (Indies); 3 Feb. 1752 (no. 10), 78 (White Russia); 18 Feb. 1754 (no. 14), 111 (isle of Gorée); *J. de Paris*, 21 Oct. 1777 (no. 294), 2 (Amazon mountains).

31. *A.-c.*, 22 Mar. 1773 (no. 12), 179. *Les numéros parisiens* (Paris: Imprimerie de la Vérité, 1788) noted that on the boulevards "there are two or three menageries where the astonished Parisian can come to contemplate for 2 sous a living tiger and a leopard in large iron cages" (no. 35: Amusemens, spectacles, page 87?).

32. See anecdote from Foire Saint Laurent in *Alm. for.* (1773), n.p., about a supposed exotic animal kept in a double-walled iron cage in a dark corner. It growled so fiercely that people were afraid to get too close. It turned out to be a person, the

son of a roofer. Nemeitz mentions people being attracted into the *tente* of a man who said he was showing a live crocodile, only to find that it was stuffed (*Séjour à Paris*, 184). A remark in the *Alm. for.* ([1775], 103) suggests that caged animals cost more to see: it noted that sieur Fassi was showing a royal tiger in a cage "for public safety (or maybe so as to charge more)."

33. [Jean de] La Fontaine, "Le singe et le léopard," in *Oeuvres complètes*, 1: *Fables, contes, et nouvelles*, ed. Jean-Pierre Collinet ([Paris]: Gallimard, 1991), 351–52 (bk. 9, fable 3).

34. These issues are discussed by Harriet Ritvo, *The Animal Estate* (Cambridge: Harvard UP, 1987), intro. (esp. 15–30); Bob Mullan and Garry Marvin, *Zoo Culture* (London: Weidenfeld & Nicolson, 1987), chap. 1; and Arnold Arluke and Clinton R. Sanders, *Regarding Animals* (Philadelphia: Temple UP, 1996), 169–86. I am borrowing Ritvo's "bad" and "good" terminology.

35. *Aff. de Prov.*, 15 Mar. 1758 (no. 11), 44.

36. Buffon, "Les Kakatoës," *HNO*, 6 (1779), 91; Mauduyt, *Encyc. méth. ois.*, 2 (1784), 145. Mauduyt explained how the trick worked: the master would suggest answers to the question; when the right one came up, he would nod his head just a tad, and the cockatoos would mimic him.

37. Duc de Croÿ, *Journal inédit*, 2:269; *A.-c.*, 20 Oct. 1766, 663 (quotation). On snakes at the fair, see also *Alm. for.* (1775), 115–16; ibid. (1777), 184; ibid (1778), 37–38; [Henri-Gabriel Duchesne and Pierre-Joseph Macquer,] *Manuel du naturaliste* (Paris: Desprez, 1771), 150.

38. Duc de Croÿ, *Journal inédit*, 2:269–70; also *A.-c.*, 16 Feb. 1767, 107 (the seal's owner is described as a [male] *maître* in the latter account). The duke described another seal in 1779 (4:150–51), and Anna Cradock saw one in 1784 (*Journal de Madame Cradock: Voyage en France [1783–1786]*, trans. and ed. Mme O. Delphin Balleyguier [Paris: Perrin, 1896], 68–69). All were described as very tame and affectionate.

39. [Abbé Ladvocat,] *Lettre sur le rhinoceros, à M***, membre de la Société royale de Londres* (Paris: Thiboust, 1749), 6.

40. Comment from Jacques de Sève, who was trying to make an illustration of the animal, in Buffon, "Seconde addition à l'article du Glouton," *HNS*, 3 (1776), 247.

41. Buffon, "L'Hyene," *HNS*, 3 (1776), 234.

42. Buffon, "Addition aux articles du Boeuf . . . ," ibid., 57–58.

43. Abbé Sauri, *Précis d'histoire naturelle* (Paris: Author, 1778), 5:79 n.

44. Buffon, "Les Orang-outangs, ou le Pongo et le Jocko," *HN*, 14 (1766), 53–54.

45. See references above on cages; also sieur Padovani's wolf, chained with a double chain (*Alm. for.* [1778], 30); a bison "which no one has ever been able to

tame" (*Aff. de Paris*, 3 Feb. 1752 [no. 10], 78); and reference to a cage ensuring "la sûreté publique," *Alm. for.* (1773), Foire Saint Laurent, Anecdote (n.p.). Cf. Ritvo on the London Zoo (*Animal Estate*, 219), and Mullan and Marvin's discussion of signs in modern zoos reading "These animals are dangerous" (*Zoo Culture*, 4–5).

46. *Alm. for.* (1775), 103–4.

47. Baboon: Buffon, "Le Papion ou Babouin proprement dit," *HN*, 14 (1766), 134–35; other fierce monkey: Daubenton, "Description de l'Ouanderou," ibid., 174.

48. *Journal de Madame Cradock*, 21 (mix of religion and amusement), 31–32 (Pentecost [quote at 32]; journal dated 23 May 1784).

49. [De Jèze,] *État . . . de Paris*, new ed., part 3, 14.

50. Letter from de G.*** to the editors, *J. de Paris*, 19 Apr. 1781 (no. 109), 441–42 (at 441). See also J. A. Dulaure, *Nouvelle description des curiosités de Paris* (Paris: Lejay, 1785), 1:534; Thomas Pennant, *Tour on the Continent 1765*, ed. G. de Beer (London, 1948), 31, quoted in Lough, *France on the Eve of Revolution*, 219; *La cour de Louis XV: Journal de voyage du comte Joseph Teleki*, ed. Gabriel Tolnai (Paris: PUF, 1943), 95–96. On the *concert spirituel* see [Pierre-Thomas-Nicolas] Hurtaut and [Pierre] Magny, *Dictionnaire historique de la ville de Paris et de ses environs . . .* (Paris: Moutard, 1779), 2:529–31. Assumptions that concertgoers were quiet and polite are probably wrong; see James Johnson, *Listening in Paris* (Berkeley: U of California P, 1995).

51. *Aff. de Paris*, 15 May 1766 (no. 38), 403, emphasis in original. *Ourvary*, or *hourvari*, was a call hunters used to summon their dogs away from a false trail. See definition in L[ouis] Liger, *Amusemens de la campagne . . .* (Paris: Prudhomme, 1734), 2:478. Often instead of the hourvari the interlude was a *peccata*, or free-for-all featuring caparisoned asses.

52. *Alm. for.* (1776), 16, 51. The *bouledogue* became a *bouledogue anglais* in 1780, when France was at war against England; see *Aff. de Paris*, 11 Mar. 1780 (no. 71), 584 and following issues. Previous notices sometimes mentioned a *bouledogue anglois* as one of the combatants, but the elevated bulldog was not routinely described as English until after this date.

53. Bulldog killer: *Aff. de Paris*, 14 Aug. 1766 (no. 63), 668; Polish bear: *J. de Paris*, 25 Dec. 1782 (no. 359), 1466; tiger: *Aff. de Paris*, 31 Mar. 1768 (no. 26), 291; 19 May 1768 (no. 39), 448; 15 Aug. 1768 (no. 63), 707–8; another tiger: ibid., 3 Nov. 1782 (no. 307), 2548; leopard: ibid., 16 Apr. 1778 (no. 31), 327; mandrill: *J. de Paris*, 27 Oct. 1782 (no. 300), 1222; *Aff. de Paris*, 27 Oct. 1782 (no. 300), 2492. Exotic animals typically appeared in special shows, but their foreign origin was not specifically noted.

54. *Alm. for.* (1778), 166.

55. Xavier Salmon, *Versailles: Les chasses exotiques de Louis XV* (Amiens: Musée de Picardie; Versailles: Musée national des châteaux de Versailles et de Tri-

anon, 1996), 170–75. Salmon does not suggest that the artist, Jean-Jacques Bachelier, attended the combat d'animaux, but he does make that suggestion for Charles Vanloo and François Desportes, who also portrayed fights or animals that often appeared in fights (130–31).

56. *J. de Paris,* 19 April 1781 (no. 109), 441–42; ibid., 21 April 1781 (no. 111), 449 (quotation); *Alm. for.* (1786), 208; ibid. (1787), 161–63.

57. *Alm. for.* (1778), 166.

58. *Journal . . . du comte Joseph Teleki,* 95–96 (2 Feb. 1761). For other accounts, see Lough, *France on the Eve of Revolution,* 219–20.

59. Peter Burke, *Popular Culture in Early Modern Europe* (New York: Harper Torchbooks, 1978), 178–243. See also Esther Cohen, "Animals in Medieval Perceptions," in *Animals and Human Society,* ed. Aubrey Manning and James Serpell (London: Routledge, 1994), 59–80, esp. 65–68; Pierre Goubert and Daniel Roche, *Les français et l'Ancien Régime,* 2: *Culture et société,* 2nd ed. (Paris: Colin, 1991), 153–54; and Daniel Roche, *La France des Lumières* (Paris: Fayard, 1993), 601. David Garrioch notes that the animal fights were not a "community sport" since they were not organized by the people (*Neighborhood and Community in Paris, 1740–1790* [Cambridge: Cambridge UP, 1986], 194). Some participation could have occurred if people brought their own dogs to fight, however, as occurred in the early nineteenth century, at least (Victor Fournel, *Le vieux Paris* [Tours: Mame & fils, 1887], 454–55).

60. Robert Isherwood has emphasized the melding of social groups in eighteenth-century Paris, where everyone enjoyed the same kinds of farcical and fantastical spectacles; even so, the nature of the enjoyment probably differed among individuals and groups. See Isherwood, *Farce and Fantasy,* esp. 54–55.

61. *Alm. for.* (1778), 170. A silly anecdote in the *Alm. for.* also suggests differential treatment. A showman who had announced that people could pay on the way out if they were happy with what they had seen gave a private tour to a nobleman after a crier cleared out the menagerie by yelling, "Place à la Noblesse." On emerging, the nobleman said he had no money. An onlooker asked him if he had been happy with the animals, and he said "No." "Then you don't owe any money," replied the onlooker. The showman then changed his spiel. *Alm. for.* (1773), Foire Saint Ovide (n.p.). Isherwood does acknowledge differential seating, but thinks it unimportant (*Farce and Fantasy,* 250).

62. *Mémoires de la baronne d'Oberkirch sur la cour de Louis XVI et la société française avant 1789,* ed. Suzanne Burkard (Paris: Mercure de France, 1970), 347–48 (the event took place in 1774 at the country home of the duchesse de Bourbon).

63. *Journal Singe,* by M. Piaud, June 1776, no. 1 (London and Paris: Cailleau), 22. This was the only issue of this short-lived journal. *Procureurs* were often thought to be greedy and ruthless.

64. *Correspondance secrète, politique & littéraire* (London: Adamson, 1787), 1:306–13 (12 April 1775) (quotes at 307, 308); notice of play: ibid., 2:3 (24 June 1775).

65. Michael R. Lynn, "Enlightenment in the Republic of Science," Ph.D. diss., Univ. of Wisconsin–Madison, 1997.

66. "Unknown": *J. de Paris*, 21 Oct. 1777 (no. 294), 2; "so rare": sieur Ruggieri [Ruggery], "Avis au Public," doc. 17 in *Catalogue de curiosités*, BnF Estampes, Yd 904. English trans. in Isherwood, *Farce and Fantasy*, 46. See Richard D. Altick, *The Shows of London* (Cambridge: Harvard UP, 1978), 36–37, for descriptions of English exhibitors appealing to naturalists.

67. *Alm. for.* (1778), 30 (emphasis in original).

68. Ruggieri, "Avis au public." Ruggieri reported that the tapir had died on its way to Paris and appeared there stuffed; Buffon wrote that the animal "lived for only a short time in Paris" and that it was drawn from life ("Du Tapir ou Maipouri," *HNS*, 6 [1782], 2).

69. *Alm. for.* (1787), 130–31.

70. Duc de Croÿ, *Journal inédit*, 2:470.

71. Jacques Roger (*Buffon* [Paris: Fayard, 1989], 357–58) mentions that Buffon visited the Saint Germain fair to view exotic animals there, Richard Altick (*Shows of London*, 37) notes that Sir Hans Sloane did the same at the Bartholomew fair in London, and Harriet Ritvo discusses menageries and naturalists in *Platypus and Mermaid*, but, overall, stage animals have received relatively little attention in works about the development of natural history.

72. Buffon, "L'Ouanderou et le Lowando," *HN*, 14 (1766), 171; "Seconde addition à l'article du Glouton," *HNS*, 3 (1776), 244–47.

73. "Le Rhinocéros," *HN*, 11 (1764), 198; "L'Élan," *IIN*, 12 (1764), 109; "L'Ocelot," *HN*, 13 (1765), 239–40; "Les Orang-outangs," *HN*, 14 (1766), 52–54; "Le Papion ou Babouin proprement dit," ibid., 134–35; "Les Kakatoës," *HNO*, 6 (1779), 91; "Addition aux articles du Boeuf . . . ," *HNS*, 3 (1776), 57–58; "Addition à l'article de la Brebis, vol. V; & à celui du Moufflon & des Brebis étrangères," ibid., 66; "Addition aux articles de la Fouine . . . & de la Zibeline," ibid., 162; "Addition à l'article de l'Hyène, de la Civette & de la Genette . . . ," ibid., 234 (hyena), 237–38 (genet); "Addition à l'article de l'Éléphant . . . ," ibid., 292–95; "Le Phoque à ventre blanc. Seconde espece," *HNS*, 6 (1782), 310–12; "Du Tapir ou Maipouri," ibid., 2; "Le Choras," *HNS*, 7 (1789), 44; "La Guenon Couronnée," ibid., 61; "Addition à l'article de l'Élan," ibid., 318; "Nouvelle addition à l'article du Tigre," ibid., 227; "Le Porc-Épic de Malaca," ibid., 303. Buffon's illustrator Jacques de Sève drew the condor; Spinacuta asked that his name be mentioned in the volume, but I have not seen it in the supplements (the main article on the condor had been published in 1770). See anonymous notes (by abbé Bexon?), "Le condor," MNIIN MS 369, f. "Oiseaux."

74. Duc de Croÿ, *Journal inédit*, 2:269; Mauduyt, *Encyc. méth. ois.*, 1 (1782), 573 (probably Spinacuta's cassowary; when the bird died, Mauduyt took its body for his cabinet); Buffon, "L'Ocelot," *HN*, 13 (1765), 239–40; "Les Kakatoës," *HNO*, 6 (1779), 91; "Addition à l'article de l'Eléphant . . . ," *HNS*, 3 (1776), 292–95; "Le Phoque à ventre blanc. Seconde espece," *HNS*, 6 (1782), 310–12.

75. Buffon, "Le Lion," *HN*, 9 (1761), 17–22; Daubenton, "Description du Lion," ibid., 27–28; "Description de l'Ours," *HN*, 8 (1760), 263–64; Buffon, "Du Loup Noir," *HN*, 9 (1761), 362–63 (at 363); Daubenton, "Description d'un Loup Noir," ibid., 364–70.

76. *Alm. for.* (1777), 54.

77. Buffon, "L'Ocelot," *HN*, 13 (1765), 239–44 (at 239). The observation about the male's superiority is given in a quotation from a letter from M. l'Escot (the ocelots' owner) to a member of the Académie des sciences (243).

78. [Jacques Christophe] Valmont de Bomare, *Dictionnaire raisonné universel d'histoire naturelle*, ed. augmented by the author (Yverdon, 1768–69), 7:416–17; [Duchesne and Macquer,] *Manuel du naturaliste*, 350; Sauri, *Précis d'histoire naturelle*, 5:74. For information taken from an exhibitor, see also Buffon, "Le Phoque à Ventre Blanc. Seconde espèce," *HNS*, 6 (1782), 310–12.

79. Martin Lister, *A Journey to Paris in the Year 1698*, ed. Raymond Phineas Stearns (repr. Urbana: U of Illinois P, 1967), 182–83.

80. [François] Le Vaillant, *Voyage de Monsieur Le Vaillant dans l'intérieur de l'Afrique* . . . (Paris: Leroy, 1790), 1:42–43.

81. *J. de Paris*, 27 Feb. 1781 (no. 58), 232–33 (excerpt from letter reprinted in *Alm. for.* [1787], 60); [Louis-Sébastien Mercier,] *Tableau de Paris*, rev. ed. (Amsterdam, 1782–83), 3:35–36. The *tarlala* is also mentioned in *Correspondance secrète*, 11:122; and [Henri-Gabriel Duchesne and Pierre-Joseph Macquer,] *Manuel du naturaliste*, 2nd ed. (Paris: Rémont, an V [1797]), 3:219–20.

82. *A.-c.*, 22 Mar. 1773, 179–84 (at 182). See Isherwood's discussion of charlatanism, *Farce and Fantasy*, 46–48. Colin Jones identified a similar pattern in the realm of medicine: the *Affiches* criticized medical charlatans but disseminated information about their remedies anyway: "The Great Chain of Buying," *American Historical Review* 101 (1996): 13–40.

83. Arlette Farge, *Fragile Lives: Violence, Power and Solidarity in Eighteenth-Century Paris*, trans. Carol Shelton (Cambridge: Harvard UP, 1993), 226–29.

84. *Alm. for.* (1773), Foire Saint Laurent (n.p.).

85. Ibid. (1777), 200; this expression, still in use, means something like, "It takes one to know one."

86. *Parallèle de Paris et de Londres*, ed. Claude Bruneteau and Bernard Cottret (Paris: Didier Érudition, 1982), 104. Mercier compares Sunday revelry in Paris with the calmer and more edifying sabbath pursuits of Londoners.

87. Dulaure, *Nouvelle description des curiosités de Paris*, 1:534; *Les numéros parisiens* (Paris: Imprimerie de la Vérité, 1788), 79. See also Bertin, "Les combats de taureaux."

88. [Jacques Christophe] Valmont de Bomare, *Dictionnaire raisonné universel d'histoire naturelle*, 4th ed., "en Suisse, chez les libraires associés" (1780), 10:23 (s.v., "rhinoceros"). New material in this edition was indicated in brackets; this comment was inserted in a description of Roman animal displays and fights. The edition was probably a pirated one, however (the legitimate fourth edition appeared in 1791), and so it is possible that the inserted comments were not by Valmont de Bomare. See entry in list of primary materials.

89. *Alm. for.* (1775), 153. I do not know what "Nations" these might be.

90. *J. de Paris*, 19 Apr. 1781 (no. 109), 442.

91. Keith Thomas, *Man and the Natural World* (New York: Pantheon, 1983), 143–91, 287–300 (at 186). See also Ritvo, *Animal Estate*, chap. 3. On sentiment against cruelty to animals in France, see Hester Hastings, *Man and Beast in French Thought of the Eighteenth Century* (Baltimore: Johns Hopkins P, 1936), part 3; Maurice Agulhon, "Le sang des bêtes: Le problème de la protection des animaux en France au XIXème siècle," *Romantisme* 31 (1981): 81–109.

92. One of the classical authors often cited as an authority was Pliny the Elder. See Pliny, *Natural History*, trans. H. Rackham (Cambridge: Harvard UP, 1938–63), 3:53 (bk. 8, chap. 29).

93. Loisel, *Histoire des ménageries*, 1:217–20. This rhinoceros became the model for Dürer's famous woodcut.

94. On the reception at Versailles, see *Mémoires du duc de Luynes sur la cour de Louis XV, 1735–1758* (Paris: Firmin-Didot frères, 1862), 9:288. The duke doubted the validity of the rumor that the captain had asked 50,000 écus (150,000 livres) for the animal. The rumored asking price doubled to 300,000 livres in Edmond-Jean-François Barbier's account (*Journal anecdotique d'un parisien sous Louis XV, 1727 à 1751*, ed. Hubert Juin [Paris: Livre club du libraire, 1963], 227).

95. For transcription of handbill, see Campardon, *Spectacles de la foire*, 2:312–13; also Laurence Berrouet and Gilles Laurendon, *Métiers oubliés de Paris* (Paris: Parigramme, 1994), 104–5.

96. Barbier, *Journal anecdotique*, 227–28. The rhinoceros was also mentioned by other contemporary chroniclers; see Franklin, *La vie privée d'autrefois: Les animaux*, 2:137–38. Barbier recorded the fees as 3 livres for first rank places, 1 livre for second, and 12 sous for third, but Campardon lists them as 24, 12, and 6 sous; he also notes that servants were required to pay, whereas usually they were allowed in free with their masters (*Spectacles de la foire*, 2:314 [see also for dramatist]). On "rhinomania," see Clarke, *Rhinoceros*, 58.

97. Buffon, "Le Rhinocéros," *HN*, 11 (1764), 174–197; Daubenton, "Descrip-

tion du Rhinocéros," ibid., 198–203. Parsons's article appeared in the *Philosophical Transactions of the Royal Society*. Buffon also included, in a footnote (179–81), the observations of a M. de Mours on the 1749 rhinoceros—all straight from the handbill. On the 1739 rhinoceros and Parsons's description, see Clarke, *Rhinoceros*, 41–46; on Oudry's painting, ibid., 64–68. On Buffon's activities in 1749, see Roger, *Buffon*, 95–96, 115–17.

98. The two most general works of natural history available at the time, "Mémoires pour servir à l'histoire naturelle des animaux" (in *Mémoires de l'Académie royale des sciences* [The Hague: Gosse & Neaulme, 1731]) and abbé Noël-Antoine Pluche's *Spectacle de la nature* (Paris, 1732–50) did not mention rhinoceroses.

99. [Ladvocat,] *Lettre sur le rhinocéros*. Loisel, *Histoire des ménageries*, 2:279 reports (with no source) that one could buy Ladvocat's booklet from the rhinoceros's keeper at the booth (no price given), as well as the poster (30 sous). Loisel, ibid., cites the *Dictionnaire de Richelet* (s.v., "Rhinocérot" *[sic]*) for identity of the author; also see Ant.-Alex. Barbier, *Dictionnaire des ouvrages anoynymes*, 3rd ed. (Paris: Daffis, 1872–79), 1:815.

100. Valmont de Bomare, *Dictionnaire . . . d'histoire naturelle* (1768–1769), 10:17–34.

101. *A.-c.*, 14 Jan. 1771 (no. 2), 23–24 (male elephant); ibid., 22 Mar. 1773 (no. 12), 182 (female elephant). Loisel (*Histoire des ménageries*, 2:281–82) states (with no source) that Louis XVI bought a female elephant from the fair in 1774 or 1775. Although this elephant may have been shown at the fair for several years (the duc de Croÿ remarked in 1775 that he was going to "look again" at the female elephant he had seen in Paris; *Journal inédit*, 3:173), I do not think that it could have been acquired for the menagerie because then there would have been two elephants there (one arrived from India in 1772; see chap. 2), and the duc de Croÿ never mentioned more than one elephant during his visits to the menagerie.

102. *Aff. de Prov.*, 6 Feb. 1771 (no. 6), 23; *A.-c.*, 14 Jan. 1771 (no. 2), 24.

103. Buffon, "Addition à l'article de l'Éléphant," *HNS*, 3 (1776), 293–95.

104. Loisel, *Histoire des ménageries*, 2:281. The print was from a drawing by Watteau.

105. [Jean-Marie Marchand,] *Mémoires de l'éléphant, écrits sous sa dictée, et traduits de l'Indien par un Suisse* (Amsterdam; Paris: Costard, 1771), 8 (observing people), 18 (bending knees), 20 (reason), 70 (mating), 77 (simple ways).

106. *A.-c.*, 8 Apr. 1771 (no. 14), 222–23. *Aff. de Prov.* was more complimentary, describing the *Mémoires* as "amusing and even instructive" (3 Apr. 1771 [no. 14], 55).

Chapter 4. The Oiseleurs' Guild

1. AN Z/1E/1166, packet IV, f. 4, subf. 2, doc. 8, testimony of Jean Baptiste Gautreau. The documents regarding this fight derive from an inquiry following the complaint against their attackers made by Bertin, Adam, and René Bourrienne, the three jurors of the oiseleurs' guild. For all documents cited from Z/1E/1166 in this chapter, only the first two or sometimes three levels (through subfolder) had been numbered by an archivist; I assigned numbers to documents according to their order in the folder.

2. Ibid., testimonies of Pierre Bernard Thuillier and Philippe Carabin. Because of conflicting stories told by the witnesses, various details remain unclear, such as who started the fight.

3. Thuillier stated that "he had given 20 sous two different times to the jurors of the oiseleurs' guild, that they had requested it in return for the liberty to sell birds at his whim, . . . that his business was inconsequential, that he sold only linnets and sparrows." A Henry Thuillier was a master oiseleur in the early eighteenth century. Pierre Thuillier, if related, may have been keeping up a family tradition on the side. For mention of Henry Thuillier (or Tuillier), see AN AD/XI/22, f. E, doc. 7 (1708), 13–14; ibid., doc. 10 (1698), list of signatories; AN Z/1E/1166, packet II, f. 2, doc. 2 (1696), list of signatories.

4. The only treatment of the guild by a historian is in Alfred Franklin's encyclopedic work on eighteenth-century life and does not use archival sources. See Franklin, *La vie privée d'autrefois: Les animaux* (Paris: Plon, 1897–99), 2:223–54; and Franklin, *Dictionnaire historique des arts, métiers et professions* (Paris: H. Welter, 1906; repr., Marseilles: Laffitte, n.d.), s.v., "oiseliers." Contemporary dictionaries usually included an entry for every guild; see, e.g., [Philippe Macquer,] *Dictionnaire portatif des arts et métiers . . .*, 2 (Paris: Lacombe, 1766), s.v. "oiseleur." Sometimes the terms *oiseleur* and *oiselier* were differentiated, the former meaning bird catcher, the latter, bird seller (see Mauduyt, *Encyc. méth. ois.*, 2 [1784], 280–81). I use *oiseleur*, which was more common, for the guild masters.

5. See note on secondary sources for literature on guilds. Membership in guilds among all workers (including day laborers) has been estimated at about 44 percent around 1700 and 55 percent around 1780: Daniel Roche, *The People of Paris*, trans. Marie Evans and Gwynne Lewis (Berkeley: U of California P, 1987), 73.

6. The statutes of 1600 were revised in 1647 and again in 1697. The king confirmed the latter in 1698 *lettres-patentes*, but they were not registered by Parlement until 1712 because of complaints by a group of bourgeois who raised and sold canaries and did not want their business interfered with by the oiseleurs. Some eighteenth-century dictionaries stated that the guild operated according to the 1647 statutes, but official documents for most of the eighteenth century refer to the

1697 statutes, which are the basis of the summary I give. See AN AD/XI/22, f. E, doc. 2 (1600 statutes), doc. 6 (1647 statutes), doc. 10 (1697 statutes), doc. 11 (registration of 1697 statutes by Parlement). The 1600 and 1697 statutes are reprinted in René de Lespinasse, *Les métiers et corporations de la ville de Paris* (Paris: Imprimerie nationale, 1886–97), 3:538–45.

7. On conflicts between oiseleurs and other merchants—mostly money changers and goldsmiths—with shops on the pont au Change, see AN AD/XI/22, f. E, doc. 1, 4–7.

8. On the *bourgeois de Paris*, see Roland Mousnier, *Les institutions de la France sous la monarchie absolue, 1598–1789*, 1 (Paris: PUF, 1974), 188–93; Joseph di Corcia, "*Bourg, Bourgeois, Bourgeois de Paris* from the Eleventh to the Eighteenth Century," *Journal of Modern History* 50 (1978): 207–33; and Sarah Maza, "Luxury, Morality, and Social Change: Why There Was No Middle-Class Consciousness in Prerevolutionary France," *Journal of Modern History* 69 (1997): 199–229. Bearers of the title bourgeois de Paris had to pay certain taxes and were accorded many of the same privileges as nobles. According to di Corcia, they were generally expected not to engage in commerce. People who did not work or who practiced professions such as surgery and law were considered to rank as bourgeois. The line between bourgeois and *marchand* (merchant), however, was not well defined.

9. Mauduyt, *Encyc. méth. ois.*, 2 (1784), 281. An anonymous nineteenth-century traveler joked that "these bird sellers, no doubt, cried often, and heartily *vive le roi, vive la reine:* for frequent coronations would be destructive to the trade": "Modern Paris, Letter 10, 2 Sept., 1807," *Literary and Philosophical Repertory* (Middlebury, Vt.) 1, no. 5 (Oct. 1813), 321–35 (at 326–27, emphasis in original).

10. AN AD/XI/22, f. E, doc. 15; AN Z/1E/1166, packet II, f. 6, subf. 3, doc. 1. See *J. de Paris*, 7 Feb. 1779 (no. 38), 150 for reports on ordinance requiring oiseleurs to supply birds and on the mass marriage ceremony.

11. [Pierre-Jean-Baptiste Nougaret,] *Les sottises et les folies parisiennes . . .* (London: Duchesne, 1781), 103.

12. From Noël d'Argonne, *Mélanges d'histoire et de littérature* (1725), 2:46, cited in Franklin, *La vie privée d'autrefois: Les animaux*, 2:252. On kings' merchants and artisans, see François Olivier-Martin, *L'organisation corporative de la France d'ancien régime* (Paris: Sirey, 1939), 252–55. I have translated *"singes, guenons & guenuches"* as "monkeys"; at the time it would have meant large and small monkeys, with and without tails.

13. Brevet d'Oiseleurs du Roy pour les Srs. Chateau père et fils, 22 Apr. 1762, AN O/1/106, fols. 132v-133r; Brevet d'Oiseleur du Roi pour Gérard Auguste Bastriès, 23 May 1782, AN O/1/126, 177–78; Brevet d'Oiseleurs du Roi pour les Srs. Chateau père et fils, 27 Jan. 1786, AN O/1/128, 14–16. Bastriès and Ange Chateau must have served simultaneously; both identified themselves in notices in *Aff. de*

Paris as oiseleur or oiselier du roi; see, e.g., 26 Feb. 1787, supp. (no. 57 bis), 574 (Chateau); 4 Mar. 1787 (no. 63), 629 (Bastriès).

14. For new regulations, see "Arrest des juges en dernier ressort des Eaux et Forêts de France . . . portant réglement pour les Oiseleurs & les Pêcheurs, 3 Septembre 1776," BHVP, *cote* (shelfmark) 131618. Records (partial?) of people who registered as oiseleurs and pêcheurs from 1781 to 1789 are in AN Z/1E/218. On the 1776 reforms, see Étienne Martin-Saint-Léon, *Histoire des corporations de métiers depuis leurs origines jusqu'à leur suppression en 1791*, 4th ed. (Paris: PUF, 1941), 512–40 (includes list of reinstated guilds); Olivier-Martin, *L'organisation corporative*, 510–41; Steven Laurence Kaplan, "Social Classification and Representation in the Corporate World of Eighteenth-Century France: Turgot's 'Carnival,'" in *Work in France*, ed. Steven Laurence Kaplan and Cynthia J. Koepp (Ithaca, N.Y.: Cornell UP, 1986), 176–228.

15. AN Z/1E/1166, packet II, f. 7, subf. b, doc. 14. If agreed to, then, this request would have given the oiseleurs free rein for confiscations.

16. For controversies over birdcages, see AN AD/XI/22, f. E, doc. 8; for conflicts with *bimbelotiers, lunetiers,* and *miroitiers,* see AN Z/1E/1166, packet IV, f. 4, subf. 1, doc. 1. The ruckus fomented by Gabriel Adam is described in ibid., packet IV, f. 4, subf. 2, doc. 7.

17. Steven Laurence Kaplan, *The Bakers of Paris and the Bread Question, 1700–1775* (Durham, N.C.: Duke UP, 1996), 83, 618 n. 6. For data on more than one hundred oiseleurs, see apps. B and C in Louise E. Robbins, "Elephant Slaves and Pampered Parrots: Exotic Animals in Eighteenth-Century France," Ph.D. diss., Univ. of Wisconsin–Madison, 1998.

18. AN Z/1E/1166, packet III, f. 2.

19. On the intermingling of home and work, see Annik Pardailhé-Galabrun, *Birth of Intimacy: Privacy and Domestic Life in Early Modern Paris,* trans. Jocelyn Phelps (Cambridge: Polity, 1991), 112–16.

20. At a 1722 meeting, nineteen masters signed their names, one signed initials, and eight made a mark (AN AD/XI/22, f. E, doc. 4, 10–11). If illiterate masters were less likely to attend meetings, then this would be an overestimate of literacy. In 1700, 75 percent of male wage earners in corporations could sign their names; for all male wage earners, the figure was about 60 percent. Roche, *People of Paris,* 203–4.

21. Half of the lower-class households in Paris had gold watches, according to a sample from the 1780s: Cissie Fairchilds, "The Production and Marketing of Populuxe Goods in Eighteenth-Century Paris," in *Consumption and the World of Goods,* ed. John Brewer and Roy Porter (London: Routledge, 1993), 228–48.

22. AN Y/15295 (Châtelet de Paris; Commissaire Claude-Etienne Prestat):

"Scellé après le décès de S. Vallin, 12 juillet 1785." I have not found inventories for any other oiseleurs.

23. AN AD/XI/22, f. E, doc. 13; AN Z/1E/1160, packet I, f. 5, subf. 2, doc. 6 (4 July 1763). The actual sequence of events is not clear. A sentence from the maître des Eaux et Forêts dated 29 Apr. 1763 forbade Vallin's acceptance as a master and made Guillaume Carcel, the oiseleur who had proposed Vallin's acceptance, pay the cost of printing posters announcing the sentence. Then on 4 July, the guild apparently agreed, reluctantly, to accept Vallin, if he would donate the 300 livres he had promised. However, on 29 July, they requested annulment of the previous request and return of Vallin's money. I do not know the immediate outcome, but by 1768 Vallin was not only a master but a juror (see AN Z/1E/1160, packet II, f. 7, subf. b, doc. 15).

24. AN Z/1E/1160, packet II, f. 6, subf. 2; AN AD/XI/22, doc. 1, 9.

25. For 1722, see AN AD/XI/22, f. E, doc. 1, 10–13. For 1775, see AN Z/1E/1166, packet IV, f. 3, subf. a, doc. 6. Neither Baveux nor Chaumont had ever served as a juror as far as I know; Baveux's name does appear on lists of masters attending meetings, but Chaumont's name appears nowhere else in the records I have seen. In both years, the guild decided that masters should contribute money rather than birds, because the birds might die on the way to Reims, but in 1775 each master was to furnish one canary to be released along with the local birds.

26. AN AD/XI/22, f. E, doc. 4, 3–4. This document is undated, but Noel Panchardy's name appears on two other documents dated 1647 and 1653. See ibid., doc. 6, 8; doc. 7, 1–2.

27. Permission was granted to René Bourienne and Henry des Champs in 1726 and to Etienne l'Albertier and Antoine Briere in 1732 to go to Picardie to buy birds (AN AD/XI/22, f. E, doc. 4, 4–5), and in 1760, a curé in Soissons certified that Nicolas Surcout had bought eight dozen birds from various people (AN Z/1E/1166, packet II, f. 4, subf. 2). This seems to have been a young man's task (for approximate ages, see Robbins, "Elephant Slaves," app. B).

28. Racine: AN AD/XI/22, f. E, doc. 7, 9–10. Briere and Rossignol: AN Z/1E/1166, packet IV, f. 4, subf. 2, doc. 2.

29. AN AD/XI/22, f. E, doc. 7 (23 May 1703), 11–13. When the jurors came to seize the illegal merchandise, Laubet and his wife threw themselves on the cages (with *menaces & injures*), breaking one, and allowing some birds to escape.

30. AN AD/XI/22, f. E, doc. 4 (ordinance of 28 Feb. 1722), 5–12. Sieur Bourrienne (probably Jean) refused to sign "even though he knows how to write." The order decreed that if they did go out of town to buy birds (presumably from other than the marchand forains) then they must bring them to the jurors for inspection and equal allotment among the masters, who would pay them according to their costs and expenses.

31. AN AD/XI/22, f. E, doc. 2, 7–12 (quote at 11).

32. On the canary fad, see [J.-C.] Hervieux [de Chanteloup], *A New Treatise of Canary-Birds* . . . (London: Lintot, 1718), 138–46, 156 (chaps. 23, 25). The book, originally published in French in 1705, reported that wealthy people would pay up to 400 livres for a fancy pair (141), but that prices had come down in the past decade and that nobles had lost interest in breeding them. See also [Macquer,] *Dictionnaire portatif des arts et métiers*, 2: 305–12.

33. AN AD/XI/22, f. E, doc. 2, 7–12.

34. AN AD/XI/22, doc. 3 (29 Dec. 1696), 14–18. The merchants described the troublemakers as *particuliers, courtiers & mâquignons d'oiseaux* (private individuals, brokers, and shady bird dealers). Since they did not specifically name oiseleurs, and since the jurors had already been notified, it is unlikely (though possible) that these people were guild members. See also ibid., 7–13, 19–25 for more legal cases involving the canary merchants. The inn where they did business seems to be gone now, but there is still an alleyway there called *Passage de la Boule Blanche*.

35. Hervieux [de Chanteloup], *New Treatise of Canary-Birds*, 126–31 (chap. 21).

36. AN AD/XI/22, f. E, doc. 9 (9 Jan. 1692), doc. 11 (6 Sept. 1712).

37. AN Z/1E/1166, packet II, f. 4, subf. 1, doc. 3. See also ibid., docs. 1, 2.

38. AN AD/XI/22, f. E, doc. 3 (10 Sept. 1684 and 6 Nov. 1686), 8, 9. The merchant's name is given as Henry Gossemaire in 1684 and Henry Hostemaire in 1686, but it is most likely the same person.

39. Gustave Loisel, *Histoire des ménageries de l'antiquité à nos jours* (Paris: Doin & fils, 1912), 2:190–93. Full text of second letter reprinted on 2:348–49. For price of desks (*bureaux*, which averaged 30 to 50 livres in 1706), see Pardailhé-Galabrun, *Birth of Intimacy*, 111.

40. Mauduyt noted that "it is not rare in times of peace to find them at the oiseleurs' shops in Paris; I have seen them there often." (*Encyc. méth. ois.*, 1 [1782], 570). For advertisements, see *Aff. de Paris*, 3 Dec. 1772 (no. 94), 979; 18 June 1778 (no. 47), 923–24; 3 Jan. 1779 (no. 3), 22; 18 Mar. 1787 (no. 77), 772. They were also listed among the *oiseaux étrangers* (foreign birds) offered for sale by Ange-Auguste Chateau, oiselier du roi, in the *A.-c.*, 26 July 1773 (no. 30), 470.

41. *Aff. de Paris*, 8 July 1788 (no. 190), 1956.

42. Ibid., 24 June 1790 supp., 1825. This was indeed a sign of new times, for fixed prices had been illegal in the old system. See Fairchilds, "Production and Marketing," 238, 247 n. 73; and Martin-Saint-Léon, *Histoire des corporations de métiers*, 467–68.

43. See, e.g. *Aff. de Paris*, 12 Oct. 1772 (no. 80), 847–48 (Chateau); 18 June 1778 (no. 47), 923–24 (Le Goubey); 4 Mar. 1786 supp. (no. 63 bis), 572 (Bastriès); 18 Feb. 1788 (no. 49), 485 (Chateau fils).

44. Mauduyt, *Encyc. méth. ois.*, 1 (1782), 486–87 (*amazone, amazone à tête*

blanche); 553 *(caille à gorge blanche)*; 570 *(cardinal)*; 657–58 *(crik, crik poudré)*; ibid., 2 (1784), 77 *(grenadin, bengalis, sénégalis, veuves)*; 138–39 *(jaco)*; 162–63 *(lori à collier)*; 223–23 *(moineau du Brésil)*; 271–72 *(oiseau royal)*; 293–94 *(padda)*; 299–300 *(pape)*; 300–301 *(papegai à bandeau rouge, p. à tête & gorge bleu, p. à ventre pourpre)*; 305–6 *(peintades panachées)*; 320 *(perruche verte de Cayenne)*; 321–22 *(perriche à tête jaune, p. pavouane)*; 326–29 *(perroquet à tête grise, p. tapiré, p. maillé)*; 331–32 *(perruche à collier couleur de rose)*; 334–35 *(perruche à tête rouge)*; 510 *(veuve dominicaine)*. For translations of species' names, see Robbins, "Elephant Slaves," app. A.

45. See Mauduyt's comments in articles cited above (the multicolored parrot is the perroquet tapiré); also Buffon, "Le Tavoua," *HNO*, 6 (1779), 240–41, where Buffon noted that the oiseleurs esteemed these rare parrots from Guyana because they learned to speak even better than grey parrots.

46. For lists of species sold by Chateau, see *A.-c.*, 4 Jan. 1768 (no. 1), 10–13; 17 Oct. 1768 (no. 42), 661–62; 26 July 1773 (no. 30), 468–73; *Aff. de Prov.*, 27 Jan. 1768 (no. 4), 15; *Aff. de Paris*, 3 Dec. 1772 (no. 94), 979. The species list from the 26 July 1773 issue of *A.-c.* is given in Robbins, "Elephant Slaves," app. D.

47. Mauduyt, *Encyc. méth. ois.*, 2 (1784), 271–72.

48. On the menagerie at rue des Postes: *Aff. de Paris*, 19 Feb. 1786 (no. 50), 452; 22 Mar. 1786 supp. (no. 81 bis), 739; 13 Sept. 1787 (no. 256), 2539–40; 2 July 1789 (no. 183), 1958 (an ad putting the house and garden up for rent). For a few examples of sales of pheasant and partridge eggs, see, for Chateau père: ibid. 12 Oct. 1772 (no. 80), 847–48; 4 Apr. 1776 (no. 28), 326; for Bastriès: ibid., 26 Sept. 1776 (no. 76), 918; 15 Apr. 1787 (no. 105), 1069; for Chateau fils: ibid., 3 Feb. 1788 (no. 34), 349; 5 Feb. 1789 supp. (no. 36 bis), 343 and refs. to menagerie, above. In 1786, the governor of the Versailles menagerie paid Chateau 105 livres for transporting an African crested crane and a white peacock from Le Havre to Versailles: see quarterly expenses for the menagerie, Apr. 1786, AN O/1/807, f. "Fournitures: Ménagerie, 1786," doc. 80.

49. Capuchin monkey: *Aff. de Paris*, 27 Jan. 1787 (no. 27), 261; green monkey: 15 June 1782 (no. 166), 1395; *Mongous:* 6 Mar. 1784 supp. (no. 66 bis), 613; *nonne* (mona monkey): 18 June 1778 (no. 47), 923–24; *Magabe* (magabey monkey?): 3 Jan. 1779 (no. 3), 22; *Indiens:* 27 Dec. 1782 (no. 361), 2994. I have not listed references simply to "monkeys" for sale.

50. Buffon, "Second addition à l'article du Glouton," *HNS*, 3 (1776), 247. Buffon identifies the oiseleur as "Saint-louis," a name I have not run across in documents pertaining to the guild.

51. *A.-c.*, 4 Jan. 1768 (no. 1), 13 (English mouse); *Aff. de Paris*, 22 Mar. 1786 supp. (no. 81 bis), 739 (goats); 27 Dec. 1782 (no. 361), 2994 (curculio).

52. Birds that Mauduyt saw only *"chez un oiseleur"* include the *caille à gorge blanche*, which he bought—the oiseleur did not know where it came from (*Encycl.*

méth. ois., 1 [1782], 552), the *lori à collier* (2 [1784], 162), the *perroquet maillé* (2, 329), and the *veuve dominicaine* (2, 510).

53. Mauduyt, *Encyc. méth. ois.*, 1 (1782), 432 (advice on food); 2 (1784), 224 *(distingué)*, 293 (Java sparrow).

54. Buffon, "Le Serin des Canaries," *HNO*, 4 (1778), 35. Usually birds lay one egg per day and they hatch one day apart. When the eggs are removed from the nest, their development is slowed (fake eggs are usually substituted); if all are re-turned to the nest when the last one is laid, then they will hatch at the same time.

55. Guéneau de Montbeillard, "Le Sizerin," ibid., 218; "Le Ministre," ibid., 86–87; "Les Bengalis et les Sénégalis, &c.," ibid., 88, 91.

56. "Industrie: Oiseaux étrangers," *A.-c.*, 26 July 1773 (no. 30), 468–73 (at 468). This description may have been written by the editors from information sup-plied by Chateau.

57. Ibid., 4 Jan. 1768 (no. 1), 10–13.

58. AN Z/1E/1166, packet II, f. 7, subf. b, doc. 8 (9 Nov. 1750).

59. Ibid., doc. 13.

60. Stoll: AD/XI/22, f. E, doc. 12; "man in red jacket": AN Z/1E/1166, packet II, f. 7, subf. b, doc. 15.

61. Transcripts of this case, from the Archives des commissaires au Châtelet de Paris, no. 3776, are in Émile Campardon, *Les spectacles de la foire* (Paris: Berger-Levrault, 1877), 1:380–81. Gernovitch, along with someone named Jacob Egger, contested the seizure, and the oiseleurs requested that the appeal be heard by officials of Eaux et Forêts rather than the police: AN Z/1E/1166, packet II, f. 7, subf. c, doc. 2. I did not see any records concerning the outcome.

62. AN Z/1E/1166, packet II, f. 7, subf. b, doc. 4. The merchant, Jean Prevost, noted that he had received license *(lettre de maîtrise)* no. 203: if the numbering began with 1 and included only bird merchants (as seems likely), then it is an indi-cation of the expansion of this area of commerce.

63. *Aff. de Paris*, 7 Feb. 1783 (no. 38), 311; 30 Dec. 1784 supp. (no. 365 bis), 3458. From 1781 to 1788, Chavanet placed up to six advertisements per year. For dates of advertisements, see Robbins, "Elephant Slaves," app. E.

64. *Aff. de Paris*, 13 Jan. 1782 (no. 31), 243. The expensive parrot was advertised 8 Jan. 1784 (no. 8), 67; and again—with no reduction in price—30 Dec. 1784 supp. (no. 365 bis), 3458.

65. Ibid., 21 Aug. 1789 (no. 233), 2407 *(perruquier);* 11 Oct. 1788 supp. (no. 285 bis), 2836 *(chapelier)*.

Chapter 5. Pampered Parrots

1. For literature on pets, see note on secondary sources.

2. *Aff. de Paris*, 7 June 1783 (no. 158), 1387. For others, see 16 May 1781 (no. 136), 1136; 20 May 1785 (no. 140), 1348–49; 8 Jan. 1788 (no. 8), 68; 11 Apr. 1788 (no. 102), 1043; 8 May 1788 (no. 129), 1316; 27 May 1788 (no. 148), 1508; 17 May 1789 (no. 137), 1473.

3. I skimmed the *Affiches* for the years 1751, 1753, 1755, 1757, 1759, 1760, 1762, 1764, 1766, 1768, 1770, 1772, 1774, 1776, 1778, 1780 (Jan.–June), 1781–89, and 1790 (Jan.–Aug.). Some numbers were missing from volumes for 1783 (1st half) and 1784 (2nd, 3rd, 4th quarters). Altogether, I found 238 for-sale ads and 198 lost-and-found notices that mentioned exotic species. On the *Affiches* and using them as a source, see Jack R. Censer, *The French Press in the Age of Enlightenment* (London: Routledge, 1994), 54–86; Colin Jones, "The Great Chain of Buying," *American Historical Review* 101 (1996): 13–40.

4. *Aff. de Paris*, 11 Jan. 1789 (no. 11), 99.

5. Ibid., 28 Oct. 1788 (no. 302), 2993. Lost items were listed in the "Annonces" column beginning in 1777. The "Effets perdus et trouvés" column first appeared in 1779.

6. Description of play "Les petites affiches," by A. Plancher-Valcour, is in L.-Henry Lecomte, *Histoire des théâtres de Paris* (Paris: Daragon, 1908), 38–39. Satire is in *Correspondance secrète, politique & littéraire* (London: Adamson, 1787), 9:205 (19 Feb. 1780), emphasis in original, as part of a ten-page fake edition of the *Annonces, affiches & avis divers*, dated 31 Dec. 1779 (ibid., 9:201–11).

7. J. M. Bechstein, *The Natural History of Cage Birds*, new ed., trans. from German (London: Groombridge & Sons, 1837), 158–59; G. T. Dodwell, *Encyclopedia of Canaries* (Hong Kong: TFH, 1976), 8–16.

8. J.-C. Hervieux [de Chanteloup], *A New Treatise of Canary-Birds* (London: Lintot, 1718), 138–41 (prices); 156–57 (how many to keep). See "Hervieux de Chanteloup," in *Biographie universelle, ancienne et moderne: Supplément*, 67 (Paris: Michaud, 1840), 152.

9. See, e.g., *Aff. de Paris*, 19 Jan. 1778 (no. 6), 88 (aviary with thirty-eight canaries, fourteen goldfinches, and two linnets); 19 June 1784 supp. (no. 171 bis), 1616 (aviary with eighty canaries of different kinds).

10. Hervieux [de Chanteloup], *New Treatise of Canary-Birds*, 53–60. The other temperaments were mischievous, base, and merry (the best).

11. *Aff. de Paris*, 17 May 1789 (no. 137), 1473; 24 Dec. 1782 supp. (no. 358 bis), 2974. The latter bird was offered along with a rosewood cage with gold-plated hardware.

12. Buffon, "Le Serin des Canaries," *HNO*, 4 (1778), 3.

13. The *Aff. de Prov.* reported in 1773 on the arrival of a pair of Amazon parrots that were very devoted to each other and suggested that parrots might one day be bred in France (1 Sept. 1773 [no. 35], 139–40). The following year it noted that a baby parrot had been born to a pair owned by a *contrôleur général des fermes du roi,* and that a member of the Académie des sciences had held it in his hand (20 July 1774 [no. 29], 115; 3 Aug. 1774 [no. 31], 123). The *A.-c.* gave detailed instructions about how to encourage parrots to mate, from the daughter of someone who had successfully bred several (17 Sept. 1774 [no. 72], 2). Buffon repeated the instructions in "Le Perroquet," *HNO,* 6 (1779), 115; Mauduyt, however, doubted that parrots would ever breed in captivity: *Encyc. méth. ois.,* 2 (1784), 325.

14. During the same period, I found for-sale ads for 117 parrots and 55 parakeets. These numbers do not include advertisements from professional bird sellers, although some of the notices were probably placed by people who maintained informal bird-selling businesses. Because of some missing issues, this sample is not complete (see n. 3 above).

15. *A.-c.,* 4 Jan. 1768 (no. 1), 10.

16. I have translated *perroquet* as parrot and *perruche* as parakeet, although the latter word was sometimes used for female parrots. The two are distinguished primarily by size, parrots being larger; technically, however, parakeets (also called long-tailed parrots) are members of the parrot family. The "green parakeet" might be the green conure.

17. [François] Salerne, *L'histoire naturelle, éclaircie dans une de ses parties principales, l'ornithologie* (Paris: Debure père, 1767), 62.

18. *Aff. de Paris,* 8 Jan. 1784 (no. 8), 67; the same parrot was listed again a year later (30 Dec. 1784 supp. [no. 365 bis], 3458). For more on Chavanet, see chap. 4.

19. Ibid., 29 Nov. 1785 (no. 333), 3188 (French and Portuguese); 28 Mar. 1784 (no. 88), 824 (French and Spanish); 3 Sept. 1785 supp. (no. 246 bis), 2371 (French and Breton); 21 Nov. 1789 supp. (no. 325 bis), 3329 ("imitant le chien, le chat, la poule, &c."); 10 Apr. 1782 (no. 100 bis), 850 (imitates canary).

20. [Pierre-Joseph Buc'hoz,] *Traité de l'éducation des animaux qui servent d'amusement à l'homme . . .* (Paris: Lamy, 1780), 204–5. Buc'hoz probably took these instructions at least in part from a work he excerpted in another of his books, Pietro Olina, *Uccellierea . . .* (Rome: Fei, 1702). (Buffon cites Olina for the recommendation that parrots be taught after dinner in "Le Jaco ou Perroquet cendré," *HNO,* 6 [1779], 104.)

21. Mauduyt, *Encyc. méth. ois.,* 2 (1784), 145–46.

22. Buffon, "Le Perroquet," *HNO,* 6 (1779), 109–10.

23. *Mémoires de la baronne d'Oberkirch sur la cour de Louis XVI et la société française avant 1789,* ed. Suzanne Burkard (Paris: Mercure de France, 1970), 399–400.

24. Mauduyt, *Encyc. méth. ois.*, 1 (1782), 494 (where to keep macaw); ibid., 2 (1784), 144, 325 (chewing on wood); 300–301 (dull parrot).

25. [Pons Augustin Alletz,] *Histoire des singes, et autres animaux curieux* . . . (Paris: Duchesne, 1752), iv.

26. In "Le Sajou," *HN*, 15 (1767), 37–38, Buffon reported that young capuchin monkeys had been born to pairs owned by Mme de Pompadour at Versailles, to Réaumur, and to a Mme de Poursel. The *Journal de Paris* (24 Nov. 1778 [no. 328], 1317) reported that a pair of *mistikis* (small monkeys) from central America belonging to the marquis de Nesle had produced young. Hennebert and Beaurieu believed that both parrots and monkeys could be bred in Europe if kept in large, warm surroundings with their native vegetation: [Jean-Baptiste-François Hennebert and Gaspard Guillard de Beaurieu,] *Cours d'histoire naturelle* (Paris: Lacombe, 1770), 2:460–61 n. From the available evidence, however, it seems to have been unusual for people to keep more than one monkey.

27. Geneviève de Malboissière, *Une jeune fille au XVIIIe siècle*, ed. le comte de Luppé (Paris: Champion, 1925), 124 (19 July 1764).

28. *Aff. de Paris*, 23 Feb. 1788 (no. 54 bis), 540; 20 July 1789 (no. 201), 2136.

29. *Mémoires de la baronne d'Oberkirch*, 159–160.

30. A few ads listed oiseaux *du Bengale*, but these were most likely *bengalis*, West African finches (probably cordons bleus or other waxbills), not birds from Bengal. See *Aff. de Paris*, 16 Feb. 1782 (no. 47), 371; 19 Apr. 1788 (no. 110 bis), 1133; 3 Jan. 1790 (no. 3), 19; 24 June 1790 supp., 1825.

31. Ibid., 19 June 1760 (no. 47), 378.

32. Buffon, "Le Hamster," *HN*, 13 (1765), 117–25; "L'Unau et L'Aï," ibid., 47–48.

33. Buffon, "Le Rat de Madagascar," *HN*, 13 (1765), 149–50. There are no solid lines between domestic, tame, and wild animals; see Jean-Pierre Digard, "Les nouveaux animaux dénatures," *Études rurales* 129–30 (1993): 169–78; Juliet Clutton-Brock, "The Unnatural World: Behavioural Aspects of Humans and Animals in the Process of Domestication," in *Animals and Human Society*, ed. Aubrey Manning and James Serpell (London: Routledge, 1994), 23–35.

34. *Aff. de Paris*, 15 Oct. 1778 (no. 81), 1516.

35. Buc'hoz described how to set up an aviary in a garden and noted that "most great lords have [aviaries] built": [Pierre-Joseph Buc'hoz,] *Les amusemens innocens* . . . (Paris: Didot le jeune, 1774), 405–6 (at 405). The *A.-c.* also mentioned the popularity of aviaries (22 Oct. 1770 [no. 43], 676–79). See descriptions of aviaries in [Luc-Vincent] Thiéry, *Guide des amateurs et des étrangers voyageurs à Paris* (Paris: Hardouin & Gattey, 1787), 2:654, 676; François Cognel, *La vie parisienne sous Louis XVI* (Paris: Calmann-Lévy, 1882), 27; *Mémoires de la baronne d'Oberkirch*, 210–11.

On window aviaries, see Alfred Franklin, *La vie privée d'autrefois: Les animaux* (Paris: Plon, 1897–99), 2:181–82.

36. *Aff. de Paris*, 24 Nov. 1760 (no. 91), 718 (macaw cage); 31 Dec. 1759 (no. 101), 804 (fountain and reservoir for flowers); 9 Dec. 1762 (no. 95), 768 (fountain); 25 May 1778 (add. à la feuille no. 41), 777 (aviary on kennel); 19 Oct. 1772 (no. 82), 860 (cavalry figures); 17 Jan. 1789 (no. 17) 147 (monkey with cabin); 22 Mar. 1781 (no. 81 bis), 680 (72-livre cage); 16 Dec. 1776, supp. (no. 98 bis), 1150 (4,000-livre cage).

37. Mauduyt, *Encyc. méth. ois.*, 2 (1784), 144, 325 (shriek when restrained). Salerne (*Ornithologie*, 62) noted that nobles in Paris let their scarlet macaws fly about, because they were tame and would return home. Several lost-and-found notices indicated that the lost parrot had a few links of a chain attached to one foot: *Aff. de Paris*, 19 Dec. 1781 (no. 353), 2909; 24 Nov. 1787 (no. 328), 3220; 29 May 1790 (no. 149), 1501.

38. [Jacques-Christophe] Valmont de Bomare, *Dictionnaire raisonné universel d'histoire naturelle*, 4th ed. (1780), 8:360 (this was probably a pirated edition [see entry in list of primary materials], and so may include text from others besides Valmont de Bomare); Buffon, "Le Jaco ou Perroquet Cendré," *HNO*, 6 (1779), 105–8; "La Mone," *HN*, 14 (1766), 260.

39. M. Mugnerot, "Le danger de la liberté," *J. de Paris*, 3 Aug. 1786 (no. 215), 889–90; reprinted in *Correspondance secrète*, 13:317–20.

40. [Hennebert and Beaurieu,] *Cours d'histoire naturelle*, 3:303 (see a similar comment about canaries: 4:1–2). Louis-Sébastien Mercier, *Tableau de Paris*, rev. ed. (Amsterdam, 1782–83), 8:335–38 (at 336).

41. *Aff. de Paris*, 15 July 1782 (no. 196), 1636–37; 19 July 1782 (no. 200), 1669.

42. Daniel Roche, *The People of Paris*, trans. Marie Evans and Gwynne Lewis (Berkeley: U of California P, 1987); Annik Pardailhé-Galabrun, *The Birth of Intimacy*, trans. Jocelyn Phelps (Cambridge: Polity, 1991); Cissie Fairchilds, "The Production and Marketing of Populuxe Goods in Eighteenth-Century Paris," in *Consumption and the World of Goods*, ed. John Brewer and Roy Porter (London: Routledge, 1993), 228–48. In after-death inventories of lower-class households, Fairchilds found an increase in pet accoutrements (e.g., birdcages and dog beds) from 1.7 percent of households in 1725 to 6.3 percent in 1785 (table 11.1, 230). The low percentages could reflect low rates of pet ownership, or perhaps notaries did not record the presence of simple cages.

43. *J. de Paris*, 23 Mar. 1787 (no. 82), 354–55; 2 Apr. 1787 (no. 93), 407.

44. Ibid., 22 Aug. 1783 (no. 234), 966–67.

45. [Pierre-Joseph] Buc'hoz, *La nature considérée sous ses différents aspects, ou Journal des trois regnes de la nature . . .* (Paris), letter 7 (28 Feb. 1782), 71. Buc'hoz mentioned the pet monkey because it caught smallpox from the little girl.

46. Malboissière, *Jeune fille au dix-huitième siècle,* 59–60, 62–64 (Brunet; quote at 60); 122, 133, 135 (Malboissière's pets); 323 (caring for Adélaïde's birds).

47. [Jean-Jacques Fillassier,] *Eraste, ou L'ami de la jeunesse . . .* (Paris: Vincent, 1773), 2:515.

48. *J. de Paris,* 22 Aug. 1783 (no. 234), 966–67.

49. Ibid., 6 June 1780 (no. 158), 647. On the practice of skimming while doing household purchases, see Sarah C. Maza, *Servants and Masters in Eighteenth-Century France* (Princeton: Princeton UP, 1983), 101–2; Cissie Fairchilds, *Domestic Enemies: Servants and Their Masters in Old Regime France* (Baltimore: Johns Hopkins UP, 1984), 26–27. Robert Darnton explores an episode of cat killing by apprentices in "Workers Revolt: The Great Cat Massacre of Rue Saint Séverin," in *The Great Cat Massacre and Other Episodes in French Cultural History* (New York: Basic, 1984), 75–104.

50. *A.-c.,* 25 Aug. 1766 (no. 34), 537. For another ad, see *Aff. de Prov.,* 23 July 1766 (no. 30), 119.

51. *J. de Paris,* 28 Nov. 1781 (no. 332), 1336; 6 Jan. 1782 (no. 6), 22–23. Réaumur, Mauduyt, and other naturalists also engaged in controversies over how best to preserve specimens. See Paul Lawrence Farber, "The Development of Taxidermy and the History of Ornithology," *Isis* 68 (1977): 550–66; P. A. Morris, "An Historical Review of Bird Taxidermy in Britain," *Archives of Natural History* 20, no. 2 (1993): 241–55.

52. Keith Thomas, *Man and the Natural World* (New York: Pantheon, 1983); Hester Hastings, *Man and Beast in French Thought of the Eighteenth Century* (Baltimore: Johns Hopkins P, 1936).

53. Daniel Roche, *La France des Lumières* (Paris: Fayard, 1993), 568. The term *conspicuous consumption* was first widely used by Thorstein Veblen, *The Theory of the Leisure Class* (1899; Harmondsworth, U.K.: Penguin, 1979), chap. 4 (68–101); on pets in this context, see 139–46. On fancy breeds of dogs and cats in nineteenth-century Britain, see Harriet Ritvo, *The Animal Estate* (Cambridge: Harvard UP, 1987), chap. 2.

54. [Mercier,] *Tableau de Paris,* 1:8–9.

55. See Roche, *La France des Lumières,* 507–20, and additional references in note on secondary sources.

56. *J. de Paris,* 18 Nov. 1787 (no. 322), 1382. Bystanders told the victim, M. Lô Leu Lin Le Long, that the man in the macaw suit was either a dance teacher or a broker, and that the one in the zebra suit was either a hairdresser or a dentist.

57. This story appears in three slightly different versions in three different collections of anecdotes by Pierre-Jean-Baptiste Nougaret: *Les milles et une folies . . .* (Amsterdam; Paris: Veuve Duchesne, 1771), 4:237–242; *Tableau mouvant de Paris*

. . . (London: Hookham; Paris: Duchesne, 1787), 3:84–85; *Paris, ou Le rideau levé*
. . . (Paris: Author, an VIII), 1:69–70.

58. A book written specifically for women is [Pierre-Joseph] Buc'hoz, *Amusemens des dames dans les oiseaux de volière* . . . (Paris: Author, 1782). The metamorphosis story is in Nougaret, *Tableau mouvant de Paris*, 2:324. The parrot woman spent all day crying "Baisez, baisez, maîtresse." Hastings gives a number of examples of attacks on women and pets in *Man and Beast*, 210–16. Emma Spary discusses the element of gender in relation to birds and natural history in "Codes of Passion: Natural History Specimens as a Polite Language in Late 18th-Century France," in *Wissenschaft als kulturelle Praxis, 1750–1900*, ed. Hans Erich Bödeker, Peter Hanns Reill, and Jürgen Schlumbohm (Göttingen: Vandenhoeck & Ruprecht, 1999), 105–35. Associations between women and animals are also discussed in Kathleen Kete, *The Beast in the Boudoir: Petkeeping in Nineteenth-Century Paris* (Berkeley: U of California P, 1994); Moira Ferguson, *Animal Advocacy and Englishwomen, 1780–1900* (Ann Arbor: U of Michigan P, 1998); Barbara Gates, *Kindred Nature: Victorian and Edwardian Women Embrace the Living World* (Chicago: U of Chicago P, 1998).

59. For male pet owners, see examples in this chapter and Franklin, *La vie privée d'autrefois: Les animaux*, 2:179–99. Of lost-and-found notices listing a contact person's name, lost parrots notices were placed by twelve women and twenty-one men; lost parakeet notices by twenty-two women and forty-six men. These numbers should be interpreted with caution since notices could have been placed by a male relative of a female pet owner.

60. "Fable," in *Le perroquet* . . . (Frankfurt am Main: 1742), week 4, 49. For a bird as the lover's stand-in, see "En envoyant un serin," *J. de Paris*, 26 Dec. 1779 (no. 360), 1469; M. Cuinet d'Orbeil, "Romance à Mlle de . . . à l'occasion d'un serin qui vint deux fois se poser sur la fenêtre où elle étoit," ibid., 31 Aug. 1782 (no. 243), 991.

61. Denis Diderot, *The Salon of 1765*, in John Goodman, ed. and trans., *Diderot on Art* (New Haven: Yale UP, 1995), 1:99.

62. "Le pour et le contre," *Almanach des muses* (1776), 73–80.

63. *Correspondance secrète* 2: 188–89 (7 Oct. 1775).

64. [Louis-Antoine de Caraccioli,] *La critique des dames et des messieurs à leur toilette* (n.p.: n.d.), 1, 3–4; *Dictionnaire critique, pittoresque et sentencieux* . . . (Lyon: Duplain, 1768), 2:226 *(perroquet)*; 3:123 *(sapajou)*. See Daniel Roche (*La culture des apparences* [Paris: Fayard, 1989], 434–40) on criticism of fashion and luxury in the *Encyclopédie*. Morag Martin discusses Caraccioli in the context of the growing criticism of cosmetics: "Consuming Beauty: The Commerce of Cosmetics in France, 1750–1800," Ph.D. diss., Univ. of California, Irvine, 1999, chap. 5.

65. [Mercier,] *Tableau de Paris*, 6: 290–91. On black servants and comparisons

to animals, see Fairchilds, *Domestic Enemies*, 146–48, 158–59; and for an analysis of images of Africans as pets, Srinivas Aravamudan, *Tropicopolitans: Colonialism and Agency, 1688–1804* (Durham, N.C.: Duke UP, 1999), chap. 1. Harriet Ritvo pointed out to me that some of the criticism of exotic pets as being frivolous probably had to do with the novelty of recreational pet keeping.

66. *Correspondance secrète*, 1:242–43 (3 Mar. 1775) (242, mispaginated as 142).

67. Nougaret, *Tableau mouvant*, 85–88. Nougaret said that the story had been presented as a play by two or three public theaters in Paris under the titles *Général Jacot* and *Le singe et la perruche*. These plays are also mentioned in a journal apparently edited by Nougaret himself, *Alm. for.* (1787), 5–6, 48–49, where the latter is attributed to the playwright Beaunoir.

68. [François] Le Vaillant, *Voyage de Monsieur Le Vaillant dans l'intérieur de l'Afrique . . .* (Paris: Leroy, 1790), 1:329.

69. [Hennebert and Beaurieu,] *Cours d'histoire naturelle*, 4:2; Caraccioli, *Dictionnaire critique*, 3:145–46.

70. *J. de Paris*, 10 Aug. 1777 (no. 222), 1–2; abbé Fromageot, *Cours d'études des jeunes demoiselles . . .* (Paris: Vincent, 1772–75), 1:xliv.

71. See Jennifer Jones, "'The Taste for Fashion and Frivolity,'" Ph.D. diss., Princeton Univ., 1991; Jones, "*Coquettes* and *Grisettes:* Women Buying and Selling in Ancien Régime Paris," in *The Sex of Things*, ed. Victoria de Grazia, with Ellen Furlough (Berkeley: U of California P, 1996), 25–53; and additional references in note on secondary sources.

72. Sarah Maza, *Private Lives and Public Affairs: The Causes Célèbres of Prerevolutionary France* (Berkeley: U of California P, 1993), 167–211.

73. On *singerie*, see H. W. Janson, *Apes and Ape Lore in the Middle Ages and the Renaissance* (London: Warburg Inst., 1952). For an analysis of monkey symbolism in Japan, see Emiko Ohnuki-Tierney, *The Monkey as Mirror* (Princeton: Princeton UP, 1987).

74. [Jean-Baptiste-Louis Gresset,] *Vairvert ou Les voyages du perroquet de la visitation de Nevers* (The Hague: Niegard, 1734). The play was put on by the Théâtre Italien in October 1790, but it "was very badly received": *J. de Paris*, 13 Oct. 1790 (no. 286), 1166. See also Jean-François Counillon, "Le XVIIIe siècle de J.-B. Gresset et 'Ver-Vert,'" *Bulletin de la Société d'émulation du bourbonnais* 67 (1995): 369–86.

75. *J. de Paris*, 20 May 1781 (no. 140), 565. Reprinted in *Almanach des muses* (1782), 240. For other examples of parrots' mechanical imitation, see *J. de Paris*, 26 Aug. 1777 (no. 238), 3; and Daniel Mornet, *Les sciences de la nature en France, au XVIIIe siècle* (Paris: Colin, 1911), 197–98.

76. M. Bérenger, "Le singe et le petit-maître," *Almanach des muses* (1786), 93–96; *Mémoires de la baronne d'Oberkirch*, 478–79; *L'échappé du Palais, ou Le Général*

Jaquot perdu (n.p., n.d.) (BHVP cote 12497). "Vers 1771" (around 1771) is penciled on the title page.

77. Jorge Martinez-Contreras pointed out, for instance, concerning Buffon's observations of a Barbary ape, that "Buffon is more concerned here with tameness *[dressage]*, or its absence, than in the spontaneous behavior of the animal": "Des moeurs des singes: Buffon et ses contemporains," in *Buffon 88*, ed. Jean Gayon (Paris: Vrin, 1992), 557–68 (at 565). On the use of anecdotes about pets, including exotic ones, in Victorian women naturalists' writings, see Gates, *Kindred Nature*, 214–35. Domestication is a topic of interest to some biologists now, as well as to anthropologists and archaeologists (e.g., Juliet Clutton-Brock, *A Natural History of Domesticated Animals* [Cambridge: Cambridge UP, 1987]), but most students of animal behavior attempt to observe the animals in their native habitats, away from human influence.

78. Mauduyt mentioned owning a Jamaican troupial (*Encyc. méth. ois.*, 1 [1782], 431); Chinese pheasant (1:433); wood duck (*canard branchu*, 1:433–34, 565); "white-throated quail" of unknown origin (1:553); black-faced munia (ibid., 2 [1784], 139); "red and purple lory" (2:164); blue-headed parrot (2:300–301); dove from Guadeloupe (2:407–8); waxbill (2:437); and whydahs (2:508).

79. See list in Gustave Loisel, *Histoire des ménageries de l'antiquité à nos jours* (Paris: Doin et fils, 1912), 2:313, to which should be added *jean-le-blanc* (*HNO*, 1 [1770], 124–29); *autour* (ibid., 234–36); *cresserelle* (ibid., 284); *hibou ou moyen duc* (ibid., 352); *effraie* (ibid., 368–69); *torcol* (*HNO*, 7 [1780], 86); *oiseau royal* (ibid., 321–23); *grand pluvier* (*HNO*, 8 [1781], 112–13), *cravant* (*HNO*, 9 [1783], 90–91). See also Jacques Roger, *Buffon* (Paris: Fayard, 1989), 279–80, 358–59.

80. Buffon, "La Mangouste," *HN*, 13 (1765), 154 n.; "Le Coaita et L'Exquima," *HN*, 15 (1767), 16–17; Daubenton, "Description de L'Unau et L'Aï," *HN*, 13 (1765), 47–48; Buffon, "Le Hamster," *HN*, 13 (1765), 117–18; "Le Cabiai," *HN*, 12 (1764), 384; "Le Calao de Malabar," *HNO*, 7 (1780), 149; "L'Ara rouge," *HNO*, 6 (1779), 189–90. On Mme de Pompadour's will: Émile Campardon, *Madame de Pompadour et la cour de Louis XV* . . . (Paris: Plon, 1867), 277.

81. Buffon, "Le Hamster," *HN*, 13 (1765), 117–18 (Montmirail); "Le Cabiai," *HN*, 12 (1764), 384 (Bouillon).

82. Buffon, "Le Jaco ou Perroquet cendré," *HNO*, 6 (1779), 105–6, 108; "Les Criks, 1: Le Crik à tête & à gorge jaune," ibid., 223–24. On Bougot and his parrot, see Henri Nadault de Buffon, ed., *Buffon: Sa famille, ses collaborateurs et ses familiers* (Paris: Veuve Renouard, 1863), 405–15.

83. Buffon, "Le Kakatoës à Huppe Jaune," *HNO*, 6 (1779), 93–95.

84. Buffon, "De la Loutre," *HNS*, 6 (1782), 283–86; "Nouvelle addition à l'article du Kinkajou," *HNS*, 7 (1789), 230.

85. Mauduyt, *Encyc. méth. ois.*, 2 (1784), 145–46, emphasis in original. Mauduyt

is referring to John Locke's theories on the difference in sensory interpretation between humans and animals.

86. Mauduyt, *Encyc. méth. ois.*, 2 (1784), 139; Buffon, "La Mone," *HN*, 14 (1766), 259–60; "Le Surikate," *HNS*, 3 (1776), 171–73; "L'Ara vert," *HNO*, 6 (1779), 195–99 (at 198). The "green macaw" was probably a chestnut-fronted or blue-winged macaw.

87. Mauduyt, *Encyc. méth. ois.*, 2 (1784), 139; letter quoted in Buffon, "Le Jaco ou Perroquet cendré," *HNO*, 6 (1779), 105–6; Salerne, *Ornithologie*, 74. Also see Mauduyt, *Encyc. méth. ois.*, 2 (1784), 324.

88. Buffon, "Discours sur la nature des animaux," *HN*, 4 (1753), 83–85 (at 84).

89. Peter Walmsley, "Prince Maurice's Rational Parrot: Civil Discourse in Locke's *Essay*," *Eighteenth-Century Studies* 28 (1995): 413–25. This account of the talking parrot, which was an addition to the fourth edition of Locke's *Essay* (1700), was included in the early-eighteenth-century French translation. [John] Locke, *Essai philosophique concernant l'entendement humain . . .*, trans. Pierre Coste, 5th ed., ed. Emilienne Naert (Amsterdam: Schreuder & Mortier le jeune, 1755; repr., Paris: Vrin, 1972), 262–63 (bk. 2, chap. 27); see also editor's introduction.

90. Julien Offray de La Mettrie, *Man a Machine*, French/English ed., ed. Gertrude Carman Bussey (La Salle, Ill.: Open Court, 1912), 28–32 (French version).

91. Mme [Louise Florence Tardieu d'Esclavelles, marquise] d'Épinay, *Les conversations d'Émilie*, ed. Rosena Davison (1st ed., 1774, 2nd ed., 1782; repr., Oxford: Voltaire Foundn., 1996), 51–52 (monkeys v. humans, kindness to animals), 94–95 (parrots), 307 (children as monkeys).

92. Mauduyt, *Encyc. méth. ois.*, 2 (1784), 324.

93. Buffon, "Le Perroquet," *HNO*, 6 (1779), 69–70 (quotation); "Discours sur la nature des animaux," 70; "Nomenclature des singes," *HN*, 14 (1766), 40. On mimicry as a mechanical effect, see also "Discours sur la nature des animaux," 87–88.

94. Buffon, "Nomenclature des singes," 41 (quotation); "Le Perroquet," 72.

95. Buffon, "L'Agami," *HNO*, 4 (1778), 497–501; Mauduyt, *Ency. méth. ois.*, 1 (1782), 428, 471.

Chapter 6. Animals in Print

1. Literacy rates for wage earners belonging to *corporations* in 1780 were 80 percent for men and 65 percent for women (figures are based on ability to sign one's name on a legal document): Daniel Roche, *The People of Paris*, trans. Marie Evans and Gwynne Lewis (Berkeley: U. of California P, 1987), 199–206. For works on publishing and reading, see note on secondary sources.

2. Anne C. Vila, *Enlightenment and Pathology* (Baltimore: Johns Hopkins UP, 1998), 104–5, 264.

3. See, e.g., Claude Labrosse, *Lire au XVIIIe siècle: La nouvelle Héloïse et ses lecteurs* (Lyon: PU de Lyon; Paris: CNRS, 1985).

4. *Fiction* was not a category in the eighteenth century, but classifiers of knowledge distinguished works of imagination from works of science or history. In d'Alembert's classification, *poetry* includes novels, drama, and poems, as well as music and painting: Jean le Rond d'Alembert, *Preliminary Discourse to the Encyclopedia of Diderot*, trans. Richard N. Schwab (Indianapolis: Bobbs-Merrill, 1963), 143–57. Following standard practice, d'Alembert placed natural history under history, not philosophy (which included theology, grammar, physics, and zoology).

5. See note on secondary sources for works on medieval bestiaries and Renaissance emblem books. On Perrault, see Jacques Barchilon and Peter Flinders, *Charles Perrault* (Boston: Twayne, 1981). On La Fontaine, see Roger Duchêne, *La Fontaine*, new ed. (Paris: Fayard, 1995); Marc Fumaroli, *Le poète et le roi* (Paris: Fallois, 1997); and Marie-Odile Sweetser, *La Fontaine* (Boston: Twayne, 1987). Two editions of collected works contain important critical material: *Oeuvres: Sources et postérité d'Ésope à l'Oulipo*, ed. André Versaille (Brussels: Complexe, 1995); *Oeuvres complètes*, 1, ed. Jean-Pierre Collinet ([Paris]: Gallimard, 1991). Perrault and La Fontaine were on opposite sides of the so-called battle of the ancients and moderns; see Aurélia Gaillard, *Fables, mythes, contes* (Paris: Champion, 1996), 81–120.

6. Florian's "De la fable" (1792), even as it declares fables indefinable, nicely lays out a variety of definitions. Essay reprinted in Jean-Noël Pascal, ed., *Les fables de Florian* (Perpignan: PU de Perpignan, 1995), 39–48. The chevalier de Jaucourt defined *fabuliste* in Diderot and d'Alembert's *Encyclopédie* (6 [Paris, 1756], 352) as someone who "announces a moral truth under the veil of fiction."

7. La Fontaine, *Oeuvres complètes*, 1:443–44 (bk. 11, fable 9).

8. Patrick Dandrey, *La fabrique des Fables*, 2nd ed. (Paris: Klincksieck, 1992), esp. 69–70, 149–51, 216–17 (cicada). Dandrey describes the rhetorical figure present in a word like *singing* as a combination of metaphorical and metonymical syllepsis (in a syllepsis, literal and figurative meanings are both in action). For a contradictory view of La Fontaine's realism, see Jacques Moussarie, "Les *Fables* de La Fontaine, un malentendu tenace," in *Fables et fabulistes*, ed. Michel Bideaux et al. (Mont-de-Marsan: Éd. InterUniv., 1992), 35–60, esp. 44–46.

9. Harriet Ritvo, "Learning from Animals," *Children's Literature* 13 (1985): 72–93. According to Ritvo, natural history writing for children in England owed more to the bestiary tradition than to the fable tradition.

10. "La perruche et le perroquet," *Aff. de Paris*, 13 Jan. 1786 (no. 13), 110; "La mère, l'enfant et les sarigues," Pascal, ed., *Les fables de Florian*, 77, 217 (bk. 2, fable 1); "Le singe et le paresseux, ou l'unau," *Almanach des muses* (1781), 149–50.

11. Jean-Noël Pascal, *Les successeurs de La Fontaine au siècle des Lumières (1715–1815)* (New York: Lang, 1995), 79–88; crayfish fable (intended to be an updated version of La Fontaine's "L'écrevisse et sa fille") at 84–86.

12. Richer, as quoted in Pascal, ibid., 20.

13. *A.-c.,* 23 Aug. 1762 (no. 34), 542–43; *J. de Paris,* 2 Apr. 1783 (no. 92), 384. The trend toward realism in depiction of animals also occurred in the visual arts; see Masumi Iriye, "Le Vau's Menagerie and the Rise of the *Animalier,*" Ph.D. diss., Univ. of Michigan, 1994.

14. Jean-Jacques Rousseau, *Émile, ou De l'éducation* (1762; Paris: Garnier frères, 1964), 110–15 (bk. 2). The fierceness of this attack has been exaggerated. Although Rousseau did criticize fables, he thought that children should not read anything at all until they were twelve, and he praised La Fontaine as reading for adults. On *Émile* and its reception, see Jean Bloch, *Rousseauism and Education in Eighteenth-Century France* (Oxford: Voltaire Foundn., 1995).

15. Robert Granderoute, "La fable et La Fontaine dans la réflexion pédagogique de Fénelon à Rousseau," *Dix-huitième siècle* 13 (1981): 335–48 (quote from Picardet at 346). One treatise on education urged that students should learn facts about natural objects first, so that the truth would sink its roots into their minds before they were exposed to La Fontaine's fables, which they would then study as literature: [Gabriel-François Coyer,] *Plan d'éducation publique* (Paris: Duchesne, 1770), 112, 184–88. Daniel Mornet mentions that Picardet recommended Buffon for children two to four years old: *Les sciences de la nature en France, au XVIIIe siècle* (Paris: Colin, 1911), 218.

16. *Aff. de Prov.,* 7 Aug. 1782 (no. 32), 127.

17. [Nicolas] Grozelier, "Le perroquet puni," *Fables nouvelles, divisées en six livres . . .* (Paris: Desaint & Saillant, 1760), 2:246–48 (bk. 11, fable 13). Statement of goal at 2:ix–x. On Grozelier, see Pascal, *Les successeurs de La Fontaine,* 111–13.

18. Nicolas-[Edmé] Restif de la Bretonne, *Lettre d'un singe aux animaux de son espèce,* ed. Monique Lebailly (Levallois-Perret: Manya, 1990). This novelette was originally published in 1781 as the continuation of vol. 3 of *La découverte australe.* See Lebailly, pref. to *Lettre d'un singe;* J. Rives Childs, *Restif de la Bretonne* (Paris: Briffaut, 1949), 278–81.

19. *Journal inédit du duc de Croÿ, 1718–1784,* ed. le vicomte de Grouchy and Paul Cottin (Paris: Flammarion, 1906–7), 4:179–80. See another mention of Pluche's book in 1:264.

20. [Noël-Antoine Pluche,] *Le spectacle de la nature, ou Entretiens sur les particularités de l'histoire naturelle . . . ,* 8 vols. (Paris, 1739–1750). See Daniel Mornet, "Les enseignements des bibliothèques privées (1750–1780)," *Revue d'histoire littéraire de la France* 17 (1910): 460; Mornet, *Les sciences de la nature,* 8–12; Caroline V. Doane, "Un succès littéraire du XVIIIe siècle: *Le spectacle de la nature* de l'abbé

Pluche," Ph.D. diss., Univ. of Paris, 1957; Stéphane Pujol, "Science et sociabilité dans les dialogues de vulgarisation scientifique au 18e siècle (de Fontenelle à l'abbé Pluche)," in *Diffusion du savoir et affrontement des idées, 1600–1770* (Montbrison: Assoc. du Centre Culturel, Montbrison, 1993), 79–96; and Dennis Trinkle, "Noël-Antoine Pluche's *Le spectacle de la nature,*" *Studies on Voltaire and the Eighteenth Century* 358 (1997): 93–134. Mme de Genlis, a writer and governess of the duc d'Orléans's children, discovered natural history when the *Spectacle de la nature* was read aloud to her as she embroidered: *Mémoires inédits de Madame la comtesse de Genlis* (Paris: Ladvocat, 1825), 1:252–53.

21. On Réaumur, see Jean Torlais, *Réaumur: Un esprit encyclopédique en dehors de "l'Encyclopédie,"* rev. ed. (Paris: Blanchard, 1961). The *A.-c.* praised Réaumur for making the history of insects interesting and useful, when until now it "seemed to have been destined only to divert the idle hours *[amuser l'oisiveté]* of a few naturalists": 8 Mar. 1774 (no. 17), 6.

22. Censor's statement quoted in *Aff. de Paris,* 9 Aug. 1764 supp., 572.

23. Éric Baratay, *L'Église et l'animal (France, XVIIe–XXe siècle)* (Paris: Cerf, 1996). According to Baratay, the focus on insects represented a shift on the part of the Roman Catholic clergy, who distanced themselves from animals like the Lamb of God and the self-sacrificing pelican in favor of more machine-like insects after 1670 as part of a turn away from a mystical religion to one that favored rational proof of God's existence.

24. [Jean-Jacques] Fillassier, *Eraste, ou L'ami de la jeunesse* . . . (Paris: Vincent, 1773). For goals of the work, see 1:vii–viii. Like many authors of such works, Fillassier lifted paragraphs from a variety of sources: he noted that he had drawn from the best writers, sometimes using their very words, which educated readers would be sure to recognize (1:viii). *Aff. de Paris* (23 May 1784, 1369–70), *Aff. de Prov.* (14 July 1773 [no. 28], 109), and *J. de Paris* (29 Mar. 1784 [no. 89], 395–96) all praised *Eraste;* the latter called it a "classic" of "indispensable utility for the education of the young."

25. Fillassier, *Eraste,* 2:512–13. Buffon also reported that condors eat young children, citing Sloane, *Trans. phil.* ("Le Condor," *HNO,* 1 [1770], 190).

26. [Louis] Cotte, *Leçons élémentaires d'histoire naturelle* (Paris: Barbou, 1784), vi. Cotte stated that he used Pluche's work as a model (iv). On Cotte, a priest and member of the Académie des sciences, see Guy Pueyo, "Les deux vocations de Louis Cotte, prêtre et météorologiste," *Bulletin des académie et société lorraines des sciences* 33, no. 4 (1994): 205–12.

27. Cotte, *Leçons élémentaires,* 15–17. On the superiority of insects, see also Fillassier, *Eraste,* 2:518.

28. Cotte, *Leçons élémentaires,* 11; [François Alexandre Aubert de La Chesnaye-

Desbois,] *Dictionnaire raisonné et universel des animaux* . . . (Paris: Bauche, 1759), 2:541 (article *"insectes"*).

29. See *J. de Paris* 13 Aug. 1777 (no. 255), 3 (letter about a dog *susceptible de sensibilité*); 10 Aug. 1783 (no. 222), 918 (a swallow that steals mud from another swallow's nest). See note on secondary sources for literature on philosophical works on the nature of animals.

30. [Pons Augustin Alletz,] *Histoire des singes, et autres animaux curieux,* . . . (Paris: Duchesne, 1752). Arab and monkey story, 51; Alletz cites De Corneille le Bruyn, *Voyage en Egypte,* vol. 2.

31. Dictionaries and encyclopedias sprouted so abundantly that reviewers sometimes sighed at the appearance of new ones (e.g., review of *Dictionnaire portatif d'histoire naturelle* in *A.-c.,* 20 Dec. 1762 [no. 51], 825). See Pierre Rétat, "L'âge des dictionnaires," in Henri-Jean Martin and Roger Chartier, eds., *Histoire de l'édition française,* 2: *Le livre triomphant* (Paris: Promodis, 1984), 186–94.

32. [Mathurin Jacques] Brisson, *Ornithologie* . . . (Paris: Bauche, 1760), 1:xv.

33. [Henri-Gabriel Duchesne and Pierre-Joseph Macquer,] *Manuel du naturaliste* (Paris: Desprez, 1771) (quote at v).

34. Mornet, *Sciences de la nature,* 248. See Robert Darnton, *The Forbidden Best-Sellers of Pre-Revolutionary France* (New York: Norton, 1996), xvii–xviii, for discussion of problems with Mornet's sample; for biographical information on Valmont de Bomare, see P. Mirault, "Notice sur la personne et les travaux de M. Valmont de Bomare. . . ," *Magasin encyclopédique, ou Journal des sciences, des lettres, et des arts* 4 (1808): 329–41; *Dictionnaire critique de biographie et d'histoire, errata et supplément pour tous les dictionnaires historiques,* ed. A. Jal, 2nd ed. (Paris: Plon, 1872), 244–45 (s.v. "Bomare, Valmont de"); "Une autobiographie inédite de Valmont de Bomare, publiée et annotée par M. E.-T. Hamy," *Bulletin du Muséum d'histoire naturelle* 12 (1906), 4–7; John G. Burke, "Valmont de Bomare," *Dictionary of Scientific Biography,* 13 (New York: Scribner's, 1976), 565–66.

35. In the 1760s, the course cost 72 livres: 30 livres for the section on minerals, 24 for plants, and 18 for animals ([de Jèze,] *État ou Tableau de la ville de Paris,* new ed. [Paris: Prault père, 1763], 192–93). The duc de Croÿ first took the course in 1765. In 1772, he noted the presence of the duc de Chartres; in 1777 he encountered many *amateurs* in the "charming" cabinet; and in 1778 he took the bishop of Tournay and the prince de Salm with him to the last lecture on animals: *Journal inédit,* 2:213–14, 3:11, 311, 4:97.

36. Geneviève de Malboissière, *Une jeune fille au XVIIIe siècle,* ed. le comte de Luppé (Paris: Champion, 1925), 56 (letter 26, 17 Nov. 1763).

37. *J. de Paris,* 28 Nov. 1777 (no. 332), 3; 15 Nov. 1778 (no. 319), 1278.

38. Duc de Croÿ, *Journal inédit,* 2:213–14. The duke met Valmont de Bomare after he had cut out from a copy of the naturalist's dictionary all of the articles on

animals and pasted them together (in the original they were interspersed with articles on plants and minerals). He then showed his compilation to Valmont de Bomare, who suggested that the duke attend his course.

39. Francisco Pelayo and Marcelo Frías, "Antonio José Cavanilles y la Historia Natural francesca: del Curso de Valmont de Bomare a la Crítica del Método de A. L. de Jussieu," *Asclepio* 47, no. 1 (1995): 197–216; *Correspondance secrète*, 10:344–45 (2 Dec. 1780).

40. The *Journal encyclopédique* lauded the first edition of the dictionary for gathering together all of the "principal facts of natural history": 1 July 1764 (5, pt. 1), 67–87 (at 67), cont. 15 July 1764 (5, pt. 2), 40–61. For other positive reviews, see ibid., 15 Feb. 1768 (2, pt. 1), 20–31 (2nd ed.); *Aff. de Paris*, 29 Mar. 1764 (no. 26), 238–39; *Aff. de Prov.*, 28 Mar. 1764 (no. 13), 50; *A.-c.*, 30 Apr. 1764 (no. 18), 285–86; and *Aff. de Prov.*, 5 June 1782 (no. 23), 91. On counterfeit 1780 edition, see entry in list of primary material.

41. [Jacques Christophe] Valmont de Bomare, *Dictionnaire raisonné universel d'histoire naturelle*, rev. ed. (Yverdon, 1768–69), 1:xiii–ix.

42. Ibid., 1:xxii–xxiii.

43. Ibid., 1:504. This story also appeared in [La Chesnaye-Desbois,] *Dictionnaire . . . des animaux*, 1:192.

44. In 1786, when Valmont de Bomare moved to the estate of Chantilly north of Paris along with his cabinet, which had been purchased by the prince de Condé, the *J. de Paris* remarked, "It is well known that the works and the lessons of this *Professeur* have particularly contributed to spreading the taste for natural history among the gens du monde": *J. de Paris*, 21 Nov. 1786 (no. 325), 1346–47.

45. *Aff. de Paris*, 13 Apr. 1768 (no. 15), 58; *J. de Paris*, 24 Oct. 1791 (no. 297), 1210; duc de Croÿ, *Journal inédit*, 2:269, 470; *Journal encyclopédique* 1 July 1764 (5, pt. 1, 67–87) (long extract from the dictionary about how to arrange and label the items in a cabinet); Malboissière, *Jeune fille au XVIIIe siècle*, 225 (letter 202, 2 May 1765), 227 (letter 204, 8 May 1765).

46. *Histoire naturelle de Pline, traduite en françois* [by Poinsinet de Sivry] (Paris: Veuve Desaint, 1771–82).

47. Although dotted with a few passages accessible to the general reader, most of the articles described dissections of exotic animals from Louis XIV's menagerie, and the work was not widely distributed or read. Geoffrey Sutton (*Science for a Polite Society* [Boulder, Colo.: Westview, 1995], 126) claims that the *Mémoires* were a popular success, but there is no evidence that they reached much beyond the world of elite salons. For their reception there, see Erica Harth, *Cartesian Women* (Ithaca, N.Y.: Cornell UP, 1992), 98–106.

48. For a few references to this very popular moniker, see *J. de Paris*, 12 Oct. 1781 (no. 285), 1148; *Aff. de Paris*, 24 June 1782 (no. 175), 1471; ibid., 28 Sept. 1784

(no. 272), 2558; Le Brun, "Ode à Monsieur de Buffon," in *Oeuvres de Ponce Denis (Écouchard) Le Brun*, ed. P. L. Ginguené (Paris: Crapelet, 1811), 1:118. [Philippe-François-Nazaire] Fabre d'Églantine referred to Buffon as *le Pline de Montbar* (Buffon lived in Montbard) and claimed credit for coining the phrase (complaining that another author stole it from him): "L'étude de la nature: Poëme à Monsieur le comte de Buffon" (London: n.p., 1783), 2. Buffon was also (infrequently) called *l'Aristote moderne* (see [Jean-Baptiste-François Hennebert and Gaspard Guillard de Beaurieu,] *Cours d'histoire naturelle* [Paris: Lacombe, 1770], 2:284).

49. On criticism of Buffon, see Mornet, *Les sciences de la nature*, 108–32; Jacques Roger, *Buffon* (Paris: Fayard, 1989), 248–69, 442–56; and, for a collection of contemporary reviews, John Lyon and Phillip R. Sloan, eds. *From Natural History to the History of Nature: Readings from Buffon and His Critics* (Notre Dame, Ind.: U of Notre Dame P, 1981).

50. *Aff. de Prov.*, 27 May 1772 (no. 22), 85–86.

51. Mornet, *Les sciences de la nature*, 217–20. Other authors such as Beaurieu and Fillassier were also recommended for children.

52. Malboissière, *Jeune fille au XVIIIe siècle*, 240 (letter 218, 25 May 1765).

53. Jean-Marc Nattier, "Le portrait de M. Bonier de la Mosson dans son cabinet" (1746), reproduced in Pierre de Nolhac, *J.-M. Nattier* (Paris: Goupil, 1905), facing 106 (the portrait has sometimes mistakenly been identified as of Buffon himself). See Frank Bourdier, "L'extravagant cabinet de Bonnier," *Connaissance des arts* (Aug. 1959): 52–61.

54. The motto was changed to *Majestati Naturae par Ingenium* (A genius equal to the majesty of nature) after someone wrote underneath the original inscription "Qui trop embrasse mal étreint" (He who embraces too much grasps poorly): Roger, *Buffon*, 472. Mme la comtesse [Fanny] de Beauharnais, "Aux incrédules: Épître envoyée à M. le comte de Buffon," *J. de Paris*, 7 Nov. 1778 (no. 311), 1245; [François-Félix Nogaret,] *Apologie de mon goût: Épître en vers sur l'histoire naturelle* (Paris: Couturier, 1771), 33.

55. "Épitre à M. le comte François d'Hartig, chambellan de l'empereur, sur la mort de M. le comte de Buffon," in Henri Nadault de Buffon, ed., *Buffon: Sa famille, ses collaborateurs, et ses familiers* (Paris: Veuve Renouard, 1863), 144–48 (at 146) (originally published in *Mercure*, 31 May 1788).

56. Buffon, "Discours sur le style," in *Oeuvres philosophiques de Buffon*, ed. Jean Piveteau (Paris: PUF, 1954), 500–504. For works on Buffon's style, see note on secondary sources.

57. *Correspondance littéraire, philosophique et critique . . .* , ed. Maurice Tourneux (Paris: Garnier frères, 1878), 4:132 (letter of 15 Aug. 1759); Jean-Jacques Rousseau, letter to Du Peyrou, 4 Nov. 1764, quoted in Roger, *Buffon*, 458.

58. Marquis de Villette, "Les Epoques," *J. de Paris*, 7 Jan. 1782 (no. 7), 25. The

Journal encyclopédique opined that "there might be other naturalists as profound and knowledgeable as Buffon" but "no one since Pliny has painted nature with such bold strokes" (1 Sept. 1764 [6, pt. 2], 60). The chevalier de Cubières and Fanny de Beauharnais also praised his artistry, exclaiming, respectively, "What fire in his tableaux! / Under his daring touch / Beautiful nature is further embellished" (de Cubières, "Épitre à M. le comte François d'Hartig," 146); and "Armed with bold brushes / he develops, explains, animates / the great tableau of the universe" (de Beauharnais, "Aux incrédules," 1245).

59. [Suzanne Curchod] Necker, *Nouveaux mélanges extraits des manuscrits de Mme Necker* (Paris: Pougens, an X [1801]), 2:129.

60. *Mémoires inédits de Madame la comtesse de Genlis*, 8:171.

61. Necker, *Nouveaux mélanges*, 1:159 (sensory word), 2:294 (bringing animals closer); [Suzanne Curchod] Necker, *Mélanges extraits des manuscrits de Mme Necker* (Paris: Pougens, an VI [1798]), 1:71–72 (linking ideas). Necker's writings are full of interesting comments about Buffon, especially his ideas about writing, apparently garnered during her conversations with him. On Necker's *mélanges*, see Geneviève Soumoy-Thibert, "Les idées de Madame Necker," *Dix-huitième siècle* 21 (1989): 357–68; Dena Goodman, "Suzanne Necker's *Mélanges*," in *Going Public: Women and Publishing in Early Modern France*, ed. Elizabeth C. Goldsmith and Dena Goodman (Ithaca, N.Y.: Cornell UP, 1995), 210–23. On her relation with Buffon, see [Gabriel Paul Othenin de Cléron,] comte d'Haussonville, *Le salon de Madame Necker d'après des documents tirés des archives de Coppet* (Paris: Calmann-Lévy, 1900), 1:304–34; Roger, *Buffon*, 495–99.

62. On authors' lack of control of their own texts, see Adrian Johns, "Natural History as Print Culture," in *Cultures of Natural History*, ed. N. Jardine, J. A. Secord, and E. C. Spary (Cambridge: Cambridge UP, 1996), 106–24.

63. The sentence was and is often misquoted as "the horse is the most noble conquest of man." See Roger, introduction to *Époques de la nature*, cxv.

64. Roger, *Buffon*, 248, 298–99.

65. *Aff. de Prov.*, 1 Mar. 1769 (no. 9), 34. Grimm rated Daubenton as a better scientist than Buffon, predicted that his reputation would outlast Buffon's, and criticized readers who preferred amusement to instruction; see *Correspondance littéraire* 3:302–3 (1 Nov. 1756); 4:132 (15 Aug. 1759); 6:22 23 (1 July 1764). On the differences between Buffon's *histoires* and Daubenton's *descriptions*, see Paul Lawrence Farber, "Buffon and Daubenton: Divergent Traditions within the *Histoire naturelle*," *Isis* 66 (1975): 63–74; Denis Reynaud, "Pour une théorie de la description au 18e siècle," *Dix-huitième siècle* 22 (1990): 347–66.

66. See note on secondary sources for references to the publishing history of the *Histoire naturelle*.

67. Georges Heilbrun, "Essai de bibliographie," in *Buffon*, ed. Muséum na-

tional d'histoire naturelle (Paris: Publications françaises, 1952), 225. For a discussion of such extracts of Rousseau's work, see Labrosse, *Lire au XVIIIe siècle*, chap. 5.

68. [Giovanni Ferry de Saint-Constant,] *Génie de M. de Buffon, par M**** (Paris: Panckoucke, 1778) (quotes at xiii, 221).

69. *Gazette du commerce*, 12 Oct. 1770 (no. 82), 735.

70. [Antoine-Hubert] Wandelaincourt, *Plan d'éducation publique* (Paris: Durand, 1777) (quotes at 22, 70) (Wandelaincourt was making a pitch for his own book, *Cours abrégé d'histoire naturelle* [Verdon: Mondon, 1778]); M. A. T. Chevignard de la Pallue, *Idée du monde, ouvrage curieux & interessant . . .*, 3rd ed. (Paris: Briand, 1788), 1:xvi–xvii (at xvii).

71. [Alexandre] Savérien, *Histoire des progrès de l'esprit humain dans les sciences et dans les arts qui en dépendent: Histoire naturelle* (Paris: Humblot, 1778); [Hennebert and Beaurieu,] *Cours d'histoire naturelle.*

72. Abbé Sauri, *Précis d'histoire naturelle . . .* (Paris: Author, 1778). See review in *J. de Paris*, 4 Dec. 1778 (no. 338), 1363–64 (1364 is mispaginated as 1366), which recommended Sauri's work as a more elementary and affordable natural history than Buffon's.

73. Ant[oine] Nic[olas] Duchesne and Aug[uste] Sav[inien] Le Blond, *Portefeuille des enfans . . .* (Paris: Mérigot, 1783–97); [Pierre-Joseph] Buc'hoz, *Histoire générale et économique des trois regnes de la nature* (Paris: Author, 1781), 8. Without liberal borrowing, often from himself, Buc'hoz would have had difficulty publishing the more than three hundred books credited to him. In a review of *Histoire naturelle des animaux, des végétaux, & des minéraux*, the *Correspondance secrète* (8:426 [6 Nov. 1779]) remarked that the "infatigable Docteur Buchoz," who had achieved celebrity through excessive mediocrity, had put nothing of his own in the book except for the wages of the engravers. Buc'hoz (1731–1807), physician to the king of Poland and later *médecin de Monsieur*, deserves more study. See Jean Sgard, ed., *Dictionnaire des journalistes (1600–1789)* (Grenoble: PU de Grenoble, 1976), 62–64.

74. Nicolas-Edmé Restif de la Bretonne, *La découverte australe par un homme-volant, ou Le dédale français* (1781; repr. Paris: France Adel, 1977). For discussion of the work, see Jacques Lacarriere's preface to reprint; Childs, *Restif de la Bretonne*, 26–28, 278–81; Charles A. Porter, *Restif's Novels . . .* (New Haven: Yale UP, 1967), 325–44; Pierre Testud, *Rétif de la Bretonne et la création littéraire* (Geneva: Droz, 1977), 213–19. See also Mark Poster, *The Utopian Thought of Restif de la Bretonne* (New York: New York UP, 1971). Patrick Graille discusses Restif's interest in hybrids in "Portrait scientifique et littéraire de l'hybride au siècle des Lumières," *Eighteenth-Century Life* 21, n.s. 2 (1997): 70–88.

75. *Aff. de Prov.*, 3 Mar. 1781 (no. 11), 43.

76. Charles E. Raven, *English Naturalists from Neckham to Ray* (Cambridge: Cambridge UP, 1947), 226. Éric Baratay identified 1670 as a turning point in writings by clerics, with earlier writers looking to literary sources for symbolic uses of animals, and later writers making more direct observations and emphasizing zoological facts: Baratay, *L'Église et l'animal.* William Ashworth documented a shift in bestiaries from 1550 to 1650 and suggested that later bestiaries presented bare facts about newly discovered species because these species had no traditional emblematic meanings: William B. Ashworth Jr., "Natural History and the Emblematic World View," in *Reappraisals of the Scientific Revolution,* ed. David C. Lindberg and Robert S. Westman (Cambridge: Cambridge UP, 1990), 303–32. Michel Foucault described an abrupt shift from an allegorical view of nature to a perspective centered on classification, and Erica Harth has claimed that science split off from fiction into a separate, objective realm at the end of the seventeenth century: Foucault, *The Order of Things* (New York: Vintage, 1973); Harth, *Ideology and Culture in Seventeenth-Century France* (Ithaca, N.Y.: Cornell UP, 1983). Erica Fudge, however, notes the persistence of fable in the new science: *Perceiving Animals* (New York: St. Martin's, 2000), chap. 4.

77. E. C. Spary, "The 'Nature' of Enlightenment," in *The Sciences in Enlightened Europe,* ed. William Clark, Jan Golinski, and Simon Schaffer (Chicago: U of Chicago P, 1999), 272–304 (at 296). Also see literature on popular science and on natural history in note on secondary sources.

78. John Bender contended that a separation of fact from fiction in the early eighteenth century was followed by a "doubly paradoxical twist" after midcentury, when fiction and science (with Buffon as the representative) each incorporated aspects of the other as a way to emphasize the difference between the genres. He focused on Buffon's use of hypothesis, however, not on his literary style or incorporation of moral lessons: "Enlightenment Fiction and the Scientific Hypothesis," *Representations* 61 (1998): 6–28. On the mingling of science and storytelling in Victorian women's writings about animals, see Barbara Gates, *Kindred Nature* (Chicago: U of Chicago P, 1998), 214–35. On moralizing in English natural history books for children, see Ritvo, "Learning from Animals."

79. J. A. Dulaure, *Nouvelle description des environs de Paris* (Paris: Lejay 1786), 1:69; Gustave Loisel, *Histoire des ménageries de l'antiquité à nos jours* (Paris: Doin & fils, 1912), 2:196–219; [François Cognel,] *La vie parisienne sous Louis XVI* (Paris: Calmann-Lévy, 1882), 104–7 (at 106–7).

80. *Aff. de Paris,* 2 May 1783 (no. 122), 1063. The seller is listed as "le sieur Desmoulins, Peintre, rue des Postes, la 3e porte-cochere à gauche par la rue de Fourcy." Nine years earlier, Desmoulins had offered for sale in the *Affiches* a cabinet containing 100 species of quadrupeds, birds, reptiles, and crustaceans, 300 insects, 300 shells and madrepores, 500 minerals, 200 specimens of wood, 200

seeds, roots, and resins, and 400 pressed plants, "all labeled with the greatest precision by an *homme de l'art.*" He gave his address as rue Ste. Croix de la Bretonnerie, vis-à-vis la rue du Puits (*Aff. de Paris*, 20 Oct. 1774 [no. 82], 860).

81. [Luc-Vincent] Thiéry, *Guide des amateurs et des étrangers voyageurs à Paris* (Paris: Hardouin & Gattey, 1787), 1:448–52 (at 452).

82. [Hennebert and Beaurieu,] *Cours d'histoire naturelle*, 1:iv, 2:320; *Aff. de Prov.*, 3 Oct. 1770 (no. 40), 158. The *Gazette du commerce* (2 Oct. 1770 [no. 82], 735–36) also gave it high praise, noting esp. that it avoided dryness by mixing in pleasant descriptions of rural pleasures.

83. Sauri, *Précis d'histoire naturelle*, 5:180; Buffon, "L'Éléphant," *HN*, 11 (1764), 2. See Harriet Ritvo, *The Animal Estate* (Cambridge: Harvard UP, 1987), 15–30, for discussion of "bad" and "good" animals in nineteenth-century English texts.

84. See D. G. Charlton, *New Images of the Natural in France* (Cambridge: Cambridge UP, 1984) and references therein.

85. *J. de Paris*, 10 Aug. 1783 (no. 222), 918.

86. Buffon, "L'Éléphant," *HN*, 11 (1764), 1–93; Buffon, "Le Tigre," *HN*, 9 (1761), 130–31.

87. Paul Sébillot, *Le folk-lore de France*, 3: *La faune et la flore* (Paris: Guilmoto, 1906), 72–73.

88. Buffon, "Discours sur la nature des animaux," *HN*, 4 (1753), 92, 94–95; "L'Éléphant," *HN*, 11 (1764), 10. On animals as moral exemplars, see also "Le Pigeon," *HNO*, 2 (1771), 523; and "Discours sur la nature des oiseaux," *HNO*, 1 (1770), 1: "This class of buoyant beings that Nature seems to have made with gaiety, can nevertheless be regarded as a tribe *[peuple]* that is serious, honest, and from which one has reason to draw moral fables and to borrow useful examples." On Buffon's moralizing, see also E. C. Spary, *Utopia's Garden: French Natural History from Old Regime to Revolution* (Chicago: U of Chicago P, 2000), chap. 3 (includes analysis of "L'histoire du lion").

89. Buffon, "Discours sur la nature des animaux," 109 (souls); ibid., 77–86 (feelings); "Du Chien," *HN*, 5 (1755) (sharing qualities). Buffon defined *sensation* as simply the agitation *(ébranlement)* of the senses, whereas *sentiment* referred to the subsequent feeling of pleasure or pain: "Les Animaux carnassiers," *HN*, 7 (1758), 11. On *sentiment*, the sentimental, and *sensibilité*, see David J. Denby, *Sentimental Narrative and the Social Order in France, 1760–1820* (Cambridge: Cambridge UP, 1994); Vila, *Enlightenment and Pathology*. On Buffon's dualism and ideas about *sentiment*, with implications for morality, see Phillip R. Sloan, "From Natural Law to Evolutionary Ethics in Enlightenment French Natural History," in *Biology and the Foundation of Ethics*, ed. Jane Maienschein and Michael Ruse (Cambridge: Cambridge UP, 1999), 52–83.

90. Buffon, "L'Éléphant," *HN*, 11 (1764), 1. Roger (*Buffon*, 370–71) suggests

that in later volumes Buffon became more prone to accord human-like traits to animals; e.g., in writing of the "intelligence" of certain species.

91. Buffon, "Premier discours de la manière d'étudier & de traiter l'Histoire Naturelle," *HN*, 1 (1749), 5–6.

92. Buffon, "Discours sur la nature des oiseaux," 68.

93. Necker, *Nouveaux mélanges*, 2:290. See also her comparison of the nightingale article to Buffon's article on the woodpecker, ibid., 2:137. Mme Necker's appreciation of his comparison of humans and animals was not without qualms: in one passage she declared the "pain" it caused her to see Buffon treat humans as animals and animals as machines (*Mélanges*, 3:318).

94. Necker, *Mélanges*, 2:339. For other examples that refer to Buffon, see *Mélanges* 1:185; 2:270; 3:138–39, 310; *Nouveaux mélanges* 1:182–83. For more general examples, see *Mélanges* 2:97–98; *Nouveaux mélanges* 1:201.

95. Buffon, "Le Cygne," *HNO*, 9 (1783), 26–28, emphasis in original.

96. "Épitre à M. le comte François d'Hartig, chambellan de l'empereur, sur la mort de M. le comte de Buffon," in Humbert-Bazile, *Buffon*, 146.

Chapter 7. Elephant Slaves

1. Buffon, "Le Cygne," *HNO*, 9 (1783), 1.

2. [Jean-Jacques Fillassier,] *Eraste, ou L'ami de la jeunesse* . . . (Paris: Vincent, 1773), 2:535; [Chambon,] *Commerce de l'Amérique par Marseille* . . . (Avignon: n.p., 1764), 2:148. English children's natural history books contained many of the same moral lessons concerning hierarchy and subordination: Harriet Ritvo, "Learning from Animals," *Children's Literature* 13 (1985): 72–93.

3. [Noël-Antoine Pluche,] *Le spectacle de la nature*, 1 (1732; repr., The Hague: Neaulme, 1739), 531–32; Buffon, "Les animaux domestiques," *HN*, 4 (1753), 170; [Jean-Baptiste-François Hennebert and Gaspard Guillard de Beaurieu,] *Cours d'histoire naturelle* . . . (Paris: Lacombe, 1770), 2:105.

4. Buffon, "Animaux du nouveau monde," *HN*, 9 (1761), 85–87: "man, in the savage state, is only a species of animal, incapable of governing others, and possessing only his individual faculties" (85); Buffon, *Les époques de la nature*, crit. ed. by Jacques Roger (Paris: Muséum national d'histoire naturelle, 1962), 217 (orig. publ. in *HNS*, 5 [1778]). See also [Guillaume Thomas François Raynal,] *Histoire philosophique et politique des établissements & du commerce des Européens dans les deux Indes*, "5th ed." (Maestricht: Dufour & Roux, 1777), 7:83 (bk. 18, chap. 18): "[In the New World], man had not subjected [animals] to his menacing voice, his terrible glance, his hand always ready to strike. He was himself a slave, and the animals were not yet. The king of nature thus knew servitude himself, before subjugating the animals."

5. For a few examples, see Buffon, "Le Zèbre," *HN*, 12 (1764), 10; "L'Agami," *HNO*, 4 (1778), 501; "De la Vigogne," *HNS*, 6 (1782), 208–12; Mauduyt, *Encyc. méth. ois.*, "Quatrième discours," 1 (1782), 427–34; *agami*, ibid., 471; *bengali*, ibid., 525; *canard huppé*, ibid., 564–65; 2 (1784), 273–75 (list of birds that should be domesticated); *ramiers*, 2:407–8; [Hennebert and Beaurieu,] *Cours d'histoire naturelle*, 2:376–77 (zebra), 2:425 (llama), 4:91–92 (colibri). This attitude continued into the nineteenth century and spawned the *Société zoologique d'acclimatation:* see Michael A. Osborne, *Nature, the Exotic, and the Science of French Colonialism* (Bloomington: Indiana UP, 1994).

6. More attacks continued after this wolf was shot: a second one was then killed, but it received less attention, and the culprit continued to be referred to in the singular, as the "bête du Gévaudan." The incident remains something of a mystery. See Richard Thompson, *Wolf-hunting in France in the Reign of Louis XV* (Lewiston: Mellen, 1991) and additional references listed in Éric Baratay and Jean-Luc Mayaud, "L'histoire de l'animal: Bibliographie," *Cahiers d'histoire* (1997, nos. 3–4): 443–80.

7. *Aff. de Paris*, 18 May 1772 (no. 39), 438.

8. *A.-c.*, 2 Dec. 1765 (no. 48), 751–53. The *Avant-coureur* followed the exploits of the beast and the hunt closely; it described the animal as a hyena escaped from a fair in the issue of 17 Dec. 1764 (no. 51), 803.

9. See George Boas, *The Happy Beast in French Thought of the Seventeenth Century* (Baltimore: Johns Hopkins P, 1933).

10. Buffon, "Les animaux domestiques," *HN*, 4 (1753), 169.

11. Buffon, "Le Cheval," *HN*, 4 (1753), 175; "Le Boeuf," ibid., 439; "Le Chameau et le Dromadaire," *HN*, 11 (1764), 228–29, 232; "Le Serin des Canaries," *HNO*, 4 (1778), 46–47; "Le Faucon," *HNO*, 1 (1770), 250–52.

12. Buffon, "De la dégénération des animaux," *HN*, 14 (1766), 317.

13. [Hennebert and Beaurieu,] *Cours d'histoire naturelle*, 1:395–97. For similar comments, see 2:1, 7, 10 (ox), 2:44–45 (sheep), 2:59–50 (goat), 2:350–52 (camel).

14. Buffon, "Le Boeuf," *HN*, 4 (1753), 440. On the cycle of life and death, ibid., 437–39; on the importance of meat for human health (at least in a temperate climate), "Les animaux carnassiers," *HN*, 7 (1758), 32–36.

15. Buffon, "Le Chameau et le Dromedaire," *HN*, 11 (1764), 228–32 (at 229).

16. Quoted in review of Rivarol, "De la nature et de l'homme," *J. de Paris*, 9 Sept. 1782 (no. 252), 1027.

17. Buffon, "Les animaux sauvages," *HN*, 6 (1756), 56.

18. Buffon, "L'Autruche," *HNO*, 1 (1770), 440.

19. [Hennebert and Beaurieu,] *Cours d'histoire naturelle*, 2:111–13.

20. Buffon, "Les Animaux Sauvages," *HN*, 6 (1756), 62.

21. [Pierre] Sonnerat, *Voyage à la Nouvelle Guinée* . . . (Paris: Ruault, 1776), 178–79.

22. [Alexandre] Savérien, *Histoire des progrès de l'esprit humain dans les sciences et dans les arts qui en dépendent: Histoire naturelle* (Paris: Humblot, 1778), 272; Nicolas-[Edmé] Restif de la Bretonne, *Lettre d'un singe aux animaux de son espèce* (1781), ed. Monique Lebailly (Levallois-Perret: Manya, 1990), 23; Buffon, "Le Serin des Canaries," *HNO*, 4 (1778), 46–47; [Hennebert and Beaurieu,] *Cours d'histoire naturelle*, 3:303 n. (parrot), 4:1–2.

23. Savérien, *Histoire des progrès de l'esprit humain*, 273; Buffon, "Le Hocco," *HNO*, 2 (1771), 380 (comparison of the pheasant to easily tamed curassow); "Le Canard," *HNO*, 9 (1783), 119.

24. [Foucher d'Obsonville,] *Essais philosophiques sur les moeurs de divers animaux étrangers* . . . (Paris: Couturier fils, 1783); this observation was mentioned in review in *J. de Paris*, 25 June 1783 (no. 176), 735.

25. Buffon, "La Macroule," *HNO*, 8 (1760), 220–21.

26. For instance, in a 1763 letter to his mother, Monsieur d'Éon (a man who masqueraded as a woman for much of his life) wrote that just as dogs and cats were "stronger, more vigorous, and more courageous in the forests than in our houses, losing these advantages when they were domesticated . . . so it is with man himself. In becoming social and the slave of the high-and-mighty, or those who ape greatness, he becomes weak, timid, and servile, while his lifestyle becomes soft and effeminate." Quoted in Gary Kates, *Monsieur d'Éon is a Woman* (New York: Basic, 1995), 148. As Kates points out, d'Éon was consciously echoing Rousseau.

27. Buffon, "Nomenclature des singes," *HN*, 14 (1766), 40; "Le Faucon," *HNO*, 1 (1770), 250–52.

28. Buffon, "L'Éléphant," *HN*, 11 (1764), 5, 16, 17 (quotes), 25–26 (description of character), 39 (scorn of Africans).

29. This contradiction in Buffon has been recognized by several historians but not thoroughly explored in relation to related cultural discourses; see Clarence J. Glacken, *Traces on the Rhodian Shore* (Berkeley: U of California P, 1967), 672–79; Jacques Roger, *Buffon* (Paris: Fayard, 1989), 310–16, 348–51, 392–98; Richard W. Burkhardt Jr., "L'idéologie de domestication chez Buffon et chez les éthologues modernes," in *Buffon 88*, ed. Jean Gayon (Paris: Vrin, 1992), 569–82, on 575–76. See also Paul Pelckmans, "When the King Learns to Do It in His Own Way: Animals in the Age of Enlightenment," in *Zoology: On (Post) Modern Animals*, ed. Bart Verschaffel and Mark Verminck (Dublin: Lilliput, 1993), 104–20. Keith Thomas has identified the juxtaposition of exploitation of and sentiment toward the natural world as a central dilemma of the modern West in *Man and the Natural World* (New York: Pantheon, 1983).

30. This comparison of England and France is based on my impressions, not

on a systematic comparative study. For some examples of criticism of "imprisoned" or "enslaved" animals in England, see Thomas, *Man and the Natural World;* Moira Ferguson, *Animal Advocacy and Englishwomen, 1780–1900* (Ann Arbor: U of Michigan P, 1998).

31. Thomas contends that the dominant attitude concerning cruelty to animals from the fifteenth through the nineteenth centuries was that "man . . . was fully entitled to domesticate animals and to kill them for food and clothing. But he was not to tyrannize or to cause unnecessary suffering" (*Man and the Natural World,* 153).

32. Mauduyt, *Encyc. méth. ois.,* 1 (1782), 427–28.

33. Buffon, "Le Canard," *HNO,* 9 (1783), 118–19; "La Perdrix Rouge," *HNO,* 2 (1771), 435–36.

34. [Hennebert and Beaurieu,] *Cours d'histoire naturelle,* 1:v. Subsequent page references given in text.

35. Mauduyt, *Encyc. méth. ois.,* 1 (1782), 472.

36. *Les oiseaux échappés, les paons et l'oiseleur, fable par l'auteur du Cri de l'honneur* (n.p., n.d.) (BHVP cote 605881) (quote at 7 n). The author of *Cri de l'honneur* (1766) is identified in *Dictionnaire des ouvrages anonymes,* 3rd ed., ed. Ant.-Alex. Barbier (Paris: Daffis, 1872–79), 1:815 as [Pierre] Barnabé Farmian de Rosoy (or Rozoi), dit Durosoi, sometime editor of the *Journal des Dames* and of the monarchist *Gazette de Paris.*

37. Jean-Jacques Rousseau, *Julie, ou La nouvelle Héloïse* (1761) (Paris: Garnier-Flammarion, 1967), 331–34, 357–58 (pt. 4, letters 10 and 11) (at 358). On the paternalistic structure of Clarens, see Michèle Duchet, *Anthropologie et histoire au siècle des Lumières* (1971; repr. Paris: Albin Michel, 1995), 149, 362–68; Jean Starobinski, *Jean-Jacques Rousseau,* trans. Arthur Goldhammer (Chicago: U of Chicago P, 1988), 97–104.

38. Glacken, *Traces on the Rhodian Shore,* 677.

39. Nancy Leys Stepan explains how such metaphors can become naturalized, or "literalized," in her analysis of analogies between blacks and women made by nineteenth- and twentieth-century scientists: "Race and Gender: The Role of Analogy in Science," *Isis* 77 (1986), 261–77.

40. *Dictionnaire historique de la langue française,* ed. Alain Rey (Paris: Robert, 1992), s.v., "esclave."

41. See Shanti Marie Singham, "Betwixt Cattle and Men: Jews, Blacks, and Women, and the Declaration of the Rights of Man," in *The French Idea of Freedom,* ed. Dale Van Kley (Stanford: Stanford UP, 1994), 114–53. The use of slavery as a metaphor for political domination is also discussed by Sue Peabody, *"There Are No Slaves in France": The Political Culture of Race and Slavery in the Ancien Regime* (New York: Oxford UP, 1996), 96. On the general problem of diluting terms by

applying them broadly, see David Brion Davis, "Reflections on Abolitionism and Ideological Hegemony," in *The Antislavery Debate*, ed. Thomas Bender (Berkeley: U of California P, 1992), 173.

42. Suzanne Curchod Necker, *Mélanges extraits des manuscrits de Mme Necker* (Paris: Pougens, an VI [1798]), 3:138–39.

43. Arthur O. Lovejoy, *The Great Chain of Being* (1936; repr., New York: Harper & Row, 1960); see also Stepan, "Race and Gender," for discussion of how the comparison between blacks and women depended on the belief in their shared inferiority. See note on secondary sources for other works discussing comparisons among subordinate human groups and animals.

44. Buffon, "Animaux du nouveaux monde," *HN*, 9 (1761), 85; "Animaux communs aux deux continens," ibid., 103–4. See also "Le Castor," *HN*, 8 (1760), 285, in which Buffon claims that the "budding societies of savage peoples" are the only sorts of human societies that can be compared to animal societies.

45. Buffon, "Nomenclature des singes," *HN*, 14 (1766), 32. Michèle Duchet explicates Buffon's ideas on the distinction between humans and animals and the differences among human races in *Anthropologie et histoire*, 229–80.

46. See note on secondary sources for works on the noble savage. On women as morally superior, see Lieselotte Steinbrügge, *The Moral Sex: Woman's Nature in the French Enlightenment*, trans. Pamela E. Selwyn (New York: Oxford UP, 1995).

47. Bernard Smith, *European Vision and the South Pacific, 1768–1850* (Oxford: Clarendon Press, 1960), 86–87, 102–9. On Marion du Fresne, see John Dunmore, *French Explorers in the Pacific*, 1 (Oxford: Clarendon Press, 1965): 166–95. On Cook's death and controversies over its interpretation, see Gananath Obeyesekere, *The Apotheosis of Captain Cook* (Princeton: Princeton UP; Honolulu: Bishop Press, 1997).

48. *Histoire philosophique . . . des deux Indes*, 1782 ed. (bk. 2, chap. 18, 258), quoted in Jean-Michel Racault, "L'effet exotique dans l'*Histoire des deux Indes* et la mise en scène du monde colonial de l'océan Indien," in Hans-Jürgen Lüsebrink and Anthony Strugnell, eds., *L'Histoire des deux Indes*, Studies on Voltaire and the Eighteenth Century, 333 (Oxford: Voltaire Foundn., 1995), 119–32, on 129. On this apostrophe, written by the passionate Diderot, see Yves Benot, *Diderot: De l'athéisme à l'anticolonialisme* (Paris: Maspero, 1970), 176–78.

49. [Raynal,] *Histoire philosophique . . . des deux Indes*, "5th ed.," 3:263–68, 356. The island of Nevis is described as another model settlement, thanks to the paternal care of its first governor: "Maybe those who are most disturbed by the destruction of the Americans and the servitude of the Africans would be somewhat consoled if Europeans were everywhere as humane as were the English on the island of Nevis" (5:209). On Paraguay, see Duchet, *Anthropologie et histoire*, 210–12, 279–80, 319–20.

50. Peabody, *"There Are No Slaves in France."* See note on secondary sources for works on antislavery sentiment.

51. Julia Douthwaite has made a similar argument to explain why novels of social criticism appealed to the very people whose customs were being criticized: "This is because the readers, all the while championing the cause of their favorite *philosophes,* were convinced in their heart of hearts that this literature could never actually affect their elite status or inherited privileges, nor provoke any substantive change in the social order": *Exotic Women: Literary Heroines and Cultural Strategies in Ancien Régime France* (Philadelphia: U of Pennsylvania P, 1992), 159.

52. Shelby T. McCloy, *The Humanitarian Movement in Eighteenth-Century France* (Lexington: U of Kentucky P, 1957); Peter Gay, *The Enlightenment,* 2 vols. (1966, 1969; repr. New York: Norton, 1977); Norman Hampson, *The Enlightenment* (London: Penguin, 1968); Robert Darnton, "George Washington's False Teeth," *New York Review of Books* (27 Mar. 1997): 34–38.

53. Richard H. Grove, *Green Imperialism* (Cambridge: Cambridge UP, 1995), 230. Grove links attitudes toward nature and society on 205–6, 230, 244–45, 481. David Arnold questions Grove's argument in *The Problem of Nature* (Oxford: Blackwell, 1996), 149–50, 183–84.

54. D. G. Charlton, *New Images of the Natural in France* (Cambridge: Cambridge UP, 1984), esp. 129–30, 196–200, 214.

55. Duchet, *Anthropologie et histoire.* William B. Cohen makes a similar point in *The French Encounter with Africans* (Bloomington: Indiana UP, 1980). Max Horkheimer and Theodor W. Adorno argued that domination and oppression took on subtler forms during the Enlightenment; see *Dialectic of Enlightenment,* trans. John Cumming (New York: Herder & Herder, 1972). On divergent views of women, see Mary Midgley, *Animals and Why They Matter* (Athens: U of Georgia P, 1983), 79. For a discussion of how indigenous people could be simultaneously regarded as noble and as barbaric, see Hayden White, "The Noble Savage Theme as Fetish," in *Tropics of Discourse* (Baltimore: Johns Hopkins UP, 1978), 183–96.

56. See works on the noble savage in note on secondary sources; Cohen, *French Encounter with Africans,* 70–73; Charlton, *New Images of the Natural,* 124; Yves Benot, *La Révolution française et la fin des colonies* (Paris: Découverte, 1987), 30. See also Pierre Pluchon, *Nègres et Juifs au XVIIIe siècle* ([Paris]: Tallandier, 1984), esp. 148–57, 281; Pluchon characterizes antislavery sentiment on the part of the philosophes as shallow and duplicitous.

57. Mad. Ymbert, "Une femme à ses serins," *Aff. de Paris,* 17 Jan. 1786 (no. 17), 150.

58. [Claude-Louis-Michel] Milscent, *Du régime colonial* (1792), repr. in *La Révolution française et l'abolition de l'esclavage,* 11 (Paris: Histoire sociale, 1968), 14. Some reformers, however, did link campaigns for emancipation and animal pro-

tection; see esp. the women profiled in Ferguson, *Animal Advocacy* (most of whom published in the nineteenth century).

59. [Abbé Sébastien-André] Sibire, *L'aristocratie négrière, ou Réflexions philosophiques et historiques sur l'esclavage et l'affranchissement des Noirs* (1789), repr. in *La Révolution française et l'abolition de l'esclavage*, vol. 2, 32–33.

60. M. Schwartz [Jean-Antoine-Nicolas de Caritat, marquis de Condorcet], *Réflexions sur l'esclavage des Nègres* (Neuchâtel, 1788), repr. ibid., vol. 6, 66–68. In 1846, editors of a French working-class magazine made a similar argument; they objected to animal-protection groups that called animals "brothers," arguing, according to Maurice Agulhon: "You say that animals are our brothers, yet you must admit that we make them work and even kill and eat them. What argument, then, can you make against white men who reduce their black brothers to slavery—since one is allowed to exploit one's 'brothers'?" The editors contended, alluding to Isidore Geoffroy Saint-Hilaire, director of the menagerie at the Jardin des plantes, that "nothing is thus more suitable for justifying slavery and the trade in so-called inferior men than this doctrine that has spread from the Jardin des plantes to the Collège de France": Agulhon, "Le sang des bêtes," *Romantisme* 31 (1981): 81–109 (at 92 n. 41).

61. David Brion Davis makes this argument in *The Problem of Slavery in Western Culture* (Ithaca, N.Y.: Cornell UP, 1966), 391–415. David J. Denby also suggests that Enlightenment sentimentalism, although it remained bound to a hierarchical system, opened up possibilities for more substantial change: *Sentimental Narrative and the Social Order in France, 1760–1820* (Cambridge: Cambridge UP, 1994), esp. 136–38.

62. Nicolas-[Edmé] Restif de la Bretonne, *Lettre d'un singe aux animaux de son espèce*, ed. Monique Lebailly (Levallois-Perret: Manya, 1990), 34 (orig. publ. 1781). On rebellion by human colonial subjects in Saint Domingue, including the influence of anticolonial writings such as *Histoire des deux Indes*, see Srinivas Aravamudan, *Tropicopolitans: Colonialism and Agency, 1688–1804* (Durham, N.C.: Duke UP, 1999), chap. 7.

Chapter 8. Vive la Liberté

1. On the meanings of liberty during the Revolution, see Mona Ozouf, "Liberty," in *A Critical Dictionary of the French Revolution*, ed. François Furet and Mona Ozouf, trans. Arthur Goldhammer (Cambridge: Belknap/Harvard UP, 1989), 716–27. On the replacement of old symbols with new, see Serge Bianchi, *La révolution culturelle de l'an II* (Paris: Aubier, 1982).

2. *J. de Paris*, 14 June 1790 (no. 165), 661. The context was a deputation to the

National Assembly by the Académie des sciences. The Académie must love liberty, the *Journal* claimed, since it is devoted to the observation of nature.

3. Isabelle Richefort, "Métaphores et représentations de la nature sous la Révolution," in *Nature, Environnement et Paysage,* ed. Andrée Corvol and Isabelle Richefort (Paris: Harmattan, 1995), 3–17.

4. M. Peuchet, "Réflexions sur le combat du taureau," *Gazette nationale, ou Le moniteur universel,* 12 Mar. 1790 (no. 71). 587–88. Reprinted (minus one paragraph) in *Actes de la Commune de Paris pendant la Révolution,* ed. Sigismond Lacroix (Paris: Cerf, 1898), 7:545–47. Peuchet, a moderate, escaped to the provinces in 1792; see *Biographie universelle ancienne et moderne,* supp. (Paris: Michaud, 1811–62), s.v. "Peuchet, Jacques." On ideas concerning the effect of bodily sensations on the mind, see Anne C. Vila, *Enlightenment and Pathology* (Baltimore: Johns Hopkins UP, 1998).

5. "Extrait d'une lettre de M. Brisset à M. Peuchet," *Gazette nationale, ou Le moniteur universel,* 18 Aug. 1790 (no. 230).

6. "Lettre de M. Bailly à M. Peuchet," *Gazette nationale, ou Le moniteur universel* 4 Aug. 1790 (no. 216). Reprinted in *Actes de la Commune de Paris,* 7:547. Bailly was executed at the Champ de Mars in 1793.

7. The notices appeared in the *J. de Paris* (14 Sept. 1790) and the *Journal de la Municipalité et des sections* (23 Sept. 1790). See *Actes de la Commune de Paris,* 7:547.

8. *Gazette national, ou Le moniteur universel,* 25 Apr. 1791 (no. 115), 211. Reprinted in *Actes de la Commune de Paris,* 7:548. According to a note in the latter publication, Peuchet was no longer the *administrateur de police* at this time.

9. *Gazette national, ou Le moniteur universel,* 12 May 1791 (no. 132), 358. Reprinted in *Actes de la Commune de Paris,* 7:548–49.

10. *Gazette nationale, ou Le moniteur universel,* 15 June 1791 (no. 166), 660–61. Reprinted in part in *Actes de la Commune de Paris,* 7:549–50; see ed. note (550) on Peuchet's distraction.

11. See 1797 letter to the *Moniteur* in Georges Bertin, "Les combats de taureaux à Paris (1781–1833)," *Revue de la Révolution* 9 (1887): 166.

12. *Actes de la Commune de Paris,* 6:438, 456 (Conseil de Ville, Commune de Paris, 8 July 1790 and 10 July 1790).

13. This summary of the contents of several declarations is provided by Sigismond Lacroix in *Actes de la Commune de Paris,* 6:479 (Eclaircissements, 10 July 1790).

14. Victor Fournel, *Le vieux Paris* (Tours: Mame et fils, 1887), 452.

15. *La chasse aux bêtes puantes et féroces . . .* (Paris: Imprimerie de la Liberté, 1789); *Chasse nouvelle aux bêtes puantes et féroces . . .* (Paris: Imprimerie de la Lanterne, 1789).

16. Annie Duprat, "Le Sage, même à cinquante ans, profite à l'école des bêtes,"

in *L'animalité*, ed. Alain Niderst (Tübingen: Narr, 1994), 131–48 (at 143) (from pamphlet entitled *Description de la ménagerie royale d'animaux vivants*). Duprat analyzes animal caricatures from 1775 to 1789.

17. Séance du 23 brumaire an II [13 Nov. 1793] and annexe A, *Procès-verbaux du Comité d'instruction publique*, ed. J. Guillaume, vol. 2 (Paris: Imprimerie nationale, 1894), 814, 816–21. The decree is also printed in *Gazette nationale, ou Le moniteur universel*, 25 Oct. 1793; and Gustave Loisel, "Histoire de la ménagerie du Muséum," *Revue scientifique* 49, no. 2 (26 Aug. 1911), 263.

18. Report from Commune de Paris of 2 prairial, an II, in *J. de Paris*, 5 prairial an II [24 May 1794], (no. 509), 2055–56.

19. Valentin Pelosse, "Imaginaire social et protection de l'animal: Des amis des bêtes de l'an X au législateur de 1850," *L'homme* 21 (1981): 5–33 and 22 (1982): 33–51. Pelosse analyzes twenty-two responses to the prize question. Maurice Agulhon argues that the fear that cruelty to animals would incite popular violence was a major factor behind passage of the Loi Grammont, which outlawed public abuse of animals, in 1850: "Le sang des bêtes: Le problème de la protection des animaux en France au XIXème siècle," *Romantisme* 31 (1981): 81–109.

20. For historical works on the founding of the new menagerie see note on secondary sources.

21. Gustave Loisel, *Histoire des ménageries de l'antiquité à nos jours* (Paris: Doin & fils, 1912), 2:255–59.

22. Paul Huot, "Les prisonniers d'Orléans," *Revue d'Alsace*, 3rd series, 4 (1868): 97–114 (at 104). The article by Huot, who makes no secret of his royalist sympathies, is quoted extensively in Loisel, *Histoire des ménageries*, 2:159–60, and the story is repeated in Bob Mullan and Garry Marvin, *Zoo Culture* (London: Weidenfeld & Nicolson, 1987), 107; and Masumi Iriye, "Le Vau's Menagerie and the Rise of the *Animalier*," Ph.D. diss., Univ. of Michigan, 1994, 194–95. Bernardin de Saint-Pierre wrote that Laimant, the head of the menagerie, had told him that a camel, several monkeys, and many birds had been taken and ultimately killed for lack of means to feed them. J.-H. Bernardin de Saint-Pierre, "Mémoire sur la nécessité de joindre une ménagerie au Jardin des plantes de Paris," in *Oeuvres complètes de J. H. Bernardin de Saint-Pierre*, new ed., ed. L. Aimé-Martin, vol. 11 (Paris: Armand-Aubrée, 1834), 395–425, 437–43 (at 402).

23. Henry Paulin Panon Desbassayns, *Voyage à Paris pendant la Révolution, 1790–1792*, ed. Jean-Claude Guillermin des Sagettes and Marie-Hélène Bourquin-Simonin (Paris: Perrin, 1985), 163. Desbassayns visited the menagerie again in January 1792, when he mentioned the rhinoceros, the lion (larger than before), and the quagga (referred to as a zebra from the Cape of Good Hope) (251); notes from a third visit, in July 1792, have become almost illegible, according to the editors, who

remark that Desbassayns recorded seeing the lion, an Asian pigeon, and the rhinoceros, "a disagreeable sight" (277).

24. Bernardin de Saint-Pierre, "Mémoire sur la nécessité." On circumstances of the memoir's publication, see Loisel, *Histoire des ménageries,* 2:162 n. 1.

25. Bernardin de Saint-Pierre, "Mémoire sur la necessité." Page references are indicated in text. On Bernardin de Saint-Pierre, see Roselyne Rey, "L'animalité dans l'oeuvre de Bernardin de Saint-Pierre," *Revue de synthèse,* ser. 4, nos. 3–4 (1992): 311–31; Phillip R. Sloan, "From Natural Law to Evolutionary Ethics in Enlightenment French Natural History," in *Biology and the Foundation of Ethics,* ed. Jane Maienschein and Michael Ruse (Cambridge: Cambridge UP, 1999), 52–83. E. C. Spary analyzes the central role of natural history in the Jacobin project to maintain public order through sentiment in *Utopia's Garden* (Chicago: U of Chicago P, 2000), chap. 5.

26. [Georges Toscan,] "Histoire du lion et du chien," *La décade philosophique, littéraire, et politique* 3, no. 18 (30 vendémiaire an III [21 Oct. 1794]), 129–38; and ibid., 3, no. 19 (10 brumaire an III [31 Oct. 1794]), 193–99. Other dog and lion stories had been popular earlier. For the story of a friendship between a dog and a lion in the Tower of London, see *Aff. de Prov.,* 25 dec. 1776 (no. 52), 208; abbé Sauri, *Précis d'histoire naturelle* (Paris: Author, 1778), 5:175–78.

27. Yves Laissus and Jean-Jacques Petter, *Les animaux du Muséum, 1793–1993* (Paris: Muséum national d'histoire naturelle, 1993), 81.

28. Printed *compte rendu* dated 10 Nov. 1792 (issue no. 42 of publication whose name is illegible), AN AJ/15/512, f. A14, subf. "Ménagerie du jardin des plantes" (quote at 120).

29. A. L. Millin, Pinel, and Alex. Brongniart, "Rapport fait à la Société d'histoire naturelle de Paris, sur la nécessité d'établir une ménagerie" (Paris: Boileau, 14 Dec. 1792).

30. Couturier to Bernardin de Saint-Pierre, 17 Jan. 1793, AN AJ/15/512, f. A14, subf. "Ménagerie du jardin des plantes," doc. 511. Also printed in E.-T. Hamy, "Les anciennes ménageries royales et la ménagerie nationale fondée le 14 brumaire an II (4 novembre 1793)," *Nouvelles archives du Muséum,* ser. 4, 5 (1893): 1–22, on 20; and Loisel, *Histoire des ménageries,* 2:162.

31. "Mémoire succinct des demandes des professeurs du Museum national d'histoire naturelle au Commissaire de l'Instruction publique," undated doc., AN AJ/15/515, f. "Instruction Publique," doc. 309 III, 1–2. There are conflicting reports about how the rhinoceros died—either by drowning in its water pool, in July 1793, or being stabbed by a saber, in September 1793; see L. C. Rookmaaker, "Histoire du rhinocéros de Versailles (1770–1793)," *Revue d'histoire des sciences* 36, 3–4 (1983): 307–18, on 311.

32. Loisel, "Histoire de la ménagerie," 262–63. For works on the founding of

the Muséum d'histoire naturelle and on science during the Revolution, see note on secondary sources.

33. Letter to Mme Suard cited in Jacques Roger, *Buffon* (Paris: Fayard, 1989), 570. Condorcet went on to say that since Buffon was a great writer and had considerable wit and some good insights, he would be able to find positive things to say about him. See Pietro Corsi, "Buffon sous la Révolution et l'Empire," in *Buffon 88*, ed. Jean Gayon (Paris: Vrin, 1992), 639–48. Spary (*Utopia's Garden*, chap. 4), however, argues that during the Revolution naturalists did not distance themselves from Buffon as much as historians have claimed they did.

34. Loisel, "Histoire de la ménagerie," 263. The number and identity of the animals brought in off the streets is unclear: see Richard W. Burkhardt Jr., "La ménagerie et la vie du Muséum," in *Le Muséum au premier siècle de son histoire*, ed. Claude Blanckaert et al. (Paris: Muséum national d'histoire naturelle, 1997), 481–508 (at 482–83 n. 3).

35. AN AJ/15/844, f. "Ménagerie." Transcription in séance du 23 brumaire an II [13 Nov. 1793], annexe A, *Procès-verbaux du Comité d'instruction publique*, 2:818; and in Hamy, "Les anciennes ménageries royales," 21.

36. "Dept. de Police contre Cochon," AN AJ/15/844, f. "Ménagerie"; Loisel, "Histoire de la ménagerie," 264.

37. Séance du 23 brumaire an II [13 Nov. 1793], annexe A, *Procès-verbaux du Comité d'instruction publique*, 2:814, 816–21. Letter from Desfontaines reprinted in Hamy, "Les anciennes ménageries royales," 22. Loisel ("Histoire de la ménagerie," 263) claims (without citation) that twenty-six animals arrived that day, but Desfontaines' letter dated Nov. 6 mentions only eleven.

38. For sieur Martin's estimates, see AN AJ/15/844, f. "Ménagerie." For summary of estimates, and those of the proprietors, see Desfontaines to Committee of Public Instruction: Séance du 23 brumaire an II [13 Nov. 1793], annexe A, *Procès-verbaux du Comité d'instruction publique*, 2:819–20. In almost all cases, the figures in the manuscript document are higher (it also includes more animals). Payment must have been slow in coming; Cochon and two other proprietors drew up a memoir requesting justice for *bons pères de famille:* ibid., 2:821. Martin's estimates were much lower than those of the owners. In a memo requesting funds, Antoine-Laurent de Jussieu remarked that the value should fall somewhere in between: see "Notes relatives a l'etablissement d'une menagerie dans le museum d'hist. naturelle," doc. n.d. (#20344), labeled de Jussieu, in AN AJ/15/844, f. "Ménagerie." Cassal, who had taken his animals to Paris from Tours following a decree issued in the latter city, may have willingly presented his animals and his services to the museum: pers. comm. from Prof. Richard W. Burkhardt Jr., based on docs. from AN AJ/15/96 and AN AJ/14/577.

39. Questions and response from Desfontaines printed in séance du 23 bru-

maire an II [13 Nov. 1793], annexe A, *Procès-verbaux du Comité d'instruction publique,* 2:818–21.

40. Loisel, "Histoire de la ménagerie," 264. On the antelope's fate, see séance du 10 floréal, an II [29 Apr. 1794], *Procès-verbaux de la Commission temporaire des arts,* ed. Louis Tuetey (Paris: Imprimerie nationale, 1912), 1:158.

41. Séance du 25 ventôse, an II [15 Mar. 1794], *Procès-verbaux de la Commission temporaire des arts,* 1:102–3; séance du 30 ventôse, an II [20 Mar. 1794], ibid., 1:104. For list of animals, see Loisel, "Histoire de la ménagerie," 264.

42. Geoffroy to "Citoyen," 26 floréal, an II, AN AJ/15/844. On Geoffroy Saint-Hilaire, see Michael A. Osborne, "Zoos in the Family," in *New Worlds, New Animals,* ed. R. J. Hoage and William A. Deiss (Baltimore: Johns Hopkins UP, 1996), 33–42.

43. See text of decree in Loisel, "Histoire de la ménagerie," 265. For other *arrêtés* of the committee from floréal and prairial an II [May and June 1794] and correspondence of Desfontaines requesting permission to buy animals, see séance du 27 floréal an II [16 mai 1794], annexe A, *Procès-verbaux du Comité d'instruction publique,* 4 (1901), 440–46.

44. An undated memoir from the professors of the museum (written after April 1794, when the animals from Versailles arrived) noted that "the throng of citizens, who crowd around to see the animals, has necessitated new expenses for barriers and [enclosures] [illegible: *parcs?*]." AN AJ/15/515, f. "Instruction publique, 2e div., 2e sec.," doc. 309 III, 2–3.

45. [Antoine-Laurent de Jussieu,] "Notes relatives à l'etablissement d'une ménagerie." The memorandum must have been written between November 1793 and April 1794 (the Versailles animals had not yet arrived); I do not know whether it was sent to the Convention.

46. For the list of articles, see annexe B, séance du 3 pluviôse, an II [22 Jan. 1794], *Procès-verbaux du Comité d'instruction publique* 3 (1897), 319–20 (see also 315).

47. "Convention Nationale: Rapport fait au nom du Comité d'Instruction Publique et des Finances, Sur le Muséum national d'histoire naturelle, Par Thibaudeau; A la séance du 21 frimaire, l'an 3." Printed pamphlet in AN AJ/15/515, f. "Règlement du Muséum, 1793" (quotes at 14). Printed in *Gazette national, ou Le moniteur universel,* 24 frimaire an III (14 Dec. 1794), 727–32. Part of text reproduced in Loisel, "Histoire de la ménagerie," 266. Antoine-Claire Thibaudeau was a moderate delegate to the Convention who served a fortnight's term as president during Year III.

48. Loisel, "Histoire de la ménagerie," 267, col. 2 n. 1 (horsemeat); Burkhardt, "La ménagerie et la vie du Muséum," 500, points out that dead menagerie animals

were an important source for studies of comparative anatomy and for the natural history collections.

49. Loisel, "Histoire de la ménagerie," 269.

50. Yves Laissus, *L'Égypte, une aventure savante avec Bonaparte, Kléber, Menou, 1798–1801* (Paris: Fayard, 1998).

51. Undated draft of letter to "Citoyen," signed "Geoffroy," AN AJ/15/742, f. An 3. See transcription in Michel Lemire, "La France et les collections du stathouder Guillaume V d'Orange," in *Een vorstelijke dierentuin/Le zoo du prince,* ed. B. C. Sliggers and A. A. Wertheim (n.p.: Walburg Inst., 1994), 119 n. 1. Many of the museum documents have been reclassified, and this one may have been moved to the AJ series from the F series identified as its location in Lemire's note.

52. This statement, probably written by Thouin, appears in an unsigned and undated doc., "Exament sommaire de la collection de quadrupedes et oiseaux qui font partie du cabinet [illegible] sur le cidevant statouder de hollande," AN AJ/15/836, f. "Stadhouder." For more on the confiscations, see Loisel, *Histoire des ménageries,* 2:35–46, 353–55; Ferdinand Boyer, "Le transfert à Paris des collections du Stathouder (1795)," *Annales historiques de la Révolution française* 43 (1971): 389–404; and articles in Sliggers and Wertheim, eds., *Le zoo du Prince.*

53. Thouin to "Citoyens collegues," 21 floréal an III [10 May 1795], AN AJ/15/836, f. "Stadhouder."

54. Sources differ on the number of animals and when they arrived. Loisel indicates eleven mammals and thirty-six birds ("Histoire de la ménagerie," 268); F. F. J. M. Pieters claims that only six mammals and eleven birds survived the trip ("La ménagerie du stathouder Guillaume V dans le domaine Het Kleine Loo à Voorburg," in *Le zoo du prince,* ed. Sliggers and Wertheim, 58).

55. On the elephants' transportation and reception, see AN AJ/15/844, f. "Transport des éléphans"; Loisel, *Histoire des ménageries,* 2:34–49; Roger Saban and Michel Lemire, "Les éléphants de la ménagerie du Stathouder Guillaume V d'Orange au Muséum d'histoire naturelle sous la Convention nationale et le Directoire," *Actes du 114e Congrès national des sociétés savantes* (Paris: CTHS, 1990): 275–300; Laissus and Petter, *Les animaux du Muséum,* 90–91; Lemire, "La France et les collections," 98–103; James H. Johnson, *Listening in Paris* (Berkeley: U of California P, 1995), 129–31; Burkhardt, "La ménagerie et la vie du Muséum," 492–95.

56. L. F. Jauffret, *Voyage au Jardin des plantes* (Paris: Houel, an VI), 126–27.

57. Loisel, "Histoire de la ménagerie," 270; Laissus and Petter, *Les animaux du Muséum,* 94–95. Richard Burkhardt suggests that "Mr. Penbrock" might have been G. Pidcock, owner of the large public menagerie in London called the Exeter 'Change (pers. comm.).

58. Richard W. Burkhardt Jr., "Unpacking Baudin," in *Jean-Baptiste Lamarck,*

1744–1829, ed. Goulven Laurent, 119e Congrès national des sociétés historiques et scientifiques, Amiens (Paris: CTHS, 1997), 497–514.

59. [Bernard-Germain-Étienne de La Ville, comte de] Lacépède, "Lettre relative aux établissemens publics destinés à renfermer des animaux vivans, et connus sous le nom de Ménageries," *La décade philosophique, littéraire et politique* 59 (20 frimaire, an IV [11 December 1795]): 449–462 (at 452 and 454). On Lacépède's vision, see Burkhardt, "La ménagerie et la vie du Muséum," which also discusses its failure to be realized.

60. Éric Baratay and Élisabeth Hardouin-Fugier, *Zoos* (Paris: Découverte, 1998), 105–8, 157–72, 205–210.

61. Delaunay to Citoyen Fourcroy, 1 brumaire an IX [Oct. 22 1800], AN AJ/15/742, f. An 8.

62. Séance du 10 pluviôse, an II [29 Jan. 1794], *Procès-verbaux de la Commission temporaire des arts,* 1:57 n. 3. The president urged creation of the menagerie despite the fact that the members (professors?) of the Muséum might not think this was the best time to do so.

63. Unsigned article in section headed "Variétés," *La décade philosophique, littéraire, et politique* 18, no. 32 (20 thermidor an VI [7 Aug. 1798]): 301–6 (at 302). A verse in a song composed for the occasion described the menagerie animals acquired through conquests (although, oddly, it did not mention those from Italy): ibid., 18, no. 31 (10 thermidor an VI [28 July 1798]): 234.

64. *Description et vente curieuse des animaux féroces mâles et femelles, de la ménagerie du cabinet d'histoire naturelle des ci-devant Jacobins . . .* (Paris: Gaulemeriti, n.d.). The references are to Jean-Baptiste Carrier, who was involved in the drownings of prisoners in Nantes, and Jean-Marie Collot d'Herbois, a former actor (thus the monkey analogy), who took part in the massacres in Lyon.

65. Jauffret, *Voyage au Jardin des plantes,* 75, 84–85, 124–26. Jauffret includes a long account of the lion and dog story (90–111). Burkhardt ("La ménagerie et la vie du Muséum," 490–91) also contrasts Cassal's attitude as described in Jauffret's book with more idealized views of the menagerie.

Epilogue

1. Voltaire, *Candide,* ed. J. H. Brumfitt (1759; Oxford: Oxford UP, 1968), 106 (chap. 19).

2. Brumfitt, introduction, ibid., 47.

3. Scholars who have discussed such symbolic uses of animals include Robert Delort, *Les animaux ont une histoire* (Paris: Seuil, 1984); Harriet Ritvo, *The Animal Estate* (Cambridge: Harvard UP, 1987); and Steve Baker, *Picturing the Beast* (Manchester: Manchester UP, 1993).

4. Of the many recent books in this field, see, e.g., Randy Thornhill and Craig Palmer, *A Natural History of Rape* (Cambridge: MIT P, 2000).

5. See especially Keith Thomas, *Man and the Natural World* (New York: Pantheon, 1983); Yi-Fu Tuan, *Dominance and Affection* (New Haven: Yale UP, 1984); Arnold Arluke and Clinton R. Sanders, *Regarding Animals* (Philadelphia: Temple UP, 1996).

6. Peter Singer, *Animal Liberation* (New York: New York Review, 1975); Mary Midgley, *Animals and Why They Matter* (Athens: U of Georgia P, 1983); James Serpell, *In the Company of Animals* (Oxford: Blackwell, 1986), esp. 171–87; Marjorie Spiegel, *The Dreaded Comparison* (New York: Mirror, 1988); Tim Ingold, "From Trust to Domination," in *Animals and Human Society,* ed. Aubrey Manning and James Serpell (London: Routledge, 1994), 1–22; Carol J. Adams and Josephine Donovan, eds., *Animals and Women* (Durham, N.C.: Duke UP, 1995).

7. Frans B. M. de Waal, "We the People (and Other Animals) . . . ," *New York Times,* 20 Aug. 1999, editorial page.

m``

PRIMARY MATERIAL

Archival Sources

Archives Nationales, Paris

AD/XI/22	Textes administratifs; Commerce et industrie, ancien régime: Médicins–oiseleurs (1402–1779). Dossier E: Oiseleurs.
AJ/15/511	Jardin du roi: Voyages et missions (1595–1793); Jardins et serres (1634–1793).
AJ/15/512	Jardin du roi: Cabinet et collections (1670–1713); Ménageries (1711–93).
AJ/15/515	Muséum: Creation et organisation, 1793–1922.
AJ/15/742	Muséum: Correspondance arrivée et départ, 1793–an VIII.
AJ/15/836	Muséum: Collections diverses, 1793–1825.
AJ/15/844	Muséum: Ménagerie, 1794–1901.
O/1/106	Secrétariat de la Maison du roi, 1762.
O/1/126	Secrétariat de la Maison du roi, 1782.
O/1/128	Secrétariat de la Maison du roi, 1786.
O/1/597	Maison du roi. Mémoires, relations, et observations présentés par divers particuliers.
O/1/807	Maison du roi. Papiers du grand-maître; Gouvernement des maisons royales—Versailles: château, parc, domaine, 1780–87.
Y/15295	Châtelet de Paris et prévôté d'île de France: Commissaires au Châtelet, office de Claude-Etienne Prestat.
Z/1E/218	Eaux et Forêts, maîtrise de Paris: Registre des oiseleurs et des pêcheurs, 1781–89.
Z/1E/1166	Eaux et Forêts, Papiers provenant des résidus des fonds judiciaires: Maître oiseleurs de Paris, 17–18 siècles.

Archives Nationales, Colonies

C/2/285 Correspondance à l'Arrivée: Compagnie des Indes: Mémoires sur
les diverses productions des comptoirs d'Asie: étoffes, soies, tein-
tures, café, thé, épices, porcelaines, animaux et curiosités, 1713–89.

C/5B/5 bis Correspondance à l'Arrivée: Cap de Bonne-Espérance, 1783–88.

C/5B/6 Correspondance à l'Arrivée: Cap de Bonne-Espérance, 1787–92.

C/5B/7 Correspondance à l'Arrivée: Cap de Bonne-Espérance, 1780–92.

C/5B/8 Correspondance à l'Arrivée: Cap de Bonne-Espérance, 1781–87:
Registre de correspondance de l'agent français au Cap.

Bibliothèque Centrale du Muséum National d'Histoire Naturelle, Paris

MS 293 Documents géographiques divers.

MS 352 Notice des oiseaux qu'on desire recevoir de Cayene vivans.

MS 369 Papiers provenant de Buffon.

MS 864 Papiers de l'abbé Bexon.

MS 1765 Relation des Indes orientales.

MS 1995 Recueil de lettres adressées par Joseph-François-Charpentier de
Cossigny à Louis-Guillaume Le Monnier de l'île de France,
1769–83.

Printed Sources

N.B. For periodicals, I have used the titles listed in Jean Sgard, *Dictionnaire des jour-
naux, 1600–1789*, 2 vols. (Paris: Universitas, 1991). For full titles and identity of
multiple publishers, please see Louise E. Robbins, "Elephant Slaves and Pampered
Parrots: Exotic Animals and Their Meanings in Eighteenth-Century France,"
Ph.D. diss., Univ. of Wisconsin–Madison, 1998.

Acts, Periodicals, and Newsletters

Actes de la Commune de Paris pendant la Révolution. Ed. Sigismond Lacroix. 15 vols.
Paris: L. Cerf, 1894–1909.

Affiches de Paris. Paris, 1751–1811. Also entitled *Annonces, affiches, et avis divers.*

Affiches de Province. Paris, 1752–85. Also entitled *Annonces, affiches, et avis divers* or
Affiches, annonces, et avis divers.

Almanach des muses. Paris, 1765–83.

Almanach forain. Paris, 1773–78, 1786–87. Also entitled *Les spectacles des Foires; Les
petits spectacles de Paris.* Because this almanac is hard to find, I list shelfmarks

(cotes) from the BnF. BnF: 1777: YF-1939; 1778: YF-1940; BnF Arsenal: 1773: GD 2740; 1774: GD 2682; 1775: GD 2682; 1776: GD 2783B; 1778: GD 2783; 1786: GD 2765; 1787: GD 2765.

Avant-coureur. Paris, 1760–73.

Buc'hoz, [Pierre-Joseph]. *La nature considérée sous ses différents aspects* . . . . Paris, 1768–1783, intermittent.

Correspondance littéraire, philosophique et critique par Grimm, Diderot, Raynal, Meister, etc. Ed. Maurice Tourneux. 16 vols. Paris: Garnier Frères, 1877–82.

Correspondance secrète, politique & littéraire. 18 vols. London: John Adamson, 1787–90. Edited reprint of *Correspondance littéraire secrète* (Paris, 1775–93) by Louis-François Métra (or Mettra) and others.

La décade philosophique, littéraire et politique. Paris, 1794–1804.

Gazette du commerce. Paris, 1763–83. Also entitled *Gazette d'agriculture, commerce, arts et finances.*

Gazette nationale, ou Le moniteur universel. Paris, 1789–99. Repr. Paris: Plon, 1858–63.

Journal de Paris. Paris, 1777–1840.

Procès-verbaux de l'Académie royale d'architecture, 1671–1793. Ed. Henry Lemonnier. 10 vols. Paris: Armand Colin, 1926.

Procès-verbaux de la Commission temporaire des arts. Ed. Louis Tuetey. 2 vols. Paris: Imprimerie nationale, 1912, 1917.

Procès-verbaux du Comité d'instruction publique de la Convention nationale. Ed. J. Guillaume. 6 vols. Paris: Imprimerie nationale, 1891–1907.

Voyage Literature

Bajon, [Bertrand]. *Mémoires pour servir à l'histoire de Cayenne, et de la Guiane françoise* . . . . 2 vols. Paris: Grangé, 1777.

Barbot, Jean. *Barbot on Guinea: The Writings of Jean Barbot on West Africa, 1678–1712.* Ed. P. E. H. Hair, Adam Jones, and Robin Law. London: Hakluyt Society, 1992.

Barrère, Pierre. *Nouvelle relation de la France equinoxiale* . . . . Paris: Piget, 1743.

Bernardin de Saint-Pierre, Jacques-Henri. "Voyage à l'île de France." In *Oeuvres complètes de J. H. Bernardin de Saint Pierre,* new ed., ed. L. Aimé-Martin, vol. 2. Paris: Armand-Aubrée, 1834.

Bosman, William. *A New and Accurate Description of the Coast of Guinea* . . . . New ed. 1705. Repr. London: Frank Cass, 1967. Orig. publ. in Dutch. Utrecht, 1704.

Bossu, [Jean Bernard]. *Nouveaux voyages dans l'Amérique septentrionale* . . . . Amsterdam: Changuion, 1777.

[Bourgeois, Nougaret]. *Voyages intéressans dans différentes colonies françaises, espag-noles, anglaises, &c*. . . . . London: n.p.; Paris: Jean-François Bastien, 1788.

Breton, père Raymond. *Relation de l'île de la Guadeloupe*. Bibliothèque d'histoire antillaise, 3. 1647. Repr. Basse-Terre: Société d'histoire de la Guadeloupe, 1978.

[Challe, Robert]. *Journal d'un voyage fait aux Indes orientales* . . . . 3 vols. Rouen: Jean Batiste Machuel le jeune, 1721.

[Chambon]. *Commerce de l'Amérique par Marseille* . . . . 2 vols. Avignon: n.p., 1764.

Charlevoix, François-Xavier de. *Journal d'un voyage fait par ordre du roi dans l'Amérique septentrionale*. Ed. Pierre Berthiaume. Montreal: P de l'Univ. de Montreal, 1994.

Chastellux, marquis de. *Travels in North-America in the Years 1780, 1781, and 1782*. Trans. "an English gentleman." 2nd ed. 2 vols. London: 1787. Repr. N.p.: Arno Press, 1968.

[Froger, François]. *Relation d'un voyage fait en 1695, 1696, & 1697, aux côtes d'Afrique, détroit de Magellan, Brezil, Cayenne, & isles Antilles* . . . . Paris, 1699.

Journal d'un voyage sur les costes d'Afrique et aux Indes d'Espagne, avec une descrip-tion particulière de la riviere de la Plata, de Buenos Ayres, & autres lieux; commencé en 1702, & fini en 1706. Amsterdam: Paul Marret, 1723.

Labat, Jean-Baptiste. *Nouvelle relation de l'Afrique occidentale* . . . . 5 vols. Paris, 1728.

La Condamine, [Charles-Marie de]. *Relation abrégée d'un voyage fait dans l'intérieur de l'Amérique méridionale* . . . . Paris: Veuve Pissot, 1745.

La Courbe, [Michel Jajolet] de. *Premier voyage du sieur de La Courbe fait à la coste d'Afrique en 1685*. Ed. Prosper Cultru. Paris: Édouard Champion, 1913.

La Pérouse, Jean-François de Galaup de. *The Journal of Jean-François de Galaup de la Pérouse, 1785–1788*. Trans. and ed. John Dunmore. 2 vols. London: Hak-luyt Society, 1995.

La Pérouse, [Jean-François de Galaup de]. *Voyage de La Pérouse autour du monde . . . rédigé par M. L. A. Milet-Mureau* . . . . Paris: Plassan, an VI [1798].

Le Page du Pratz. *Histoire de la Louisiane* . . . . 3 vols. Paris: De Bure l'aîné, 1758.

Léry, Jean de. *Histoire d'un voyage faict en la terre du Brésil (1578)*. 2nd ed. (1580). Ed. Frank Lestringant. Paris: Librairie générale française, livre de poche, 1994.

Le Vaillant, [François]. *Voyage de Monsieur Le Vaillant dans l'intérieur de l'Afrique, par le cap de Bonne-Espérance, dans les années 1780, 81, 82, 83, 84, et 85*. 2 vols. Paris: Leroy, 1790.

Loyer, Godefroy. *Relation du voyage du royaume d'Issyny, Côte d'Or, païs de Guinée en Afrique* . . . . Paris, 1714. Repr. in *L'établissement d'Issiny, 1687–1702*, ed. Paul Roussier, 109–235. Paris: Larose, 1935.

Moreau de Saint-Méry, Médéric Louis Élie. *Description topographique, physique,* civile, politique et historique de la partie française de l'isle Saint Domingue. New ed. Philadelphia, 1797. Ed. Blanche Maurel and Étienne Taillemite. Paris: Société de l'histoire des colonies françaises, 1958.

Pernetty, Dom [Antoine-Joseph]. *Histoire d'un voyage aux isles Malouines, fait en 1763 & 1764* . . . . New ed. 2 vols. Paris: Saillant & Nyon, 1770.

Sonnerat, [Pierre]. *Voyage à la Nouvelle Guinée* . . . . Paris: Ruault, 1776.

———. *Voyage aux Indes orientales et à la Chine* . . . . 2 vols. Paris: Author; Froulé, 1782.

[Thibault de Chanvalon, Jean Baptiste]. *Voyage à la Martinique* . . . . Paris: Cl. J. B. Bauche, 1763.

Tibierge, "Journal du sieur Tibierge, principal commis de la compagnie de Guynée sur le vaisseau 'Le pont d'or' au voyage de l'année 1692." In *L'établissement d'Issiny, 1687–1702*, ed. Paul Roussier, 51–69. Paris: Larose, 1935.

Wimpffen, [François Alexandre Stanislaus, Baron de]. *Voyage à Saint-Domingue, pendant les années 1788, 1789, et 1790.* Paris: Chocheris, an 5 [1799].

Descriptions of and Guides to Paris

[Dezallier d'Argenville, Antoine-Joseph]. *Voyage pittoresque des environs de Paris* . . . . Paris: De Bure l'aîné, 1755.

Dulaure, J. A. *Nouvelle description des curiosités de Paris* . . . . 2 vols. Paris: Lejay, 1785.

———. *Nouvelle description des environs de Paris* . . . . 2 vols. Paris: Lejay, 1786.

Hurtaut, [Pierre-Thomas-Nicolas], and [Pierre] Magny. *Dictionnaire historique de la ville de Paris et de ses environs* . . . . 4 vols. Paris: Moutard, 1779.

[Jèze, de]. *État ou Tableau de la ville de Paris.* New ed. Paris: Prault père, 1763.

[Le Rouge, George-Louis]. *Curiosités de Paris . . . et des environs.* New ed. 2 vols. Paris, 1778.

Mercier, Louis-Sébastien. *Parallèle de Paris et de Londres: Un inédit de Louis-Sébastien Mercier.* Ed. Claude Bruneteau and Bernard Cottret. Paris: Didier Érudition, 1982.

[Mercier, Louis-Sébastien]. *Tableau de Paris.* Rev. ed. 8 vols. Amsterdam, 1782–83.

Nemeitz, J[oachim] C[hristoph]. *Séjour de Paris* . . . . Leiden: Jean Van Abcoude, 1727.

Thiéry, [Luc-Vincent]. *Guide des amateurs et des étrangers voyageurs à Paris* . . . . 2 vols. Paris: Hardouin & Gattey, 1787.

Barbier, Edmond-Jean-François. *Journal anecdotique d'un parisien sous Louis XV, 1727 à 1751*. Ed. Hubert Juin. Paris: Livre club du libraire, 1963.

Cognel, François. *La vie parisienne sous Louis XVI*. Paris: Calmann-Lévy, 1882.

Cradock, [Anna Francesca]. *Journal de Madame Cradock: Voyage en France (1783–1786)*. Trans. and ed. Mme O. Delphin Balleyguier. Paris: Perrin, 1896.

Croÿ, Emmanuel, duc de. *Journal inédit du duc de Croÿ, 1718–1784*. Ed. le vicomte de Grouchy and Paul Cottin. Paris: Ernest Flammarion, 1906–7.

Hézecques, comte de. *Page à la cour de Louis XVI: Souvenirs du comte d'Hézecques*. Ed. Emmanuel Bourassin. N.p.: Tallandier, 1987.

Lister, Martin. *A Journey to Paris in the Year 1698*. Ed. Raymond Phineas Stearns. Repr. Urbana: U of Illinois P, 1967.

Luynes, Charles Philippe d'Albert, duc de. *Mémoires du duc de Luynes sur la cour de Louis XV, 1735–1758*. 17 vols. Paris: Firmin-Didot frères, 1862.

Malboissière, Geneviève de. *Une jeune fille au XVIIIe siècle: Lettres de Geneviève de Malboissière à Adélaïde Méliand, 1761–1766*. Ed. le comte de Luppé. Paris: Édouard Champion, 1925.

Oberkirch, Henriette-Louise de Waldner de Freundstein, baronne d'. *Mémoires de la baronne d'Oberkirch sur la cour de Louis XVI et la société française avant 1789*. Ed. Suzanne Burkard. Paris: Mercure de France, 1970.

Panon Desbassayns, Henry Paulin. *Voyage à Paris pendant la Révolution, 1790–1792: Journal inédit d'un habitant de l'île Bourbon*. Ed. Jean-Claude Guillermin des Sagettes and Marie-Hélène Bourquin-Simonin. Paris: Perrin, 1985.

Perrault, Charles. *Charles Perrault: Memoirs of My Life*. Ed. and trans. Jeanne Morgan Zarucchi. Columbia: U of Missouri P, 1989.

Réaumur, René-Antoine Ferchault de. *Correspondance inédite entre Réaumur et Abraham Trembley*. Ed. Maurice Trembley. Geneva: Georg, 1943.

Sabran, [Françoise-Éléanore de Jean de Manville], comtesse de. *Correspondance inédite de la comtesse de Sabran et du chevalier de Boufflers, 1778–1788*. 2nd ed. Paris: E. Plon., 1875.

Teleki, Joseph. *La cour de Louis XV: Journal de voyage du comte Joseph Teleki*. Ed. Gabriel Tolnai. Paris: PUF, 1943.

Encyclopedias, Dictionaries, Treatises, and Philosophical Works

Bacon, Francis. *Essays and New Atlantis*. Roslyn, N.Y.: Walter J. Black, 1942.

Chevignard de la Pallue, M. A. T. *Idée du monde, ouvrage curieux & interessant . . . .* 3rd ed. 3 vols. Paris: Briand, 1788.

Diderot, Denis. *Salons,* vol. 3: *1765.* Ed. Jean Seznec. 2nd ed. Oxford: Clarendon Press, 1979.

L'encyclopédie, ou Dictionnaire raisonné des sciences, des arts et des métiers. Ed. Denis Diderot and Jean le Rond d'Alembert. 35 vols. Paris and Neuchâtel, 1751–80. (I use the Pergamon Press compact edition [Elmsford, N.Y., 1985] but cite the original volume and page numbers.)

La Mettrie, Julien Offray de. *Man a Machine.* French/English ed. Ed. Gertrude Carman Bussey. La Salle, Ill.: Open Court, 1912.

Locke, [John]. *Essai philosophique concernant l'entendement humain . . . .* 5th ed. Trans. [Pierre] Coste. Amsterdam/Leipzig: J. Schreuder & Pierre Mortier le jeune, 1755. Repr., ed. Emilienne Naert. Paris: J. Vrin, 1972.

[Macquer, Philippe]. *Dictionnaire portatif des arts et métiers. . . .* Paris: Lacombe, 1766.

Milscent, [Claude-Louis-Michel]. *Du régime colonial.* 1792. Repr. in *La Révolution française et l'abolition de l'esclavage,* vol. 11. Paris: Éditions d'histoire sociale, 1968.

Necker, [Suzanne Curchod]. *Mélanges extraits des manuscrits de Mme Necker.* 3 vols. Paris: Charles Pougens, an VI [1798].

———. *Nouveaux mélanges extraits des manuscrits de Mme Necker.* 2 vols. Paris: Charles Pougens, an X [1801].

[Raynal, Guillaume Thomas François]. *Histoire philosophique et politique des établissements & du commerce des Européens dans les deux Indes.* "5th ed." 7 vols. Maestricht: Jean-Edme Dufour & Philippe Roux, 1777.

Savérien, [Alexandre]. *Histoire des progrès de l'esprit humain dans les sciences et dans les arts qui en dépendent: Histoire naturelle.* Paris: Humblot, 1778.

Schwartz, M. [Jean-Antoine-Nicolas de Caritat, marquis de Condorcet]. *Réflexions sur l'esclavage des Nègres.* Neuchâtel, 1788. Repr. in *La Révolution française et l'abolition de l'esclavage,* vol. 6. Paris: Éditions d'histoire sociale, 1968.

Sibire, [abbé Sébastien-André]. *L'aristocratie négrière, ou Réflexions philosophiques et historiques sur l'esclavage et l'affranchissement des Noirs.* 1789. Repr. in *La Révolution française et l'abolition de l'esclavage,* vol. 2. Paris: Éditions d'histoire sociale, 1968.

Children's Books and Books on Education

Berquin, [Armand]. *L'ami des enfants et des adolescents.* New ed. 2 vols. Paris: Didier, 1855. Repr. New York: Roe Lockwood and Son, n.d.

Cotte, [Louis]. *Leçons élémentaires d'histoire naturelle . . . .* Paris: J. Barbou, 1784.

[Coyer, Gabriel-François]. *Plan d'éducation publique.* Paris: Duchesne, 1770.

Duchesne, Ant[oine] Nic[olas], and Aug[uste] Sav[inien] Le Blond. *Portefeuille des enfans* . . . . Paris: Mérigot, 1783–97.

Épinay, Mme [Louise Florence Tardieu d'Esclavelles, marquise] d'. *Les conversations d'Émilie.* 2nd ed., 1782. Ed. Rosena Davison. Studies on Voltaire and the Eighteenth Century, 342. Oxford: Voltaire Foundation, 1996.

[Fillassier, Jean-Jacques]. *Eraste, ou L'ami de la jeunesse* . . . . 2 vols. Paris: Vincent, 1773.

Fromageot, abbé. *Cours d'études des jeunes demoiselles* . . . . 8 vols. Paris: Vincent, 1772–75.

Wandelaincourt, [Antoine-Hubert]. *Cours abrégé d'histoire naturelle.* Verdun: Mondon; Paris: Delalain, 1778.

———. *Plan d'éducation publique.* Paris: Durand, 1777.

Literary and Satirical Works

Beauharnais, Mme la comtesse [Fanny] de. "Aux incrédules: Épître envoyée à M. le comte de Buffon." *Journal de Paris,* 7 Nov. 1778 (no. 311), 1245.

[Caraccioli, Louis-Antoine de]. *La critique des dames et des messieurs à leur toilette.* N.p.: n.d.

———. *Dictionnaire critique, pittoresque et sentencieux* . . . . 3 vols. Lyon: Benoît Duplain, 1768.

[Cerfvol, de, P.-J.-B. Nougaret, and J. H. Marchand]. *Le Radoteur* . . . . 2 vols. [Paris]: Jean-François Bastien, 1777.

La chasse aux bêtes puantes et féroces . . . . Paris: Imprimerie de la Liberté, 1789.

Chasse nouvelle aux bêtes puantes et féroces . . . . Paris: Imprimerie de la Lanterne, 1789.

Cubières, Chevalier de. "Épitre à M. le comte François d'Hartig, chambellan de l'empereur, sur la mort de M. le comte de Buffon." In Henri Nadault de Buffon, ed., *Buffon: Sa famille, ses collaborateurs, et ses familiers: Mémoires par M. Humbert-Bazile, son secrétaire* (Paris: Veuve Renouard, 1863), 144–48.

Description et vente curieuse des animaux féroces mâles et femelles, de la ménagerie du cabinet d'histoire naturelle des ci-devant Jacobins . . . . Paris: Gaulemeriti, n.d.

L'échappé du Palais, ou Le Général Jaquot perdu. N.p., n.d. BHVP cote 12497. *"Vers 1771"* penciled on title page.

Fabre d'Églantine, [Philippe-François-Nazaire]. "L'étude de la nature: Poëme à Monsieur le comte de Buffon." London: n.p., 1783.

[Ferry de Saint-Constant, Giovanni]. *Génie de M. de Buffon, par M***.* Paris: Panckoucke, 1778.

Florian, [Jean-Pierre de Claris]. *Les fables de Florian.* Ed. Jean-Noël Pascal. Perpignan: PU de Perpignan, 1995.

[Gresset, Jean-Baptiste-Louis]. *Vairvert, ou Les voyages du perroquet de la visitation de Nevers. Poëme heroi-comique.* The Hague: Guillaume Niegard, 1734.

Grozelier, [Nicolas]. *Fables nouvelles . . . .* Paris: Desaint & Saillant, 1760.

Journal Singe. Par M. Piaud, no. 1. London: n.p.; Paris: Cailleau, June 1776.

La Fontaine, [Jean de]. *Oeuvres: Sources et postérité d'Ésope à l'Oulipo.* Ed. André Versaille. Brussels: Complexe, 1995.

————. *Oeuvres complètes,* vol. 1: *Fables, contes, et nouvelles.* Ed. Jean-Pierre Collinet. [Paris]: Gallimard, 1991.

————. *Oeuvres complètes,* vol. 2: *Oeuvres diverses.* Ed. Pierre Clarac. [Paris]: Gallimard, 1958.

Le Brun, Ponce Denis Écouchard. "Ode à monsieur de Buffon." In *Oeuvres de Ponce Denis (Écouchard) Le Brun,* ed. P. L. Ginguené. Paris: Imprimerie de Crapelet, 1811.

Liger, L[ouis]. *Amusemens de la campagne . . . .* Paris: Claude Prudhomme, 1734.

————. *La nouvelle maison rustique . . . .* 4th ed. 2 vols. Paris: Veuve Prudhomme, 1736.

[Marchand, Jean-Marie]. *Mémoires de l'éléphant, écrits sous sa dictée, et traduits de l'Indien par un Suisse.* Amsterdam; Paris: J. P. Costard, 1771.

[Mercier, Louis-Sébastien]. *L'an deux mille quatre cent quarante. Rêve s'il en fût jamais.* London: n.p., 1772.

Nogaret, Félix. *Apologie de mon goût: Épître en vers sur l'histoire naturelle.* Paris: Couturier père et fils, 1771.

[Nougaret, Pierre-Jean-Baptiste]. *Les milles et une folies.* 4 vols. Amsterdam; Paris: Veuve Duchesne, 1771.

Nougaret, Pierre-Jean-Baptiste. *Paris, ou Le rideau levé: . . . .* 3 vols. Paris: Author; Desenne, an VIII.

————. *Les sottises et les folies parisiennes; . . . .* London; Paris: Veuve Duchesne, 1781.

————. *Tableau mouvant de Paris . . . .* 3 vols. London: Thomas Hookham; Paris: Veuve Duchesne, 1787.

Les numéros parisiens. Paris: Imprimerie de la Vérité, 1788.

Les oiseaux échappés, les paons et l'oiseleur, fable par l'auteur du Cri de l'honneur. N.p., n.d. BHVP cote 605881.

Le perroquet, ou Mélange de diverses pièces intéressantes pour l'esprit et pour le coeur. Frankfurt am Main, 1742.

Restif de la Bretonne, Nicolas-Edmé. *La découverte australe par un homme-volant, ou Le dédale français.* Paris: France Adel, 1977. Orig. publ. Leipzig, 1781.

————. *Lettre d'un singe aux animaux de son espèce.* Ed. Monique Lebailly. Levallois-Perret: Manya, 1990. Orig. publ. Leipzig, 1781.

Rousseau, Jean-Jacques. *Émile, ou De l'éducation*. 1762. Paris: Garnier frères, 1964.

———. *Julie, ou La nouvelle Héloïse*. 1761. Paris: Garnier-Flammarion, 1967.

Natural History (broadly defined)

[Alletz, Pons Augustin]. *Histoire des singes, et autres animaux curieux* . . . . Paris: Duchesne, 1752.

[Bennett, Edward]. *The Tower Menagerie* . . . . London: Robert Jennings, 1829.

Bernardin de Saint-Pierre, Jacques-Henri. "Mémoire sur la nécessité de joindre une ménagerie au Jardin des plantes de Paris." In *Oeuvres complètes de J.-H. Bernardin de Saint-Pierre*, new ed., ed. L. Aimé-Martin, vol. 11: 395–425, 437–43. Paris: Armand-Aubrée, 1834.

Brisson, [Mathurin Jacques]. *Ornithologie* . . . . 6 vols. Paris: Cl. Jean Baptiste Bauche, 1760.

Buc'hoz, [Pierre-Joseph]. *Amusemens des dames dans les oiseaux de volière* . . . . Paris: Author, 1782.

———. *Histoire générale et économique des trois regnes de la nature*. Paris: Author, 1781.

[Buc'hoz, Pierre-Joseph]. *Les amusemens innocens, contenant le Traité des oiseaux de voliere, . . . Traduit en partie de l'ouvrage italien d'Olina, & mis en ordre d'après les avis des plus habiles oiseleurs*. Paris: P. Fr. Didot le jeune, 1774.

———. *Traité de l'éducation des animaux qui servent d'amusement à l'homme* . . . . *Par M****. Paris: Lamy, 1780.

Buffon, Georges-Louis, comte de. *Les époques de la nature*. Critical ed. by Jacques Roger. Paris: Éditions du Muséum national d'histoire naturelle, 1962. Orig. publ. *HNS*, vol. 5 (1778).

———. *Histoire naturelle, générale et particulière; Histoire naturelle des oiseaux; and Supplément à l'Histoire naturelle*. See list of abbreviations.

[Duchesne, Henri-Gabriel, and Pierre-Joseph Macquer]. *Manuel du naturaliste: Ouvrage dédié à M. de Buffon*. Paris: Desprez, 1771.

———. *Manuel du naturaliste* . . . . 2nd ed. 4 vols. Paris: Rémont, an V (1797).

[Foucher d'Obsonville]. *Essais philosophiques sur les moeurs de divers animaux étrangers* . . . . *extraits des voyages de M***. en Asie*. Paris: Couturier fils; Veuve Tilliard, 1783.

[Hennebert, Jean-Baptiste-François, and Gaspard Guillard de Beaurieu]. *Cours d'histoire naturelle* . . . . 7 vols. Paris: Lacombe, 1770.

Hervieux [de Chanteloup, J.-C.]. *A New Treatise of Canary-Birds* . . . . London: Bernard Lintot, 1718. Orig. publ. as *Nouveau traité des serins de canarie*, 1705.

Jauffret, L. F. *Voyage au Jardin des plantes*. Paris: Ch. Houel, an VI.

[La Chesnaye-Desbois, François Alexandre Aubert de]. *Dictionnaire raisonné et universel des animaux* . . . . 4 vols. Paris: Claude-Jean-Baptiste Bauche, 1759.

[Ladvocat, abbé]. *Lettre sur le rhinocéros, à M***, membre de la Société royale de Londres*. Paris: Thiboust, 1749.

Mauduyt, *Encyc. méth. ois.* See list of abbreviations.

"Mémoires pour servir à l'histoire naturelle des animaux." In *Mémoires de l'Académie royale des sciences*, vol. 1. The Hague: P. Gosse & I. Neaulme, 1731.

Millin, A. L., Pinel, and Alex. Brongniart. "Rapport fait à la Société d'histoire naturelle de Paris, sur la nécessité d'établir une ménagerie." Paris: Boileau, 14 Dec. 1792.

Pliny the Elder. *Natural History.* Trans. H. Rackham. 10 vols. Cambridge: Harvard UP, 1938–63.

Pluche, abbé [Noël-Antoine]. *Le spectacle de la nature* . . . . The Hague: Jean Neaulme, 1735–?. Orig. publ. Paris, 1732–50.

Salerne, [François]. *L'histoire naturelle, éclaircie dans une de ses parties principales, l'ornithologie* . . . . Paris: Debure père, 1767.

Sauri, abbé. *Précis d'histoire naturelle* . . . . 5 vols. Paris: Author, 1778.

Valmont de Bomare, [Jacques Christophe]. *Dictionnaire raisonné universel d'histoire naturelle*, Ed. enlarged by author. 12 vols. Yverdon, 1768–69.

———. *Dictionnaire raisonné universel d'histoire naturelle*, 4th ed. "En Suisse, chez les libraires associés." 12 vols. 1780. Pirated edition. See correspondence between Valmont de Bomare and a Genevan publisher concerning counterfeit edition: *J. de Paris*, 22 Feb. 1780 (no. 53), 223; 15 Mar. 1780 (no. 75), 311; 23 Mar. 1780 (no. 83), 341–42; 1 Oct. 1780 (no. 275), 1114; and announcements in *Aff. de Paris* 28 May 1782 (no. 148), 1254–55; *Aff. de Province*, 5 June 1782 (no. 23), 91.

This note gives an indication of the secondary sources this study draws on and may serve as a guide to the relevant literature. It is not comprehensive on either count. Many references cited in notes are not listed below, and I do not include general synthetic historical works. Except for Buffon, biographical information appears only in the notes.

General Works

Material culture in preindustrial Europe and North America has been a lively area of scholarship. See Arjun Appadurai, ed., *The Social Life of Things: Commodities in Cultural Perspective* (Cambridge: Cambridge UP, 1986); Annik Pardailhé-Galabrun, *The Birth of Intimacy*, trans. Jocelyn Phelps (Cambridge: Polity Press, 1991); John Brewer and Roy Porter, eds., *Consumption and the World of Goods* (London: Routledge, 1993); Dominique Poulot, "Une nouvelle histoire de la culture matériale?" *Revue d'histoire moderne et contemporaine* 44 (1997): 344–57; and many works by Daniel Roche, the latest of which is *A History of Everyday Things: The Birth of Consumption in France, 1600–1800*, trans. Brian Pearce (Cambridge: Cambridge UP, 2000).

For science in the Enlightenment, see Thomas L. Hankins, *Science and the Enlightenment* (Cambridge: Cambridge UP, 1985); J. V. Golinski, "Science in the Enlightenment," *History of Science* 24 (1986): 411–24; Thomas Broman, "The Habermasian Public Sphere and 'Science in the Enlightenment,'" *History of Science* 36 (1998): 123–49; William Clark, Jan Golinski, and Simon Schaffer, eds., *The Sciences in Enlightened Europe* (Chicago: U of Chicago P, 1999). On the Paris Academy of Sciences, see Alice Stroup, *A Company of Scientists: Botany, Patronage, and Community at the Seventeenth-Century Parisian Royal Academy of Sciences* (Berkeley: U of California P, 1990); Roger Hahn, *The Anatomy of a Scientific Institution: The Paris Academy of Sciences, 1666–1803* (Berkeley: U of California P, 1971).

Both French and American scholars have begun to pay more serious attention to popular science. See Robert Darnton, *Mesmerism and the End of the Enlighten-*

ment in France (Cambridge: Harvard UP, 1968); Andreas Kleinert, "La science qui se vulgarise et la science qui se fait: A propos d'un article d'Hélène Metzger sur la littérature scientifique française au XVIIIe siècle," in *Nature, histoire, société: Essais en hommage à Jacques Roger,* ed. Claude Blanckaert et al. ([Paris]: Klincksieck, 1995), 321–26; Geoffrey V. Sutton, *Science for a Polite Society: Gender, Culture, and the Demonstration of Enlightenment* (Boulder, Colo.: Westview, 1995); Alain Niderst, ed., "La diffusion des sciences au XVIIIe siècle," special issue of *Revue d'histoire des sciences* 44, no. 3–4 (1991); "Animalité et anthropomorphisme dans la diffusion et la vulgarisation scientifiques," special issue of *Revue de synthèse,* ser. 4, no. 3–4 (1992); Michael R. Lynn, "Enlightenment in the Republic of Science: The Popularization of Natural Philosophy in Eighteenth-Century France," Ph.D. diss., U of Wisconsin–Madison, 1997; Gilles Chabaud, "Entre sciences et sociabilités: Les expériences de l'illusion artificielle en France à la fin du XVIIIe siècle," *Bulletin de la Société d'histoire moderne et contemporaine* (1997, no. 3–4): 36–49. On approaches to popular science, see Roger Cooter and Stephen Pumphrey, "Separate Spheres and Public Places: Reflections on the History of Science Popularization and Science in Popular Culture," *History of Science* 32 (1994): 237–67; Jonathan R. Topham, "Beyond the 'Common Context': The Production and Reading of the Bridgewater Treatises," *Isis* 89 (1998): 233–62.

Synthetic works on changing attitudes toward nature in the West include Clarence J. Glacken, *Traces on the Rhodian Shore: Nature and Culture in Western Thought from Ancient Times to the End of the Eighteenth Century* (Berkeley: U of California P, 1967); Carolyn Merchant, *The Death of Nature: Women, Ecology, and the Scientific Revolution* (1980; paperback, San Francisco: Harper and Row, 1983); Keith Thomas, *Man and the Natural World: A History of the Modern Sensibility* (New York: Pantheon, 1983); Richard H. Grove, *Green Imperialism: Colonial Expansion, Tropical Island Edens, and the Origins of Environmentalism, 1600–1860* (Cambridge: Cambridge UP, 1995); Lorraine Daston and Katharine Park, *Wonders and the Order of Nature, 1150–1750* (New York: Zone Books, 1998). Some important older works on ideas about nature in France are Daniel Mornet, *Le sentiment de la nature en France de J.-J. Rousseau à Bernardin de Saint-Pierre* (1907; repr. New York: Burt Franklin, n.d.); Mornet, *Les sciences de la nature en France, au XVIIIe siècle* (Paris: Armand Colin, 1911); Jean Ehrard, *L'idée de nature en France dans la première moitié du XVIIIe siècle* (1963; repr. Paris: Albin Michel, 1994); Geoffroy Atkinson, *The Sentimental Revolution: French Writers of 1690–1740* (Seattle: U of Washington P, 1965); Robert Lenoble, *Esquisse d'une histoire de l'idée de nature* (Paris: Albin Michel, 1969). On the "preromantic" period, see Christopher Thacker, *The Wildness Pleases: The Origins of Romanticism* (London: Croom Helm; New York: St. Martin's, 1983); D. G. Charlton, *New Images of the Natural in France: A Study of Euro-*

pean Cultural History, 1750–1800 (Cambridge: Cambridge UP, 1984); Roy Porter, "The New Taste for Nature in the Eighteenth Century," *Linnean* 4, no. 1 (1988): 14–30.

Much of the work on animals in France has focused on philosophical issues: on the seventeenth and eighteenth centuries, see George Boas, *The Happy Beast in French Thought of the Seventeenth Century* (Baltimore: Johns Hopkins P, 1933); Hester Hastings, *Man and Beast in French Thought of the Eighteenth Century* (Baltimore: Johns Hopkins P, 1936); Leonora Cohen Rosenfield, *From Beast-Machine to Man-Machine: Animal Soul in French Letters from Descartes to La Mettrie* (New York: Oxford UP, 1940; rcpr. New York: Octagon Books, 1968); Virginia Dawson, *Nature's Enigma: The Problem of the Polyp in the Letters of Bonnet, Trembley, and Réaumur* (Philadelphia: American Philosophical Society, 1987); Elizabeth Anderson, intro. to Charles-Georges Le Roy, *Lettres sur les animaux*, ed. Elizabeth Anderson (Oxford: Voltaire Foundation 1994), 1–71. For a vast survey from Plato to Derrida, see Elisabeth de Fontenay, *Le silence des bêtes: La philosophie à l'épreuve de l'animalité* (Paris: Fayard, 1998).

For histories of the animals themselves, see Alfred Franklin, *La vie privée d'autrefois: Les animaux, du XVe au XIXe siècle*, 2 vols. (Paris: Plon, 1897–99); Robert Delort, *Les animaux ont une histoire* (Paris: Seuil, 1984); and a special issue of *Cahiers d'histoire* (1997, nos. 3–4) that includes several interesting articles and an extensive bibliography and historiography by Éric Baratay and Jean-Luc Mayaud. For pets and menagerie and show animals, see below.

Works on animals in England include Thomas, *Man and the Natural World;* Harriet Ritvo, *The Animal Estate: The English and Other Creatures in the Victorian Age* (Cambridge: Harvard UP, 1987); Moira Ferguson, *Animal Advocacy and Englishwomen, 1780–1900* (Ann Arbor: U of Michigan P, 1998); Barbara Gates, *Kindred Nature: Victorian and Edwardian Women Embrace the Living World* (Chicago: U of Chicago P, 1998); Erica Fudge, *Perceiving Animals: Humans and Beasts in Early Modern English Culture* (New York: St. Martin's, 2000).

Many anthropologists and sociologists have explored the symbolic and cultural meanings of animals; see esp. Emiko Ohnuki-Tierney, *The Monkey as Mirror: Symbolic Transformations in Japanese History and Ritual* (Princeton: Princeton UP, 1987); R. G. Willis, ed., *Signifying Animals: Human Meaning in the Natural World* (London: Unwin Hyman, 1990); Steve Baker, *Picturing the Beast: Animals, Identity, and Representation* (Manchester: Manchester UP, 1993); Arnold Arluke and Clinton R. Sanders, *Regarding Animals* (Philadelphia: Temple UP, 1996); Elizabeth Atwood Lawrence, *Hunting the Wren: Transformation of Bird to Symbol: A Study in Human-Animal Relationships* (Knoxville: U of Tennessee P, 1997).

On domestication and domestic animals, see Juliet Clutton-Brock, *A Natural*

History of Domesticated Animals (Cambridge: Cambridge UP, 1987); Jean-Pierre Digard, *L'homme et les animaux domestiques: Anthropologie d'une passion* (Paris: Fayard, 1990).

Chapter 1. Live Cargo

For overseas trade in all of Europe, see Michel Devèze, *L'Europe et le monde à la fin du XVIIIe siècle* (Paris: Albin Michel, 1970); Ralph Davis, *The Rise of the Atlantic Economies* (Ithaca, N.Y.: Cornell UP, 1973); Frédéric Mauro, *L'expansion européenne, 1600–1870*, 4th ed. (Paris: PUF, 1996) (includes an extensive bibliography); Liliane Hilaire-Pérez, *L'expérience de la mer: Les Européens et les espaces maritimes au XVIIIe siècle* (Paris: Seli Arslan, 1997). For a summary of French trade, see Paul Butel, "France, the Antilles, and Europe in the Seventeenth and Eighteenth Centuries: Renewals of Foreign Trade," in *The Rise of Merchant Empires: Long-Distance Trade in the Early Modern World, 1350–1750*, ed. James D. Tracy (Cambridge: Cambridge UP, 1990), 153–73. Much of the best work on French trade is in histories of the major trading cities; see references in works just cited. The big books on the East India trade are Louis Dermigny, *La Chine et l'Occident: Le commerce à Canton au XVIIIe siècle, 1719–1833*, 3 vols. (Paris: SEVPEN, 1964); and Philippe Haudrère, *La Compagnie française des Indes au XVIIIe siècle*, 4 vols. (Paris: Librairie de l'Inde, 1989); see also Claude Nières, ed., *Histoire de Lorient* (Toulouse: Privat, 1988). For a comparative perspective, see Holden Furber, *Rival Empires of Trade in the Orient, 1600–1800* (Minneapolis: U of Minnesota P, 1976).

For the French slave trade, see Robert Louis Stein, *The French Slave Trade in the Eighteenth Century: An Old Regime Business* (Madison: U of Wisconsin P, 1979); Pierre Pluchon, *La route des esclaves, négriers et bois d'ébène au XVIIIe siècle* (Paris: Hachette, 1981); Serge Daget, *La traite des Noirs: Bastilles négrières et velléités abolitionnistes* ([Rennes]: Ouest-France Université, 1990); Jean-Michel Deveau, *La traite rochelaise* (Paris: Karthala, 1990) (daily life on a slave ship); Deveau, *La France au temps des négriers* (Paris: France-Empire, 1994). For Atlantic slavery in general, see Robin Blackburn, *The Making of New World Slavery* (London: Verso, 1997).

Several general works provide overviews of French colonial efforts: Herbert Ingram Priestley, *France Overseas through the Old Régime: A Study of European Expansion* (New York: D. Appleton-Century, 1939); Jean Meyer et al., *Histoire de la France coloniale des origines à 1914* (Paris: Armand Colin, 1991); Pierre Pluchon, *Histoire de la colonisation française*, 1: *Le premier empire colonial: Des origines à la Restauration* (Paris: Fayard, 1991); Philippe Haudrère, *L'empire des rois, 1500–1789* (Paris: Denoël, 1997). On the French in India, see S. P. Sen, *The French in India*,

1763–1816 (Calcutta: Firma K. L. Mukhopadhyay, 1958); Sudipta Das, *Myths and Realities of French Imperialism in India, 1763–1783* (New York: Peter Lang, 1992).

The most complete work on French voyages of exploration is Numa Broc, *La géographie des philosophes: Géographes et voyageurs français au XVIIIe siècle* (Paris: Ophrys, 1975). For biographies of naturalists abroad, see Alfred Lacroix, *Figures de savants*, vols. 3 and 4 (Paris: Gauthier-Villars, 1938). See also P. Fournier, *Voyages et découvertes scientifiques des missionnaires naturalistes français à travers le monde pendant cinq siècles: XVe à XXe siècles* (Paris: Paul Lechevalier & fils, 1932); P. Huard and M. Wong, "Les enquêtes scientifiques françaises et l'exploration du monde exotique aux XVIIe et XVIIIe siècles," *Bulletin de l'École française d'Extrême-Orient* 52 (1964): 143–56 (includes list of voyages); John Dunmore, *French Explorers in the Pacific*, 2 vols. (Oxford: Clarendon Press, 1965–69); Yves Laissus, "Les voyageurs naturalistes du Jardin du roi et du Muséum d'histoire naturelle: Essai de portrait-robot," *Revue d'histoire des sciences* 34 (1981): 259–317; Jean Lescure, "L'épopée des voyageurs naturalistes aux Antilles et en Guyane," in *Voyage aux îles d'Amérique* (Paris: Archives nationales, 1992), 59–70; Marie-Noëlle Bourguet, "Voyage, mer et science au XVIIIe siècle," *Bulletin de la Société d'histoire moderne et contemporaine* (1997, 1–2): 39–56; Bourguet, "Voyage et histoire naturelle (fin XVIIe siècle–début XIXe siècle)," in *Le Muséum au premier siècle de son histoire*, ed. Claude Blanckaert et al. (Éditions du Muséum national d'histoire naturelle, 1997), 163–96. On specific voyages, see this chapter's notes.

On voyage accounts, see Michèle Duchet, *Anthropologie et histoire au siècle des Lumières* (1971; repr. Paris: Albin Michel, 1995), 65–136; Roger Chartier, "Les livres de voyage," in *Histoire de l'édition française*, 2: *Le livre triomphant, 1660–1830*, ed. Henri-Jean Martin and Roger Chartier (Paris: Promodis, 1984), 216–17; Pierre Berthiaume, *L'aventure américaine au XVIIIe siècle: Du voyage à l'écriture* (Ottawa: P de l'Univ. d'Ottawa, 1990); Mary Louise Pratt, *Imperial Eyes: Travel Writing and Transculturation* (New York: Routledge, 1992).

For a general history of Western (esp. French) attitudes toward the exotic, see André Bourde, "Histoire de l'exotisme," in *Encyclopédie de la Pléiade, Histoire des Moeurs*, 3: *Thèmes et systèmes culturels*, ed. Jean Poirier (Paris: Gallimard, 1991), 598–701. Also see works on natural history and natural history cabinets listed below. On "exotic" peoples, see Gilbert Chinard, *L'Amérique et le rêve exotique dans la littérature française au XVIIe et au XVIIIe siècle* (Paris: E. Droz, 1934); Daniel Droixhe and Pol-P. Gossiaux, eds., *L'homme des Lumières et la découverte de l'Autre* (Bruxelles: Éditions de l'Univ. de Bruxelles, 1985); G. S. Rousseau and Roy Porter, eds., *Exoticism in the Enlightenment* (Manchester: Manchester UP, 1990); Julia V. Douthwaite, *Exotic Women: Literary Heroines and Cultural Strategies in Ancien Régime France* (Philadelphia: U of Pennsylvania P, 1992); Anthony Pagden, *European En-*

counters with the *New World: From Renaissance to Romanticism* (New Haven: Yale UP, 1993); Claude Blanckaert, "Postface: Les archives du genre humain," in Duchet, *Anthropologie et histoire* (1995 ed.), 565–608.

There has been little work on the importation of exotic animals in the eighteenth century other than literature on menageries (see below) and Jean-Bernard Lacroix, "L'approvisionnement des ménageries et les transports d'animaux sauvages par la Compagnie des Indes au XVIIIe siècle," *Revue française d'histoire d'outre-mer* 65 (1978): 153–79. Liliane Bodson, ed., *Les animaux exotiques dans les relations internationales: Espèces, fonctions, significations* (Liège: U de Liège, 1998) includes one article on nineteenth-century France. On nineteenth-century animal importations, see Michael A. Osborne, *Nature, the Exotic, and the Science of French Colonialism* (Bloomington: Indiana UP, 1994); Nigel Rothfels, "Bring 'Em Back Alive: Carl Hagenbeck and the Exotic Animal and People Trades in Germany, 1848–1914," Ph.D. diss., Harvard Univ., 1994; Elizabeth Anne Hanson, "Nature Civilized: A Cultural History of American Zoos, 1870–1940," Ph.D. diss., Univ. of Pennsylvania, 1996.

Chapter 2. The Royal Menagerie

The foundational work on menageries is Gustave Loisel, *Histoire des ménageries de l'antiquité à nos jours,* 3 vols. (Paris: Octave Doin et fils; Henri Laurens, 1912). See also R. J. Hoage and William A. Deiss, eds., *New Worlds, New Animals: From Menagerie to Zoological Park in the Nineteenth Century* (Baltimore: Johns Hopkins UP, 1996); and Éric Baratay and Élisabeth Hardouin-Fugier, *Zoos: Histoire des jardins zoologiques en occident (XVIe–XXe siècle)* (Paris: Éditions La Découverte, 1998). On the Versailles menagerie, see Gérard Mabille, "La ménagerie de Versailles," *Gazette des Beaux-Arts* 83 (1974): 5–36; Masumi Iriye, "Le Vau's Menagerie and the Rise of the *Animalier:* Enclosing, Dissecting, and Representing the Animal in Early Modern France," Ph.D. diss., Univ. of Michigan, 1994.

Works on menageries in other European countries include Wilma George, "Alive or Dead: Zoological Collections in the Seventeenth Century," in *The Origins of Museums: The Cabinet of Curiosities in Sixteenth- and Seventeenth-Century Europe,* ed. Oliver Impey and Arthur MacGregor (Oxford: Clarendon Press, 1985), 179–87; *Hendrik Engel's Alphabetical List of Dutch Zoological Cabinets and Menageries,* 2nd, enlarged ed., by Pieter Smit with A. P. M. Sanders and J. P. F. van der Veer (Amsterdam: Rodopi, 1986); F. F. J. M. Pieters and M. F. Mörzer Bruyns, "Menagerieën in Holland in de 17e en 18e eeuw," *Holland, regionaal-historisch Tijdschrift* 20, no. 4/5 (1988): 195–209; Sally Festing, "Menageries and the Landscape Garden," *Journal of Garden History* 8, no. 4 (1988): 104–17 (includes an inventory of English menageries to 1800); B. C. Sliggers and A. A. Wertheim, eds.,

Een vorstelijke dierentuin: De menagerie van Willem V/Le zoo du prince: La ménagerie du stathouder Guillaume V (n.p.: Walburg Instituut, 1994). See also Ritvo, *Animal Estate;* Bob Mullan and Garry Marvin, *Zoo Culture* (London: Weidenfeld & Nicolson, 1987); Robert W. Jones, "'The Sight of Creatures Strange to our Clime': London Zoo and the Consumption of the Exotic," *Journal of Victorian Culture* 2, no. 1 (1997): 1–26; Randy Malamud, *Reading Zoos: Representations of Animals and Captivity* (New York: New York UP, 1998).

Chapter 3. Fairs and Fights

For traveling menageries and animals as popular entertainment, see Émile Campardon, *Les spectacles de la foire,* 2 vols. (Paris: Berger-Levrault, 1877); Victor Fournel, *Le vieux Paris: Fêtes, jeux et spectacles* (Tours: Alfred Mame et fils, 1887); Robert Isherwood, *Farce and Fantasy: Popular Entertainment in Eighteenth-Century Paris* (Oxford: Oxford UP, 1986); Loisel, *Histoire des ménageries,* 2:275–84. On roving rhinoceroses, see L. C. Rookmaaker, *Bibliography of the Rhinoceros: An Analysis of the Literature on the Recent Rhinoceroses in Culture, History and Biology* (Rotterdam: A. A. Balkema, 1983); T. H. Clarke, *The Rhinoceros from Dürer to Stubbs, 1515–1799* (London: Sotheby's Publications, 1986).

On fairs and boulevard entertainment in general, see also Arthur Heulhard, *La foire Saint-Laurent: Son histoire et ses spectacles* (Paris, 1878; repr. Geneva: Slatkine, 1971); Martine de Rougement, *La vie théâtrale en France au XVIIIe siècle* (1988; repr. Geneva: Slatkine, 1996), 261–78; Lynn, "Enlightenment in the Republic of Science."

On animal fights, see Georges Bertin, "Les combats de taureaux à Paris (1781–1833)," *Revue de la Révolution* 9 (1887): 160–68; Fournel, *Le vieux Paris,* 449–55; "Les courses de taureaux en France sous l'ancien régime," *Archives historiques, artistiques et littéraires* 1 (1889–90): 29–30; "Interdiction des combats de taureaux par le Département municipal de police," in *Actes de la Commune de Paris pendant la Révolution,* ed. Sigismond Lacroix (Paris: L. Cerf, 1898), 7:544–50; Loisel, *Histoire des ménageries,* 2:282–84; and Isherwood, *Farce and Fantasy,* 209–11.

Chapter 4. The Oiseleurs' Guild

Very little literature exists on commerce in exotic animals in the eighteenth century or earlier. See G. Musset, "Les collectionneurs de bêtes sauvages (1047–1572)," *Bulletin du Muséum d'histoire naturelle* (1902): 242–43; E.-T. Hamy, "Le commerce des animaux exotiques à Marseille à la fin du XVIe siècle," *Bulletin du Muséum d'histoire naturelle* 7 (1903): 316–18.

On *corporations* in ancien-régime France, see François Olivier-Martin, *L'organ-*

isation corporative de la France d'ancien régime (Paris: Recueil Sirey, 1939); Émile Coornaert, *Les corporations en France avant 1789,* 7th ed. (Paris: Gallimard, 1941). On guilds, see René de Lespinasse, *Les métiers et corporations de la ville de Paris,* 3 vols. (Paris: Imprimerie nationale, 1886–97); Étienne Martin-Saint-Léon, *Histoire des corporations de métiers depuis leurs origines jusqu'à leur suppression en 1791,* 4th ed. (Paris: PUF, 1941); Steven L. Kaplan, "The Luxury Guilds in Paris in the Eighteenth Century," *Francia* 9 (1981): 257–98; Michael Sonenscher, *Work and Wages: Natural Law, Politics, and the Eighteenth-Century French Trades* (Cambridge: Cambridge UP, 1989); Jacques Revel, ed., "Corps et communautés d'ancien régime," spec. issue, *Annales ESC* (Mar.-Apr. 1988, no. 2); Abel Poitrineau, *Ils travaillaient la France: Métiers et mentalités du XVIe au XIXe siècle* (Paris: Armand Colin, 1992); Steven L. Kaplan, *The Bakers of Paris and the Bread Question, 1700–1775* (Durham, N.C.: Duke UP, 1996); Carolyn Sargentson, *Merchants and Luxury Markets: The Marchands Merciers of Eighteenth-Century Paris* (London: Victoria and Albert Museum with J. Paul Getty Museum, 1996); Robert Fox and Anthony Turner, eds., *Luxury Trades and Consumerism in "Ancien Régime" Paris: Studies in the History of the Skilled Workforce* (Aldershot, U.K.: Ashgate, 1998).

Chapter 5. Pampered Parrots

Alfred Franklin *(Les animaux)* collected stories on exotic pets in ancien-régime France, and Hester Hastings discusses ideas about pets in *Man and Beast,* 206–16. On pets in nineteenth-century France, see Kathleen Kete, *The Beast in the Boudoir: Petkeeping in Nineteenth-Century Paris* (Berkeley: U of California P, 1994).

For other works on the history of pet keeping, see Yi-Fu Tuan, *Dominance and Affection: The Making of Pets* (New Haven: Yale UP, 1984); James Serpell, *In the Company of Animals: A Study of Human-Animal Relationships* (Oxford: Basil Blackwell, 1986); Ritvo, *The Animal Estate;* Digard, *L'homme et les animaux domestiques;* Liliane Bodson, ed., *L'animal de compagnie: ses rôles et leurs motivations au regard de l'histoire* (Liège: U de Liège, 1997). On animals and sympathy in eighteenth-century England, see Jonathan Lamb, "Modern Metamorphoses and Disgraceful Tales," *Critical Inquiry,* special issue, fall 2001.

On criticism of luxury, see Sarah Maza, "Luxury, Morality, and Social Change: Why There Was No Middle-Class Consciousness in Prerevolutionary France," *Journal of Modern History* 69 (1997): 199–229; Morag Martin, "Consuming Beauty: The Commerce of Cosmetics in France, 1750–1800," Ph.D. diss., Univ. of California, Irvine, 1999. On women, consumption, and morality, see Jennifer Jones, "*Coquettes* and *Grisettes:* Women Buying and Selling in Ancien Régime Paris," in *The Sex of Things: Gender and Consumption in Historical Perspective,* ed. Victoria de Grazia, with Ellen Furlough (Berkeley: U of California P, 1996), 25–53; Sylvana

Tomaselli, "The Enlightenment Debate on Women," *History Workshop Journal* 20 (1985): 101–24; Tjitske Akkerman, *Women's Vices, Public Benefits: Women and Commerce in the French Enlightenment* (Amsterdam: Het Spinhuis, 1992); Lieselotte Steinbrügge, *The Moral Sex: Woman's Nature in the French Enlightenment*, trans. Pamela E. Selwyn (New York: Oxford UP, 1995).

Chapter 6. Animals in Print

Much excellent work exists on the history of reading and publishing. For eighteenth-century France, see Robert Darnton, *The Literary Underground of the Old Regime* (Cambridge: Harvard UP, 1982); Jean Sgard, "La multiplication des périodiques," in *Histoire de l'édition française* 2:198–205; Roger Chartier, *Lectures et lecteurs dans la France d'ancien régime* (Paris: Seuil, 1987); Daniel Roche, *Les républicains des lettres: Gens de culture et Lumières au XVIIIe siècle* (Paris: Fayard, 1988); Robert Darnton, "What Is the History of Books?" and "First Steps toward a History of Reading," both in *The Kiss of Lamourette: Reflections in Cultural History* (New York: W. W. Norton, 1990); James Smith Allen, "From the History of the Book to the History of Reading: Review Essay," *Libraries and Culture* 28 (1993): 319–26; Jack R. Censer, *The French Press in the Age of Enlightenment* (London: Routledge, 1994); Darnton, *The Forbidden Best-Sellers of Pre-Revolutionary France* (New York: W. W. Norton, 1996). Daniel Mornet's *Les origines intellectuelles de la Révolution française (1715–1787)* (Paris: Armand Colin, 1933) is stimulating for its consideration of the relation between reading and revolution. Also see articles in *Revue française d'histoire du livre*.

On passionate reading, see Robert Darnton, "Readers Respond to Rousseau: The Fabrication of Romantic Sensitivity," in *The Great Cat Massacre and Other Episodes in French Cultural History* (New York: Basic Books, 1984), 215–56; Anne Vincent-Buffault, *Histoire des larmes* (Paris: Rivages, 1986); Jean Marie Goulemot, *Ces livres qu'on ne lit que d'une main: Lectures et lecteurs de livres pornographiques au XVIIIe siècle* (Aix-en-Provence: Alinéa, 1991).

On medieval bestiaries, see Florence McCulloch, *Medieval Latin and French Bestiaries*, Studies in the Romance Languages and Literatures 33 (Chapel Hill: U of North Carolina P, 1960); Willene B. Clark and Meradith T. McMunn, eds., *Beasts and Birds of the Middle Ages: The Bestiary and Its Legacy* (Philadelphia: U of Pennsylvania P, 1989). On emblem books, see William B. Ashworth Jr., "Emblematic Natural History of the Renaissance," in *Cultures of Natural History*, ed. Jardine et al., 17–37. Éric Baratay discusses religious animal symbolism in *L'Église et l'animal (France, XVIIe–XXe siècle)* (Paris: Éditions du Cerf, 1996). On Aldrovandi, see Paula Findlen, *Possessing Nature: Museums, Collecting, and Scientific Culture in Early Modern Italy* (Berkeley: U of California P, 1994). For a discussion of *vraisemblance*, see Bernard Tocanne, *L'idée de nature en France dans la seconde moitié du XVIIe siècle: Contribution à l'histoire de la pensée classique* ([Paris]: Klincksieck, 1978); and, on truth, fiction, and science, Erica Harth, *Ideology and Culture in Seventeenth-Century France* (Ithaca, N.Y.: Cornell UP, 1983). On natural history in early modern England, see Charles E. Raven, *English Naturalists from Neckham to Ray: A Study of the Making of the Modern World* (Cambridge: Cambridge UP, 1947).

Several authors, most notably Jean-Noël Pascal, have begun to explore the neglected area of eighteenth-century fables. See Pascal, ed., *La fable au siècle des Lumières, 1715–1815: Anthologie des successeurs de La Fontaine, de La Motte à Jauffret* (Saint-Étienne: Publications de l'Univ. de Saint-Étienne, 1991); Pascal, "Les successeurs de La Fontaine et la poétique de la fable, de La Motte à Florian," in *Fables et fabulistes: Variations autour de La Fontaine,* ed. Michel Bideaux et al. (Mont-de-Marsan: Éditions InterUniversitaires, 1992), 171–201; Pascal, *Anthologie des fabulistes français de La Fontaine au Romantisme* (Étoile-sur-Rhône: Nigel Gauvin, 1993); Pascal, ed., *Les fables de Florian* (Perpignan: PU de Perpignan, 1995); Pascal, *Les successeurs de La Fontaine au siècle des Lumières (1715–1815)* (New York: Peter Lang, 1995). See also Roseann Runte, "The Paradox of the Fable in Eighteenth-Century France," *Neophilologus* 61 (1977): 510–17; Runte, "From La Fontaine to Porchat: The Bee in the French Fable," *Studies in Eighteenth-Century Culture* 18 (1988): 78–89; and Friederike Hassauer, *Die Philosophie der Fabeltiere* (Munich: Wilhelm Fink Verlag, 1986). On La Fontaine, see notes to chapter 6.

On eighteenth-century education, see Dominique Julia, "Les recherches sur l'histoire de l'éducation en France au siècle des Lumières," *Histoire de l'éducation* 1 (1978): 17–38; François Lebrun, Marc Venard, and Jean Quéniart, *Histoire générale de l'enseignement et de l'éducation en France,* 2: *De Gutenberg aux Lumières* (Paris: Nouvelle libraire de France, 1981); Dominique Julia, "Livres de classe et usages pédagogiques," in *Histoire de l'édition française,* 2:468–97; Jean Bloch, *Rousseauism and Education in Eighteenth-Century France* (Oxford: Voltaire Foundation, 1995); Marcel Grandière, *L'idéal pédagogique en France au dix-huitième siècle* (Oxford: Voltaire Foundation, 1998). Several dozen educational treatises are collected in the microfiche series "La réforme de l'enseignement au siècle des Lumières," ed. Dominique Julia. On children's literature in France, see Mornet, *Sciences de la nature,* 216–25; François Caradec, *Histoire de la littérature enfantine* (Paris: Albin Michel, 1977).

For eighteenth-century natural history, see Michel Foucault, *The Order of Things: An Archaeology of the Human Sciences* (New York: Vintage, 1973); Wolf Lepenies, "De l'histoire naturelle à l'histoire de la nature," *Dix-huitième siècle* 11 (1979): 175–84; Jacques Roger, "The Living World," in *The Ferment of Knowledge: Studies in the Historiography of Eighteenth-Century Science,* ed. G. S. Rousseau and Roy Porter (Cambridge: Cambridge UP, 1980), 255–83; Phillip R. Sloan, "Natural

History, 1670–1802," in *Companion to the History of Modern Science*, ed. R. C. Olby et al. (London: Routledge, 1990), 295–313; articles in N. Jardine et al., *Cultures of Natural History*; Harriet Ritvo, *The Platypus and the Mermaid and Other Figments of the Classifying Imagination* (Cambridge: Harvard UP, 1997); E. C. Spary, *Utopia's Garden: French Natural History from Old Regime to Revolution* (Chicago: U of Chicago P, 2000). Daniel Mornet's classic *Les sciences de la nature* is still a superb source. For Linnaeus, see Lisbet Koerner, *Linnaeus: Nature and Nation* (Cambridge: Harvard UP, 1999).

On natural history cabinets in France, see Yves Laissus, "Les cabinets d'histoire naturelle," in *Enseignement et diffusion des sciences en France au XVIIIe siècle*, ed. René Taton (Paris: Hermann, 1964), 659–712 (with a list of close to two hundred in Paris alone); Antoine Schnapper, *Le géant, la licorne, la tulipe: Collections françaises au XVIIe siècle* (Paris: Flammarion, 1988); Krzysztof Pomian, *Collectors and Curiosities: Paris and Venice, 1500–1800*, trans. Elizabeth Wiles-Portier (Cambridge: Polity Press, 1990); Katie Whitaker, "The Culture of Curiosity," in *Cultures of Natural History*, ed. Jardine et al., 75–90. On the Jardin du roi, see Yves Laissus, "Le Jardin du roi," in *Enseignement et diffusion des sciences*, ed. Taton, 287–341; Spary, *Utopia's Garden*.

The definitive intellectual biography of Buffon is Jacques Roger, *Buffon: Un philosophe au Jardin du roi* (Paris: Fayard, 1989) (Eng. trans., 1997). For access to the literature on Buffon, see E. Genet-Varcin and Jacques Roger, "Bibliographie de Buffon," in *Oeuvres philosophiques de Buffon*, ed. Jean Piveteau (Paris: PUF, 1954), 513–70; Marie-Françoise Lafon, "Bibliographie de Buffon (1954–1991)," in *Buffon 88*, ed. Jean Gayon (Paris: J. Vrin, 1992), 691–743; John H. Eddy Jr., "Buffon's *Histoire naturelle*: History? A Critique of Recent Interpretations," *Isis* 85 (1994): 644–61; Phillip R. Sloan, "Buffon Studies Today," *History of Science* 32 (1994): 469–77. For a collection of some of Buffon's discourses and contemporary reviews, see John Lyon and Phillip R. Sloan, eds., *From Natural History to the History of Nature: Readings from Buffon and His Critics* (Notre Dame, Ind.: U of Notre Dame P, 1981). On the publishing history of the *Histoire naturelle*, see Georges Heilbrun, "Essai de bibliographie," in *Buffon*, ed. Muséum national d'histoire naturelle (Paris: Publications françaises, 1952), 225–37; Paul-Marie Grinevald, "Les éditions de l'*Histoire naturelle*," in *Buffon 88*, ed. Gayon, 631–37; L. C. Rookmaaker, "J. N. S. Allamand's Additions (1769–1781) to the *Nouvelle Édition* of Buffon's *Histoire naturelle* Published in Holland," *Bijdragen tot de Dierkunde* 61, no. 3 (1992): 131–62.

Only a few works since Mornet's *Les sciences de la nature en France* have paid significant attention to Buffon's style or popularity; see Charles Bruneau, "Buffon et le problème de la forme," in *Oeuvres philosophiques de Buffon*, 491–99; Jacques Roger, introduction to Buffon, *Les époques de la nature*, critical ed. by J. Roger

(Paris: Éditions du Muséum national d'histoire naturelle, 1962), cxiv–cxxvii; Otis E. Fellows and Stephen F. Milliken, *Buffon* (New York: Twayne, 1972), 148–70; Michel Espagne, "'Le style est l'homme même': *A priori* esthétique et écriture scientifique chez Buffon et Winckelmann," in *Leçons d'écriture: Ce que disent les manuscrits*, ed. Almuth Grésillon and Michaël Werner (Paris: Lettres modernes, Minard, 1985), 51–67; Pietro Corsi, "Buffon sous la Révolution et l'Empire," in *Buffon 88*, ed. Gayon, 639–48; Jeff Loveland, "Rhetoric and Science in Buffon's Natural History," Ph.D. diss., Duke Univ., 1994; Spary, *Utopia's Garden*, chap. 3.

Chapter 7. Elephant Slaves

On the noble savage, see Gilbert Chinard, *L'Amérique et le rêve exotique dans la littérature française au XVIIe et au XVIIIe siècle* (1913; repr. Paris: E. Droz, 1934); Bernard Smith, *European Vision and the South Pacific, 1768–1850: A Study in the History of Art and Ideas* (Oxford: Clarendon Press, 1960); Hélène Clastres, "Sauvages et civilisés au XVIIIe siècle," in *Histoire des idéologies*, 3: *Savoir et pouvoir du XVIIIe au XXe siècle*, ed. François Châtelet (n.p.: Hachette, 1978), 209–28; Hayden White, "The Noble Savage Theme as Fetish," in *Tropics of Discourse: Essays in Cultural Criticism* (Baltimore: Johns Hopkins UP, 1978), 183–96.

On racism and French views of Africans, see Roger Mercier, *L'Afrique Noire dans la littérature française: Les premières images (XVIIe–XVIIIe siècles)* (Dakar: Publications de la section de langues et littératures, Univ. de Dakar, 1962); William B. Cohen, *The French Encounter with Africans: White Response to Blacks, 1530–1880* (Bloomington: Indiana UP, 1980); Pierre Pluchon, *Nègres et Juifs au XVIIIe siècle: Le racisme au siècle des Lumières* ([Paris]: Tallandier, 1984); Pierre H. Boulle, "In Defense of Slavery: Eighteenth-Century Opposition to Abolition and the Origins of a Racist Ideology in France," in *History from Below: Studies in Popular Protest and Popular Ideology in Honour of George Rudé*, ed. Frederick Krantz (Montreal: Concordia Univ., 1985), 221–41. See also works on the "exotic," above.

On anticolonialism and antislavery movements, see Russell Jameson, *Montesquieu et l'esclavage: Étude sur les origines de l'opinion antiesclavagiste en France au XVIIIe siècle* (Paris: Hachette, 1911); Carl Ludwig Lokke, *France and the Colonial Question: A Study of Contemporary French Opinion, 1763–1801* (1932; repr. New York: Octagon Books of Farrar, Straus & Giroux, 1976); Edward Derbyshire Seeber, *Anti-Slavery Opinion in France during the Second Half of the Eighteenth Century* (1937; repr. New York: Burt Franklin, 1971); Duchet, *Anthropologie et histoire;* David Brion Davis, *The Problem of Slavery in the Age of Revolution, 1770–1823* (Ithaca, N.Y.: Cornell UP, 1975); Yves Benot, *La Révolution française et la fin des colonies* (Paris: La Découverte, 1987); Thomas Bender, ed., *The Antislavery Debate: Capitalism and Abolitionism as a Problem of Historical Interpretation* (Berkeley: U of

California P, 1992); Shanti Marie Singham, "Betwixt Cattle and Men: Jews, Blacks, and Women, and the Declaration of the Rights of Man," in *The French Idea of Freedom: The Old Regime and the Declaration of Rights of 1789*, ed. Dale Van Kley (Stanford: Stanford UP, 1994), 114–53; Srinivas Aravamudan, *Tropicopolitans: Colonialism and Agency, 1688–1804* (Durham, N.C.: Duke UP, 1999); and works on colonialism and slavery listed above.

On Raynal, see Yves Benot, *Diderot: De l'athéisme à l'anticolonialisme* (Paris: François Maspero, 1970), 162–259; Michèle Duchet, *Diderot et l'Histoire des deux Indes, ou L'écriture fragmentaire* (Paris: A. G. Nizet, 1978); Hans-Jürgen Lüsebrink and Manfred Tietz, eds., *Lectures de Raynal: L'Histoire des deux Indes en Europe et en Amérique au XVIIIe siècle*, Studies on Voltaire and the Eighteenth Century 286 (Oxford: Voltaire Foundation, 1991); Hans-Jürgen Lüsebrink and Anthony Strugnell, eds., *L'Histoire des deux Indes: réécriture et polygraphie*, Studies on Voltaire and the Eighteenth Century, 333 (Oxford: Voltaire Foundation, 1995); Aravamudan, *Tropicopolitans*.

On the history of comparisons between subordinate human groups and animals, see Mary Midgley, *Animals and Why They Matter* (Athens: U of Georgia P, 1983); Harriet Ritvo, "Border Trouble: Shifting the Line between People and Other Animals," *Social Research* 62 (1995): 481–500. On slaves and animals, see David Brion Davis, *The Problem of Slavery in Western Culture* (Ithaca, N.Y.: Cornell UP, 1966); Marjorie Spiegel, *The Dreaded Comparison: Human and Animal Slavery*, 2nd ed. (New York: Mirror Books, 1988); Karl Jacoby, "Slaves by Nature? Domestic Animals and Human Slaves," *Slavery and Abolition* 15 (1994): 89–99; on women and animals, see Sylvana Tomaselli, "The Enlightenment Debate on Women," *History Workshop Journal* 20 (autumn 1985): 101–24; Londa Schiebinger, *Nature's Body: Gender in the Making of Modern Science* (Boston: Beacon, 1993). See also Nancy Leys Stepan, "Race and Gender: The Role of Analogy in Science," *Isis* 77 (1986), 261–77; Singham, "Betwixt Cattle and Men." Jonathan Lamb explores boundary crossing between humans, animals, and things in "Modern Metamorphoses and Disgraceful Tales."

Chapter 8. Vive la Liberté

On science during the revolutionary period and the early nineteenth century, see Hahn, *Anatomy of a Scientific Institution;* Charles Coulston Gillispie, *Science and Polity in France at the End of the Old Regime* (Princeton, N.J.: Princeton UP, 1980); Pietro Corsi, *The Age of Lamarck: Evolutionary Theories in France, 1790–1830*, trans. Jonathan Mandelbaum, rev. ed. (Berkeley: U of California P, 1988); Nicole and Jean Dhombres, *Naissance d'un pouvoir: Sciences et savants en France (1793–1824)* (Paris: Payot, 1989); Keith Michael Baker, "Science and Politics at the End of the Old Regime," in *Inventing the French Revolution: Essays on French Political Culture in the Eighteenth Century* (Cambridge: Cambridge UP, 1990), 153–66; Michael A. Osborne, "Applied Natural History and Utilitarian Ideals: 'Jacobin Science' at the Muséum d'Histoire Naturelle, 1789–1870," in *Re-creating Authority in Revolutionary France*, ed. Bryant T. Ragan Jr. and Elizabeth A. Williams (New Brunswick, N.J.: Rutgers UP, 1992), 125–43.

On the founding of the Muséum d'histoire naturelle, see E.-T. Hamy, "Les derniers jours du Jardin du roi et la fondation du Muséum d'histoire naturelle," in *Centenaire de la fondation du Muséum national d'histoire naturelle* (Paris: Imprimerie nationale, 1893); Camille Limoges, "The Development of the Muséum d'Histoire Naturelle of Paris, c. 1800–1914," in *The Organization of Science and Technology in France, 1808–1914*, ed. Robert Fox and George Weisz (Cambridge: Cambridge UP, 1980), 211–40; Dorinda Outram, "New Spaces in Natural History," in *Cultures of Natural History*, ed. Jardine et al., 249–65; Claude Blanckaert et al., eds., *Le Muséum au premier siècle de son histoire* (Paris: Éditions du Muséum national d'histoire naturelle, 1997); Spary, *Utopia's Garden*.

On the founding of the menagerie at the Jardin des plantes, see E.-T. Hamy, "Les anciennes ménageries royales et la ménagerie nationale fondée le 14 brumaire an II (4 novembre 1793)," *Nouvelles archives du Muséum*, 4e série, 5 (1893): 1–22; Gustave Loisel, "Histoire de la ménagerie du Muséum," *Revue scientifique* 49, no. 2 (1911): 262–77; Loisel, *Histoire des ménageries;* Osborne, "Applied Natural History and Utilitarian Ideals"; Yves Laissus and Jean-Jacques Petter, *Les animaux du Muséum, 1793–1993* (Paris: Muséum national d'histoire naturelle, 1993); Richard W. Burkhardt Jr., "La ménagerie et la vie du Muséum," in *Le Muséum au premier siècle de son histoire*, ed. Blanckaert, 481–508; Baratay and Hardouin-Fugier, *Zoos*.

Académie royale des sciences, 39, 44–45, 170

acclimatization. *See* domestication

Adam, Gabriel (oiseleur), 100, 101

Adanson, Michel, 244 n. 48

advertisements, 116, 117, 120, 123, 126, 128–29, 130, 132, 139, 180

Affiches de Paris (newssheet), 80, 86, 116, 120, 123, 126–29, 137–38, 204. *See also* advertisements; lost-and-found notices

Affiches de Province (newssheet), 62, 91, 97, 162, 172, 175, 177, 181

Africa, 13–14, 24, 27, 47, 54; animals from, 20, 22, 23–24, 27–29, 39, 43, 47, 52–60, 88, 117, 120, 126, 130, 226. *See also* slave trade; *particular locations*

Africans, 49, 195, 201, 202. *See also* blacks; slavery; slave trade

Alexandria, 21, 22

Alletz, Pons Augustin *(Histoire des singes, et autres animaux curieux)*, 165

Almanach forain (journal), 85, 86, 87, 89, 91, 93, 282 n. 67

Amazon parrot, 29, 119, 123, 126, 128

animal displays: audiences at, 85–87; at fairs, 74–78; naturalists and, 87–89; during Revolution, 211–12

animal exhibitors: criticisms of, 91–92; at fairs, 72–80, 88; during Revolution, 220–21; selling birds, 119–20

animal fights: at court, 37–38, 259–60 n. 7; criticism of, 92–93; naturalists and, 88–90; in Paris, 71–72, 80–85; during Revolution, 207–10

animal-human relationship, 4, 6, 151–53, 174, 183–84, 200–201, 203, 204–5, 218. *See also* domination of animals; friendship; pets; sympathy toward animals

animal rights, 6, 235

animal slavery: Buffon on, 2, 190–95, 196, 199–200; criticized, 190–95; Hennebert and Beaurieu on, 190, 192, 197; and menagerie, 219, 224, 228; Mercier on, 134; as metaphor, 5–6, 198–200, 203, 204–5, 233, 235; proposals to alleviate, 195–98

animals. *See individual species*

animals, exotic, definition of, xiii

animals, wild. *See* wild animals

antelope (hartebeest), 217, 222

Antilles. *See* Guadeloupe; Martinique; Saint Domingue; West Indies

antislavery movement, 202–5

Aristotle, 170, 175

armadillo, 68

Astley (performer), 145, 149

attachment, animal-human, 151–52. *See also* sympathy toward animals

Australia, 227

Avant-coureur (journal), 1, 61, 91–92, 97, 99, 118, 161

aviaries, 28, 43, 105, 110, 113, 125, 132, 134, 198–99

baboon, 52, 72, 80

Bachelier, Jean Jacques, 84

Bacon, Francis, 61

badger, 52

Bailly, Jean Sylvain, 209

Bajon, Bertrand, 35

Barbot, Jean, 28–29

Barrère, Pierre, 247 n. 70

Bastriès, Gérard Auguste, 107, 117, 120

Baudin, Nicolas, 19, 225, 227

Baudouin, Mlle (taxidermist), 139

Bazin, Gilles-Augustin, 163–64

bear, 45, 71, 83, 85, 88, 89, 91

Beauharnais, Fanny de, 173

Beaurieu. See *Cours d'histoire naturelle*

Beauvau, prince de, 24

beaver, 181, 184, 192, 193

Beliardy, abbé, 26–27

Bernardin de Saint-Pierre, Jacques-Henri, 22, 35, 203, 213, 215, 223; memoir on menagerie, 215–19

Bertin, François (oiseleur), 100, 101

Bexon, Gabriel-Léopold, 172

Bible, 187

bird merchants, traveling *(marchands forains)*, 102, 104–5, 112, 113–15

bird sellers. *See* oiseleurs

birds. *See individual species*

birds, as symbols of liberty, 105, 106, 107–8, 134, 193, 198, 204, 207, 208

bison, 88

blacks: monkey compared to, 131; servants, 143, 155. *See also* Africans; slavery; slave trade

boar, 71, 83

Bonnier de la Mosson, Joseph, 172

Bordeaux, 13

Bosman, William, 28

Boufflers, chevalier de, 23–24

Bougainville, Louis-Antoine de, 9–11, 17, 26

Bouillon, duc de, 149, 150

boulevards, 71, 88, 119, 131

bourgeois de Paris, 105, 114–15

Bourgogne, duchesse de, 43, 45

Bourrienne family (oiseleurs), 109, 119

Brazil, 10, 11, 13, 20, 25, 30, 34, 117

breeding: of canaries, 113, 114; of monkeys, 130; of parrots, 126. *See also* domestication

Brisson, Mathurin Jacques, 166, 176

Buc'hoz, Pierre-Joseph, 177

buffalo, 72, 79

Buffon, George-Louis Leclerc, comte de: and animal fights, 88–89; and animal slaves, 2, 190–95, 196, 199–200; on beavers, 181, 192; Bernardin de Saint-Pierre on, 216; on bustards, 66; on camels, 191; on canaries, 125, 192–93; and captivity, 36, 65–67, 199–200; cited by animal exhibitors, 87; cited by Cassal, 230; and compilers, 175–77; Condorcet on, 220; criticism of, 165, 170, 203, 220; on dogs, 153–55, 181, 190; on elephants, 2, 97–99, 153, 181–83, 193–95; and fables, 161, 182–85; and fairs, 81, 88; on falcons, 190, 193; *Histoire naturelle*, 2, 5, 51, 66–67, 166, 169–75, 227; on horses, 174; on humans, 151–52, 153, 183, 188, 201; on hummingbirds, 36; on insects,

182–83; and menagerie at Montbard, 65, 67; and menagerie at Versailles, 65–67; on monkeys, 153, 181; and oiseleurs, 117–18; on ostriches, 192; on oxen, 191; on parrots, 129, 153; on partridges, 196; and pets, 134, 149–55; and readers, 150–51, 172–75, 184–85; reputation, 5; and rhinoceros, 95; and sociopolitical commentary, 199–200; and style, 5, 173–75; on swans, 184–85, 186; on tigers, 182; on trumpeters, 153–55; on zebras, 53, 62
bull, 72, 83, 85
bulldog, 82, 83
bullfights, 72, 80, 83. *See also* animal fights
bunting *(pape)*, 27, 111, 117, 118
bustard, 66

Cabinet du roi, 19, 61–62, 66, 149, 181
cages, 30–31, 79–80, 132–34, 228. *See also* captivity; liberty
Calabar, 28
camel, 47, 64, 91, 191, 226
Camper, Petrus, 63, 95
Canada, 13, 89
canary: as pets, 122, 124–25, 136, 139, 192–93; in poems, 134, 204; trade in, 101, 102, 113–15; and women, 143, 145, 151
Canary Islands, 113, 124, 145
Cape of Good Hope, 21, 35, 49, 54, 56, 58, 62, 72
captivity, 36, 192–93, 198, 216, 224, 228, 230. *See also* animal slavery; cages; liberty
capybara, 150
caracal, 65
Caraccioli, Louis-Antoine de, 143
cardinal, 11, 27, 115–16

Caribbean. *See* West Indies
Cassal, Félix, 222, 226, 230
cassowary, 54, 58, 74, 88, 230
Castriès, maréchal de, 24, 51, 52, 55–58
cat, 129, 145, 151, 192
Cayenne, Guyana, 30, 35
Ceylon, 72, 88
chaffinch, 111
chain of being, 200–201
Challe, Robert, 11, 33
Champs de Mars, 210–11
Chantilly, estate of, 179–80, 214, 222
charlatans, 91–92
Charlemagne, 94
Charlevoix, François-Xavier de, 27
Chartres, duc de, 86, 288 n. 35
Chasses exotiques, 46
Chastellux, marquis de, 19
Chateau, Ange, 106–7, 117, 120
Chateau, Ange-Auguste, 2, 101, 106–7, 117–18, 120, 126
Chevalier, Jean-Baptiste, 49
Chevignard de la Pallue, M. A. T., 176
Chevreau (*intendant* of île de France), 52, 54, 55
children: books for, 160, 161–62, 164, 176–77, 230; and pets, 136–38, 142, 146, 151, 152
Chimay, princesse de, 131
chimpanzee, 33, 36, 79
China, 16, 19
civet, 29, 214, 221
coati, 169
Cochinchina, 22, 23
cockatoo, 10, 11, 22, 79, 88, 126, 128, 129, 130, 134, 150, 151, 230
Colbert, Jean-Baptiste, 20, 43
colonies: criticism of, 202; trade with, 12–16
combat d'animaux. See animal fights

combat du taureau. See bullfights
Committee of Public Safety, 213, 222
commodities, animals as, 123, 140–41,
    143, 145, 232. See also trade, animal
Compagnie de la Guyane, 14
Compagnie des Indes, 14, 16, 19, 20, 22
Compagnie du Sénégal, 14, 20, 22
concert spirituel, 80
Condé, prince de, 115–16, 179–80, 214
condor, 47, 68, 164
Condorcet, Jean-Antoine-Nicolas de
    Caritat, marquis de, 170, 205, 220
coot, 193
cordon bleu, 118. See also finch
coronation, 105–6, 107, 111
Correspondance secrète (newsletter), 63,
    168
Cossigny, Jean-François Charpentier
    de, 23
Cotte, Louis, 164–65
cougar, 79, 92
Cours d'histoire naturelle (Hennebert
    and Beaurieu), 176, 181, 190, 192,
    193, 196–97
Courtivron, marquis de, 150
Cradock, Anna Francesca, 80
crane: African crowned, 117, 274 n. 48;
    demoiselle, 43, 44, 45
crocodile, 43
Croÿ, Emmanuel, duc de: at fair, 74,
    79, 87–88, 169; at menagerie, 62–65;
    and Pluche, 163; and Valmont de
    Bomare, 168, 169
Cubières, chevalier de, 173, 185
curlew, 11

Dandrey, Patrick, 159–60
Daubenton, Louis Jean-Marie: and
    animal fights, 89–90; and Histoire
    naturelle, 170–72, 175; memoir on

importing birds, 29–30; and
    menagerie, 65–66; and pets, 149; and
    rhinoceros, 95
deer, 72, 83, 84; axis, 56; Tibetan musk,
    49
de Gennes, captain, 20
de La Borde (médecin du roi), 155
de La Motte, Houdar, 161
Delaunay (director of menagerie),
    226, 228, 229
d'Entrecasteaux, chevalier, 17, 19, 26,
    32
de Robien (president of Parlement of
    Brittany), 149
Desfontaines, René, 215, 221–22
Desmoulins (taxidermist), 180–81
Dictionnaire raisonné universel d'histoire
    naturelle (Valmont de Bomare), 92–
    93, 97, 118, 166–67, 168–69, 170, 176
Diderot, Denis, 142, 201, 202, 203
Dieppe, 25, 113
dog, 11, 89, 124, 129; in animal fights,
    82–85; friend of lion, 217, 218; as
    pet, 31, 141, 145, 150, 151, 153, 155,
    197–98; servility of, 195; superiority
    of, 153, 155; taming of, 190
domestic animals, 188, 190–91, 195, 197
domestication, 29–31, 149, 188, 193–98,
    217, 218, 223; of hummingbirds, 35–
    36; of monkeys, 130; of vicuñas,
    26–27
domination of animals, 152, 187–80
Duchesne, Antoine Nicolas (Porte-
    feuille des enfans), 176
Duchesne, Henri-Gabriel (Manuel du
    naturaliste), 89, 166
Duchet, Michèle, 203
duck, wood, 283 n. 78
Dürer, Albrecht, 96
Dutch East India Company, 54

eagle, 20, 214, 222

East Indies, 11, 16, 19, 25, 39, 117, 225

Eaux et Forêts, 104, 107, 108, 112, 114, 119

education: and fables, 161–63; and menagerie, 224, 228; and natural history texts, 172, 176; and pet keeping, 145–46

eland, 54

elephant: Buffon on, 2, 97–98, 153, 181–83, 193–95; confiscated from Dutch, 225–27; death at sea, 19; and menagerie, 37, 43, 45, 50–51, 52, 62, 63, 64, 65; shown in Paris, 1, 88, 93–94, 97–99; symbolism of, 235

elk, 88

emblem books, 158–59, 182

Encyclopédie, 5, 60–61

Encyclopédie méthodique. See Mauduyt de la Varenne

Endric, Martin, 72–74

England, 6, 140, 160, 195

entrepreneurs. *See* animal exhibitors

environmentalism, 203

Épinay, Louise Florence Tardieu d'Esclavelles, marquise d', 152

exotic, definition of, xiii

exotic animals, definition of, xiii

exotic commodities, 141

extinct animals, 58, 192

fables, 159–62; and Buffon, 182–85; definition of, 179; and natural history, 160–62, 177–82

fairs, 69–71; and audiences, 85–87; and charlatans, 91–92; elephant at, 97–99; and exhibits, 72–81; and naturalists, 87–92, 169. *See also* animal displays; animal exhibitors; *specific fairs*

falcon, 190, 193

Falkland Islands, 9

Farge, Arlette, 92

Ferry de Saint-Constant, Giovanni (*Génie de M. de Buffon*), 175

Festival of Federation, 210–11

fiction, 158–63

fights. *See* animal fights

Fillassier, Jean-Jacques, 164, 187

finch, 27, 29, 116, 131

French Revolution. *See* Revolution, French

friendship: animal-human, 151; between animals, 150–51, 217, 218; between humans, 22–23, 145, 151

Froger, François, 33–34

Fromageot, abbé, 146

Fromageot de Verrax (taxidermist), 139

Galam: birds from, 120; white monkey of, 36, 193

Gambia, governor of, 33

gazelle, 19, 43, 54, 226

genet, 72, 88

Génie de M. de Buffon (Ferry de Saint-Constant), 175

Genlis, Mme de, 173

Geoffroy Saint-Hilaire, Étienne, 222, 225

Gévauden, bête du, 188–89

gifts, animals as, 9–10, 20–24, 49–50, 216, 226

Gold Coast, 28, 120

goldfinch, 111, 112, 124, 125, 134, 136

grenadier, 117

Gresset, Jean-Baptiste-Louis, 148

Greuze, Jean-Baptiste, 142

Grimm, Friedrich Melchior, 170, 173, 291 n. 65

Grove, Richard, 203
Grozelier, Nicolas, 162
Guadeloupe, 15, 27, 35, 283 n. 78
guan, crested, 35
Guéneau de Montbeillard, Philibert, 117, 118, 172, 184
guilds, 101, 103–4, 107. *See also* oiseleurs
Guinea, 27, 28–29, 130
guineafowl, 22, 32, 43, 52, 110, 117
guinea pig, 197
Guyana, 27–28, 29–30, 31, 35, 130, 155, 274 n. 45

Haiti. *See* Saint Domingue
hamster, 132, 149
hedgehog, 197
Hennebert, Jean-Baptiste-François, and Gaspard Guillard de Beaurieu (*Cours d'histoire naturelle*), 176, 181, 190, 192, 193, 196–97
Henri III, 25
Hervieux de Chanteloup, J.-C., 114, 24–25, 273 n. 32
Histoire des progrès de l'esprit humain (Savérien), 176, 192, 193
Histoire des singes, et autres animaux curieux (Alletz), 165
Histoire naturelle, génerale et particulière. See Buffon; Daubenton
Histoire naturelle des oiseaux. See Buffon; Guéneau de Montbeillard
Histoire philosophique et politique des deux Indes, 202
Holland, 51, 54, 56, 62, 72; animals confiscated from, 225–26, 230
hornbill, 150
horse, 24, 83, 84, 174, 190, 197, 205
Hottentots (Khoikhoi), 201–2
humanitarianism, 203

humans. *See* animal-human relationship; attachment, animal-human; domination of animals; friendship; sympathy toward animals
hummingbird, 19, 31, 35–36, 217
Huot, Paul, 214
hyena, 52, 79, 88, 188–89, 226, 230

île de Bourbon, 16
île de France, 16, 52, 217
India, 16, 21, 39, 47, 49–50, 60, 98, 202
Indians, American, 35, 188, 201
indigenous people, 32, 67, 235; owning, trapping, and trading animals, 10, 27, 35, 243 n. 29, 247 nn. 69, 70; views of, 27, 210
insects, 163–65, 173, 182–83
Italy, confiscation of animals from, 229–30

Jacobins, 214, 230
jaguar, 9, 219
Jardin des plantes. *See* menagerie at Jardin des plantes
Jardin du roi, 67, 173, 212
Joseph II (Holy Roman emperor), 64
Journal de Paris, 80, 161, 206; letters to, 91, 93, 136, 138–40, 141, 145–46, 165, 182
Jussieu, Antoine-Laurent de, 223

Kamchatka, 20
kangaroo, 19
Khoikhoi (Hottentots), 201–2
kinkajou, 79, 88, 117, 150

Labat, Jean-Baptiste, 36
Lacépède, Bernard de, 227–28, 229
La Chesnaye-Desbois, François Alexandre Aubert de, 165

La Condamine, Charles-Marie de, 34
La Courbe, Michel Jajolet de, 22, 32
Lacroix, Jean-Bernard, 49
Ladvocat, abbé *(Lettre sur le rhinocéros)*, 95, 97
La Fontaine, Jean de: "Le corbeau et le renard," 161; fables of, 158–62; illustration of fair booths, 77–78; and menagerie at Chantilly, 180; and menagerie at Versailles, 44; in natural history text, 176, 181; "Le singe et le léopard," 76–78; "Le souris et le chat-huant," 159. *See also* fables
La Mettrie, Julien Offray de, 152
La Motte, Antoine Houdar de, 161
La Pérouse, Jean-François de Galaup, comte de, 11, 17, 20, 26, 32, 34
Laudonnière, René de, 245–46 n. 59
Lauriston, Jean Law de, 49
Le Blond, Auguste Savinien *(Portefeuille des enfans)*, 176
Le Havre, 21, 26, 113, 115, 126
Leleu, sieur (proprietor of animal fights), 71
Le Monnier, Louis-Guillaume, 23
lemur, 117, 245 n. 57
leopard: and animal fights, 71, 83; at Champs de Mars, 211; at fair, 76–77, 91; and menagerie at Jardin des plantes, 230; and menagerie at Versailles, 58; and menagerie at Vincennes, 37, 45; in town square, 25
Léry, Jean de, 11, 12, 35
Lettre sur le rhinocéros, 95, 97
Le Vaillant, François, 19, 91, 145
Levant, 20, 21
Le Vau, Louis, 38, 40
Lévi-Strauss, Claude, 4
liberty, 36, 134, 191, 196, 199, 204–5, 206–7. *See also* cages; captivity

linnet, 111, 125
lion: and animal fights, 38, 71, 80, 84, 88–89; benevolence of, 181, 217, 218; at Champs de Mars, 211; and dog friend, 217, 218; at an inn, 25; and menagerie at Jardin des plantes, 222, 226; and menagerie at Versailles, 47, 60, 215, 216–17, 219, 222; and menagerie at Vincennes, 37, 38, 45; and naturalists, 88–89; owned by exhibitor, 73; taming, 219
Lister, Martin, 91
literacy, 109, 157
literature, animal, 156–58; compilations, 175–77; fables, 158–63; *Histoire naturelle*, 169–75; moral lessons in, 177–79, 181–85; natural theology and philosophy, 163–66; reference works, 166–69; travel narratives, 34–36
llama, 31
Locke, John, 152
Lorient, 16, 21, 23, 57, 58
loris, 117
lost-and-found notices, 123, 126, 132, 136–38, 142
Louisiana, 13, 27
Louis XIV, 20, 37–40, 43–46
Louis XV: and animal paintings, 46–47, 48, 83, 84; criticism of, 60–61, 148–49; and menagerie, 20–21, 38–39, 45–47, 49–50
Louis XVI: criticism of, 186, 198, 212; and menagerie, 20, 21, 38–39, 47, 50–52
Louzardy, Bernard, 222, 225, 229
lovebird, 28
luxury: menagerie as, 216, 218–29; pets as, 136, 140, 145
Luynes, Charles Philippe d'Albert, duc de, 47

Luzerne, comte de, 60
lynx, 226

macaque *(ouanderou)*, 80, 81, 88
macaw, 120, 126, 130, 132, 134, 141, 150, 151
Macquer, Pierre-Joseph *(Manuel du naturaliste)*, 89, 166
Madagascar, 16, 22, 25, 117, 130, 132
Malacca, 33, 88
Malboissière, Geneviève de, 130, 136, 167, 169, 172
Malouines Islands, 9
Malvinas Islands, 9
mandrill, 52, 71, 83, 222, 226
Manuel du naturaliste (Duchesne and Macquer), 89, 166
Marchini, C. Dominique, 221
Marie Antoinette, 106
marmoset, 132
Marsan, comtesse de, 132
Marseille, 15, 21, 22, 25, 72
marten, 88
Martin, sieur, 71, 72, 88–89, 222, 224
Martinique, 15, 27
Mascarene Islands, 16, 52
Maudave, comte de, 34
Mauduyt de la Varenne, Pierre-Jean-Claude: on domestication, 31, 65, 196; and fairs, 88; and importing birds, 29, 31, 65; and menagerie, 65; and oiseleurs, 106, 116–18; on parrots, 129, 130, 153; and pets, 129, 130, 149, 151; on trumpeters, 31, 155, 197–98
Mauritius (île de France), 16, 52, 217
Mediterranean, 13, 20, 21, 39, 43
Mégisserie, quai de la, 101, 109, 110
Méliand, Adélaïde, 136, 167
Mémoires de l'éléphant, 98–99, 162

Mémoires pour servir à l'histoire naturelle des animaux, 45, 170
menagerie, Dutch, 51, 54, 225
menagerie at Cape Town, 54, 56
menagerie at Chantilly, 179–80, 214
menagerie at Jardin des plantes: arrival of animals, 220–22; competing visions of, 227–30; establishment of, 213, 220, 223–24; keepers, 221–22, 228–29, 230; proposals for, 215–20; war booty, 224–27
menagerie at Montbard (Buffon's), 65, 67
menagerie at Versailles, 37–40, 43, 47; to 1750, 40–47; acquisition of animals, 20–22, 49–60; Bernardin de Saint Pierre on, 216–17; Buffon on, 66–67; criticism of, 60–62, 65–67, 218–19; design of, 40–43; naturalists and, 63–67; revival of, 51–52; during Revolution, 212, 214–15, 222; visitors to, 44, 60–67; zebra quest, 52–60
menagerie at Vincennes, 37, 38, 45
Mercier, Louis-Sébastien, 61–62, 91, 92, 134, 141, 143
mimicry, 125, 126, 129, 146–49, 152–53
Mississippi, 28
mongoose, 149
monkey: Buffon on, 153, 181, 193, 201; and captivity, 36, 192; at Champs de Mars, 211; domesticating, 197; eaten, 11, 12; at fair, 68, 72–74, 76, 78, 80–81, 88; as gifts, 20, 22, 24; in literature, 161, 162–63, 165, 205; and menagerie at Jardin des plantes, 221; and mimicry, 146–47, 148–49, 152–53; and naturalists, 149, 150, 151; and oiseleurs, 110, 115–16, 117; as pets, 128, 130–31, 132, 134, 135, 136, 141,

143–46, 149, 150, 151; resemblance to humans, 131, 201; in satire, 212, 286; on ships, 17, 32–34; trade in, 25–29; and women, 143–46. *See also* baboon; chimpanzee; mandrill; orangutan

Montevideo, 9

Montmirail, marquis de, 132, 149, 150

Montmorency, duc de, 180

moral lessons: based in nature, 160, 179, 181–82; in fables, 158–60; in natural history books, 89, 163–64, 179, 181–85; provided by animals, 217, 228

Mother Goose Tales, 158

Mout, Douwe, 94–95

Mozambique, 58

Mukerji, Chandra, 40

munia, black-faced, 283 n. 78

Museum national d'histoire naturelle, 58, 219, 220

Native Americans, 35, 188, 201

native people. *See* indigenous people

natural history: and Buffon, 169–75, 183–84; course, 167–68; definition of, xiii–xiv; and fables, 160–62, 177–82; and fairs, 69, 87, 91; and menagerie, 39, 47, 65, 216, 219; and oiseleurs, 117–18; and pets, 149–50; popularity of, 4–5, 16, 156–57; popular texts, 175–77, 181, 203; reference works, 166–69; as substitute for pets, 146; theological approach, 163–66. *See also* Buffon; naturalists; *titles of specific works*

natural history cabinet, 166, 215, 224; of Bonnier de la Mosson, 172; confiscated from Holland, 225; definition of, xiv; of duc de Monmorency,

180–81; of prince de Condé, 179–80; of Valmont de Bomare, 166–69, 180; and voyagers, 19. *See also* Cabinet du roi

natural theology, 163–66, 187

naturalism in fables, 159–62

naturalists: and animal fights, 88–90; at fairs, 87–92; and menagerie 40, 63–67, 216, 217; and oiseleurs, 117–18; and pets, 149–55; and voyages, 16–19

naturalization. *See* domestication

nature: attitudes toward, 3–4, 156; as basis for morality, 160, 179, 181–82, 187; domination of, 187–88; and freedom, 206

Necker, Suzanne Curchod, 174, 184, 199–200

nightingale, 125, 184

noble savage, 98, 201–2

Nogaret, Félix, 173

North America, 13, 19, 27, 117

Nouveau traité des serins de Canarie, 114, 124–25

Oberkirch, Henriette-Louise de Waldner de Freundstein, baronne d', 51, 131

ocelot, 68, 88, 89

oiseleurs (bird sellers): and bourgeois de Paris, 105, 114–15; and exotic species, 112–20; and native species, 111–12; and naturalists, 117–18; shop inventory, 109–11; workings of guild, 101–11

Opéra Comique, 70

opossum, 19, 161

Orange, prince of, 51, 54, 56

"orang-outang," 79, 212, 230

orangutan, 193

ornithology, 29, 31, 117

ostrich: as gift, 24; harassed by humans, 192; and menagerie at Cape Town, 54; and menagerie at Jardin des plantes, 226; and menagerie at Versailles, 20, 21, 43, 52, 57; *Ostrich hunt* (Vanloo), 48; as pet, 25; shown in Paris, 68; Valmont de Bomare on, 169

otter, 150

ouanderou (macaque), 80, 81, 88

Oudry, Jean-Baptiste, 76–78, 95

Ourika, 245 n. 53

Ovid, 43

oxen, 190, 191, 197

paca, 31

Pacific Islands, 26, 34

paintings, 43, 46–47, 50–51, 83

panther, 46, 52, 180, 212, 226

pape (bunting), 27

parakeet: in fable, 11, 161; as gifts, 22, 24; lost, 137–38; and oiseleurs, 101, 110, 115, 116–17, 120; as pets, 126, 128, 136, 140, 142; in satire, 145; trade in, 27–29

Parlement, Paris, 104

parrot: Buffon on, 129, 153; eaten, 11, 12; edibility of, 11; in fable, 11, 161; as gifts, 22–24; lost, 137–38; and mimicry, 11, 126, 129, 130, 146, 148, 152–53; and oiseleurs, 110, 115, 116–17, 119–20; as pets, 123, 125–30, 132, 133, 134, 136, 150, 193; in satires, 141–42, 143, 145; on ships, 9–12, 32, 35; trade in, 25–30; *Ver-vert*, 148; and women, 142–43, 145, 151. *See also* Amazon parrot; cockatoo; macaw

Parsons, James, 95

partridge, red, 196

peacock, 117, 198, 214

pelican, 47, 57, 64, 75, 158

penguin, 192

Penthièvre, duc de, 131–32

Percheron (French agent), 54–58

Pernetty, Antoine-Joseph, 9, 10, 19

Perrault, Charles, 158

Perrault, Claude, 44, 45

petit-maître, 146–48

pets: advertised, 123, 126, 128–29, 130; and cages, 132–39; as commodities, 123, 140–41, 143, 145; criticism of, 140–49; deaths of, 139–40; and naturalists, 149–55; and owners, 136–38, 149–52; prices of, 124, 126, 128, 130–31, 132–33; on ships, 32–34; types of, 124–32; and women, 142–46, 151

Peuchet, Jacques, 207, 209–10

pheasant, 117, 193; Chinese, 110, 283 n. 78; golden, 46

philosophes, 60, 62, 140

Picardet, H. C., 162

Pliny, 95, 170, 175, 176, 182

Pluche, Noël-Antoine, 163, 166, 188

Poivre, Pierre, 203

Poix, prince de, Philippe-Louis-Marc-Antoine de Noailles, 24, 51, 52, 54, 57, 58, 60, 62

polar bear, 71, 75, 79–80, 83, 88, 169, 221, 222

political criticism, 6, 98–99, 186–87, 198, 233

Pompadour, Mme de, 46, 150, 278 n. 26

Pondicherry, 16, 49, 150

porcupine, 21, 25, 58, 60, 230

Portefeuille des enfans, 176

Portugal, king of, 43, 45, 94

Précis d'histoire naturelle (Sauri), 89, 176, 181

Príncipe Island, 28, 29
Provence, comte de, 110

quagga (striped horse), 55, 58, 59, 215, 222
quail, 107, 110, 283 n. 78

racoon, 91
Raincy, estate of, 222
ram, 72, 88, 228
Raynal, Guillaume Thomas François, 202, 203
reading: and Buffon, 150, 170–75, 184; growth of, 4–5; and natural history, 156–58, 203
Réaumur, René-Antoine Ferchault de, 19, 163, 166, 170, 176, 278 n. 26
redpoll, 118
Reims, 105, 111
religion. *See* natural theology
religious holidays, 70, 80, 83, 85, 104
Renaissance, 158–59, 179, 182
Restif de la Bretonne, Nicolas-Edmé, 162, 177, 178, 192, 205
Réunion (île de Bourbon), 16
Revolution, American, 36, 51, 120
Revolution, French, 60, 169, 206–7; and animal displays, 210–12; and animal fights, 207–10, 259–60; and birds, 208; and menagerie at Chantilly, 214; and menagerie at Jardin des plantes, 213, 215–30; and menagerie at Versailles, 214–15, 222
rhinoceros: Bernardin de Saint Pierre on, 216–17; and menagerie at Versailles, 49–50, 51, 62–65, 215, 216–17, 219; observed by duc de Croÿ, 62–65; in Paris (1749), 73, 75, 79, 88, 93–97; during Revolution, 215, 219
Ritvo, Harriet, 4

Rivarol, Antoine de, 191
Rouen, 113
Rouillé (navy minister), 47
Rousseau, Jean-Jacques, 145, 161, 173, 181, 198–99, 200, 201
Ruggieri, sieur (exhibitor), 87, 88

Sabran, Françoise-Éleanore de Jean de Manville, comtesse de, 23–24
sailors, 10, 26, 27, 32, 33
Saint Claire fair, 69, 70
Saint Domingue, 13, 15, 23, 28, 58, 202–3
Saint Germain Fair, 68–70, 73, 75, 79, 83, 86, 91, 94, 97
Saint Laurent fair, 69, 70, 81, 123
Saint Martin, sieur (proprietor of animal fights), 71, 72, 88–89, 222, 224
Saint Ovide fair, 69, 70–71, 74, 89, 92
Salerne, François, 151
satires and caricatures, 61, 86–87, 212, 230
Sauri, abbé, 89, 176, 181
Savérien, Alexandre, 176, 192, 193
science, 5, 17, 156, 160, 179, 216, 220, 234; definition of, xiii. *See also* Académie royale des sciences; natural history
screamer *(kamichi)*, 173–74
seal, 47, 73, 74, 79, 87, 88
Senegal, 22–24, 27–28, 30, 60, 118, 120, 126, 130
sentiment, 183
serval, 58, 65, 222
servants, 138–39, 143, 199, 200
servants, animal, 195–98
Seven Years War, 39, 49, 60
ship captains, 21, 32, 55, 56, 58
show people. *See* animal exhibitors
Siam, king of, 43

siskin, 111, 118
slavery, 200, 202–5. *See also* animal
    slavery
slave trade, 14, 20, 27–29, 120
sloth, 132, 149, 161
snake, 79, 88
Société d'histoire naturelle de Paris,
    219
solitaire, 36
Sonnerat, Pierre, 18, 19, 26, 192
Sonnini de Manoncourt, Charles-
    Nicolas-Sigisbert, 150, 251 n. 106
Souillac, vicomte de, 52, 54, 55, 57, 58
South America, 26, 34, 88, 117, 126
sparrow, 101; Java, 22, 117, 118, 131
Spary, E. C., 179
Spectacle de la nature (Pluche), 163,
    166, 188
Spinacuta, Laurent, 68, 74, 88
spoonbill, 24
squirrel, 110, 117; flying, 27, 118, 128
starling, 57, 112
style, literary, 5, 173–75
suricate, 151
swamphen, 20, 21, 24, 43, 58
swan, 117, 184–85
Swiss bird merchants, 113–14
sympathy toward animals, 5–6, 140,
    189–93, 232–33, 234

Table de Marbre, 104, 114
tapir, 87, 88
taxidermy, 139–40
Teleki, Joseph (Hungarian count), 62,
    83–85
Thibaudeau, Antoine-Claire, 223–24
Thomas, Keith, 140
Thouin, André, 215, 225
tiger: and animal fights, 37, 71, 80, 83;
    Buffon on, 182; as caricature, 212,

230; at Champs de Mars, 211; at
Chantilly, 214; at fair, 68, 76–77, 79,
88; and menagerie at Jardin des
plantes, 226; and menagerie at Versailles, 21, 47, 49, 50, 65–66; and
menagerie at Vincennes, 37, 45; in
natural history cabinet, 180; taming
of, 193, 219, 226
"tigre" (jaguar), 9, 52
Tippoo Saib, 226, 229
Torres, Antonio de, 25
tortoise, 20
toucan, 31, 62
Toulon, 72
trade, animal, 25–31, 100–121, 124–31
trade, colonial, 12–16
trade, slave, 14, 20, 27–29, 120
transporting animals, 21–22, 29–31,
    34–36, 50, 55–57, 73
troupial, 30, 283 n. 78
trumpeter, 30, 31, 153–55, 197–98
Tunis, 72, 226
turtledove, 34
Tyrol, 113

Vallière, duchesse de, 129
Vallin, Antoine, 109–11, 119–20
Valmont de Bomare, Jacques
    Christophe, 166–70, 176, 180; and
    fairs, 89; and parrot, 134; and rhi-
    noceros, 97
van de Graaff, Cornelis Jacob, 56
Vanloo, Charles André, 48, 264 n. 55
Versailles, palace and grounds, 40, 46,
    60–61
Versailles menagerie. *See* menagerie at
    Versailles
Ver-vert (Gresset), 148
veuve des Indes, 132
Victoire, abbé de la, 115–16

vicuña, 26–27
Villette, marquis de, 173
Vincennes menagerie, 37, 38, 45
violence, 85, 93, 207–10, 213, 218
Voltaire, 231
voyage accounts, 10–11, 16, 33–35
voyages of exploration, 16–17

Waal, Frans B. M. de, 235
Wandelaincourt, Antoine-Hubert, 176
warbler, 184
waxbill, 117, 283 n. 78

West Indies, 13–16, 27–29, 117
whydah, 117, 283 n. 78
wild animals, 76, 79–80, 188–89, 191–95, 197, 207, 210–13, 230
Wimpffen, François Alexandre, Stanis-laus, baron de, 26, 33
wolf, 71, 83, 85, 188–89; black, 89–90
women: and pets, 142–46, 151; and sub-ordinate status, 200–201, 235

zebra, 49, 52–58, 61, 62, 141, 197, 225, 227